THE ORGANIC CHEMISTRY OF BORON

THE ORGANIC CHEMISTRY
OF BORON

W. GERRARD

Head of Department of Chemistry, Mathematics, Biology and Geology
The Northern Polytechnic, London

1961

ACADEMIC PRESS

LONDON AND NEW YORK

ACADEMIC PRESS INC. (LONDON) LTD.
17 OLD QUEEN STREET
LONDON, S.W.1

547
G 32 o

U.S. Edition published by

ACADEMIC PRESS INC.
111 FIFTH AVENUE
NEW YORK 3, NEW YORK

42,143

Library of Congress Catalog Number 60–16984

Printed in Great Britain by
John Wright and Sons Ltd., The Stonebridge Press, Bristol

PREFACE

THE organic chemistry of boron is in a sense new chemistry; it affords unlimited fields for study at all levels of ability and experience.

The purpose of this monograph is to describe the reactions and techniques used in the organic chemistry of boron, to mention actual, suggested and potential industrial applications, and to pave the way to a more general better acquaintance with this chemistry. It is probable that the whole of organic chemistry will now come under review relating to the influence boron might have on the properties of critical linkages in organic compounds containing it.

The marvellous potentialities of the organic chemistry of boron have only just emerged. Revelations have been slow in coming; but during the past three or four years the gush of information has been really remarkable. It is therefore fitting and timely that a monograph on the subject should be presented to the more general reader, especially as the chemistry has up to the present been in the hands of a somewhat restricted number of research schools.

Interest is quickening in the extent to which ordinary reactions may be applied in relation to the fission susceptibility of boron–carbon, boron–hydrogen, boron–oxygen, boron–nitrogen, and boron–halogen bonds.

This upsurge of interest was caused mainly by the production and harnessing of diborane as a commercial chemical, the search for inorganic and semi-inorganic polymers applicable for high-temperature duty, by the revelation of the potentialities of boron trichloride as a reagent in organic chemistry, and the possible biological activity of organic boron compounds, particularly with reference to neutron absorption by B^{10} which is present in ordinary boron to the extent of about 20 per cent.

"Diborane" has long been in the text-books and college examination papers, and much learned discussion at the highest levels has related to the peculiar valency features of the compound. It was for many years prepared and handled only by the most tedious and specialized techniques. Then a technically workable method was discovered, and wide avenues for development suddenly appeared. About 1948 boron trichloride was difficult to obtain commercially and one made it as required in one's own laboratory; now it is readily available in cylinders. Diborane can be easily made in the usual laboratories for use as a reagent either *in situ* or for reaction in an adjoining vessel. It is probably not far from the truth to state that the primary starting materials in the practice of the organic chemistry of boron are boric acid, diborane, and boron trichloride.

Although there are scattered references to esters of boric acid in the first half of the nineteenth century, techniques which are still not outmoded are associated with Frankland (1860), Schiff (1867), Michaelis (1880), and Khotinsky and Melamed (1909).

The author is grateful to Dr. M. J. Frazer for his help with reference checks and to Mr. E. F. Mooney and Mr. H. A. Willis for their co-operation with the section on infra-red spectroscopy.

January, 1961   W. GERRARD

CONTENTS

Preface v

Chapter I. Reactivity of Alkyl and Aryl Groups 1

Chapter II. Esters of Boric Acid 5
    I. Preparation 5
    II. Properties of Alkyl and Aryl Borates 10
        A. Physical Properties 10
        B. Chemical Properties 11
    III. Dihydric Alcohol Systems 16
    IV. Polyhydroxy–Boric Acid Polymers 19
    V. Trends 20
    VI. Analysis 21

Chapter III. Boron Trichloride—Alcohol, Phenol and Ether Systems . 22
    I. Boron Trichloride—Alcohol Systems 22
        A. Alkyl Dichloroborinates 26
        B. Chloroboronates 30
        C. Alkenyl and Alkynyl Compounds . . . 31
        D. Interaction of Boron Trichloride and Ethylene
           Glycol 32
    II. Boron Trichloride—Phenol Systems 34
        A. Monohydric Phenols 34
        B. Dihydric and Polyhydric Phenols . . . 35
    III. Boron Trichloride—Ether Systems 36
        A. Cyclic Ethers 40
        B. Alkyl Chloroalkyl Ethers 42
        C. Methanolysis 44
        D. Unsaturated Ethers 44
        E. Dioxygen Cyclic Ethers 48

Chapter IV. The Boron Chemistry of Carboxylic Acids, Esters and
        Carbonyl Compounds 51
    I. Acids and Esters 51
    II. Carbonyl Compounds 54
        A. Aldehydes 54
        B. Ketones 57

Chapter V. The Attachment of One Hydrocarbon Group to Boron . 58

   I. Direct Formation of Alkyl- or Arylboron Dihalides . 58

      A. Properties of Alkyl- or Arylboron Dihalides . 59

   II. Attachment by Grignard Reagent 62

   III. Attachment by Organo–lithium Reagents . . . 63

   IV. Boronic Acids 64

   V. Properties of Boronic Acids 67

      A. Acid Strength 67

      B. Substitution in the Aromatic Nucleus . . 67

      C. Anhydride Formation 68

      D. Formation and Properties of Esters . . . 71

      E. Formation and Properties of Halogeno-Esters . 73

   VI. Vinylboronic Acid 75

   VII. Applications of Boronic Acids 75

   VIII. Summary 76

      A. C—B Bond Cleavage 77

      B. Hydrolytic C—B Bond Cleavage in the Presence
of Metal Salts 79

Chapter VI. The Attachment of Two Hydrocarbon Groups to Boron 81

   I. Attachment of Alkyl Groups 81

      A. Properties of the Dialkylboron Halides . . 84

   II. Attachment of Aryl Groups 86

   III. Ethanolamine Esters 90

   IV. Stabilization by Chelation 91

Chapter VII. The Attachment of Three Hydrocarbon Groups to Boron 92

   I. Formation of Trialkyl- and Triarylborons . . . 92

      A. Aluminium Alkyls 95

      B. Cyclic Boron–Carbon Compounds . . . 95

   II. Properties of Trisubstituted Borons 96

      A. Co-ordination with Bases 98

   III. Suggested Applications of Trisubstituted Borons . . 99

   IV. Polymerization by Trialkylborons 100

   V. Mutual Replacement of Groups in Trialkylborons . 102

**Chapter VIII. Special Tetravalent Boron Compounds Derived from
Triphenylboron** 108

   I. Introduction 108

   II. Tetraarylboron Salts 109

   III. Sodium Tetraphenylboron as an Analytical Reagent . 111

   VI. Tetraphenylboron Derivatives of Amides . . . 112

Chapter IX. Oxidative Fission of Carbon–Boron Bonds .   .   . 114

Chapter X. Hydrido-compounds of Boron 118
    I. Introduction 118
   II. Boron Hydrides 118
      A. Historical Development 118
      B. Preparative Reactions 121
      C. Structure and Properties 122
      D. Alkylated and Arylated Boranes   .   .   . 124
      E. Complexes of Decaborane 128
      F. High-energy Fuels 129
  III. Metal Borohydrides 130
      A. Introduction 130
      B. Preparation 130
      C. Properties of Sodium Borohydride   .   .   . 132
  IV. Borohydrides as Reducing Agents   134
      A. The Effect of Metal Ions and Solvents   .   . 134
      B. The Effect of Alkoxy Substituents on the Reduc-
         ing Properties 139
      C. Application of Borohydrides as Reducing Agents 141
      D. Hydroboration of Olefins 152
      E. Preparation of Borate Esters using Sodium Boro-
         hydride 158
   V. Suggested Applications of Hydrido Compounds of
      Boron 158

Chapter XI. Boron–Nitrogen Compounds 161
    I. Co-ordination Compounds derived from Diborane   . 161
   II. Co-ordination of Trialkyl and Triarylborons with Nitro-
      gen Bases 164
  III. Mono-aminoborons "Borazenes"   166
  IV. Borazines (XBNY), Borazole Derivatives .   .   . 174
   V. Borazole Polymers 182
  VI. Interaction of Boron Halides and Nitriles .   .   . 183
 VII. Interaction of Boron Trihalides and Amides   .   . 185

Chapter XII. Boron–Phosphorus Compounds   187

Chapter XIII. Certain Boron–Sulphur Compounds   .   .   . 194

Chapter XIV. Miscellaneous Ring Systems 197

Chapter XV. Other Boron–Halide Systems 209
    I. Introduction 209
    II. Reactions with Boron Trifluoride 209
    III. Reactions with Diboron Tetrachloride . . . 213
    IV. Reactions with Diboron Tetrafluoride . . . 217
    V. Reactions with Dichloroborane 217
    VI. Reactions with Boron Tribromide 218
    VII. Boron–Iodine Systems 220

Chapter XVI. The Infra-red Spectra of Boron Compounds . . 223
    I. Introduction 223
    II. Characteristic Group Frequencies 223
        A. B–O Frequencies 223
        B. B–Aryl Frequencies 225
        C. B–Methyl Frequencies 226
        D. B–Cl Frequencies 226
        E. B–N Frequencies 229
        F. B–H Frequencies 231
    III. Co-ordination of the Carbonyl Group to Boron . . 231

Appendix 233

References 253

Author Index 283

Subject Index 297

Chapter I

REACTIVITY OF ALKYL AND ARYL GROUPS

WITHOUT detailed reference to the following and other related points, descriptions in the organic chemistry of inorganic non-metal halides are at the best misleading, and often totally false. As the formation of alkyl chloride is significant in a considerable field of the organic chemistry of boron it is necessary to draw attention to the factors involved, and for reasons which will appear later, the silicon system is a good reference model.

In the organic chemistry of inorganic non-metal halides the degree of reactivity of the alcoholic or phenolic carbon atom, for convenience henceforth called the *central* carbon atom, is of first importance. There are three centres of reactivity involving two bonds in a hydroxy-compound, three involving two bonds in an inorganic ester, and two involving one bond in an inorganic non-metal (E) halide (X). Electron release to the carbon atom from groups attached to it will increase electron density on the atom of oxygen, and by implication its nucleophilic power. Such an effect will increase the reactivity of the C—O bond for mechanisms tending to form a carbonium cation; but will

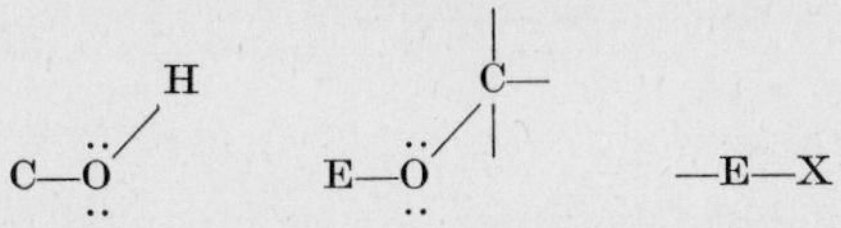

$$
\begin{array}{ccc}
& H & C- \\
& / & / | \\
C-\overset{\cdot\cdot}{O} & E-\overset{\cdot\cdot}{O} & —E—X \\
\cdot\cdot & \cdot\cdot &
\end{array}
$$

decrease the reactivity of the O—H bond for mechanisms involving the removal of a proton. A pull of electron density away from the oxygen atom by atoms or groups attached to the central carbon will have the reverse effect. Steric factors, however, can intervene in a somewhat complicated way; hindrance to SN2 (end-on) approach to the central carbon atom may have little to do with approaches to the oxygen atom.

The first reaction in the sequence occurring in alcohol-hydrogen chloride systems is an association with hydrogen chloride which may be formulated according to precise degree of separation of H and Cl. The solubility of hydrogen chloride in organic compounds containing oxygen has been correlated with electron density on oxygen, and an idea of this correlation may be obtained from Gerrard *et al.* (1956, 1959, 1960).

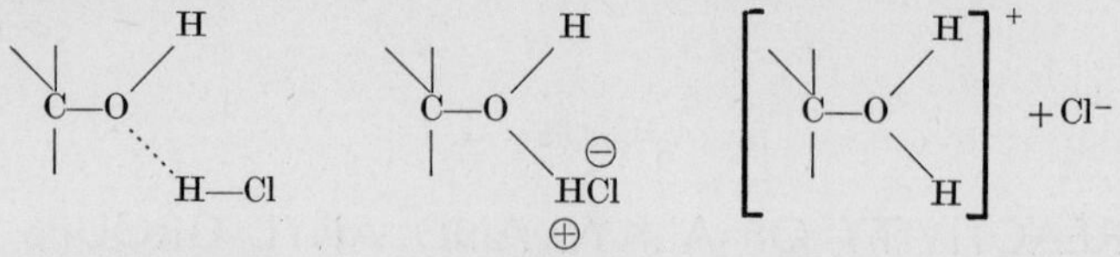

Hydrogen bond   Ion pair

TABLE I

Solubility of HCl in Moles per Mole of ROH at 10°C

Substance	Solubility
CH_3OH	0.857
CH_3CH_2OH	0.950
$(CH_3)_2CHOH$	1.030
$CH_3CH_2(CH_3)CHOH$	1.048
CH_2ClCH_2OH	0.53
$CHCl_2CH_2OH$	0.17
CCl_3CH_2OH	0.09

The solubility in *t*-butanol is evidently greater than in *s*-butanol, but intervention of the second stage prevents accurate measurement. The second stage of the overall sequences is the breaking of the C—O bond and the formation of alkyl chloride. For alcohols of what may be termed ordinary reactivity this stage is slow, and probably involves mainly an SN2 or end-on replacement on carbon.

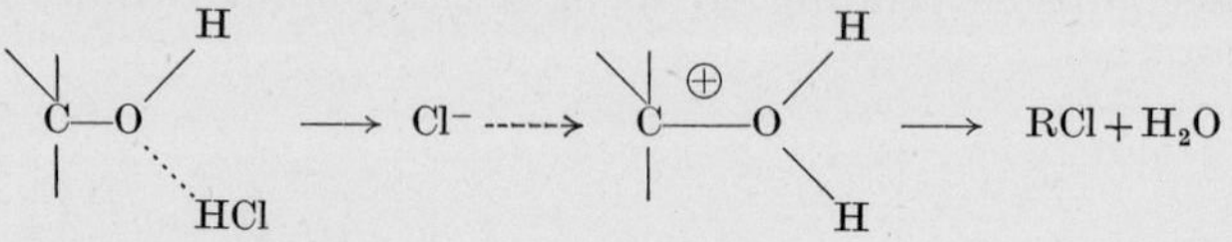

With secondary alcohols this may be accompanied to some extent by the SN1 or SNi mechanism.

SN1   Part of product may be rearranged

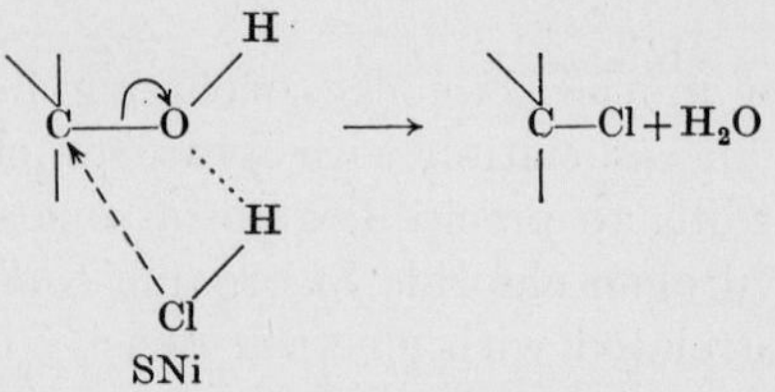

SNi

With tertiary alcohols such as *t*-butanol, the second stage is quick, and presumably involves mainly the SN1 mechanism. For the chloro-ethyl alcohols the second stage is slow, and is indeed extremely so for 2,2,2-trichloroethanol. It has very low probability of occurrence in the examples of the simpler phenols, simply because the benzene ring is usually not susceptible to nucleophilic attack.

Now although a decrease in electron density on the oxygen atom decreases the reactivity of the C—O bond in the sense mentioned, the effect is in general opposite for the O—H bond, and for this reason alone phenols and the chloroethanols might be expected to be well to the fore in systems involving attachment to oxygen by the severance of the O—H bond. The simplest formulation of the primary inter-action of a hydroxy-compound and an inorganic non-metal halide is as a four-centre broadside approach of the reactants involving only one transition state, as suggested by Gerrard (1939, 1940, 1944, 1945, 1946, 1951). It may be visualized that the driving force is the nucleophilic approach of oxygen to the electrophilic non-metal, mutual enhance-ment of the existing dipoles on approach bringing the four centres into the same transition state. A more complicated formulation involves a

$$
\begin{array}{ccc}
& \text{H} & \\
& \diagup \uparrow & \\
\text{R—O} & \text{Cl} & \longrightarrow \quad \text{R—O} + \text{HCl} \\
\downarrow \diagdown & & \diagdown \\
\text{—Si} & & \text{SiCl}_3 \\
\diagup \diagdown & &
\end{array}
$$

nucleophilic end-on replacement on the non-metal, followed by another nucleophilic replacement on hydrogen.

$$
\text{R—O:} \dashrightarrow \text{Si—Cl} \longrightarrow \left[\text{R—O—Si—} \right]^{+} + \text{Cl}^{-}
$$

$$
\text{Cl}^{-} \dashrightarrow \text{H—O} \oplus \longrightarrow \text{HCl} + \text{ROSiCl}_3
$$

For many alcohols and phenols this process is quick, although there are examples such as trichloro-*t*-butanol (chloretone) and 1,1,1,3,3,3-hexachloropropan-2-ol for which the rate is very small.

Keeping for the moment to the silicon system, one can understand that the electrophilic power of silicon in the alkoxytrichlorosilane is reduced by the replacement of chlorine by alkoxyl. It is to be expected that the next replacement to give $(RO)_2SiCl_2$ will be slower than the first, and so on. The fall-off in rate, however, is much greater than this, because association between hydrogen chloride and the alcohol, ROH_2Cl, decreases considerably the nucleophilic power of the alcoholic oxygen atom (Gerrard *et al.*, 1952; Currell *et al.*, 1960). Whereas tri-*n*-butoxychlorosilane will react readily with *n*-butanol, it does not appear to do so with *n*-$BuOH_2Cl$.

With alcohols such as *t*-butanol, in which the C—O bond has considerably more, but the O—H bond very much less reactivity than in the example of *n*-butanol, which for convenience is referred to as of ordinary reactivity, the replacement reactions depicted above are comparatively slow. But the decomposition of the alcohol–hydrogen chloride is comparatively quick. What evidence is available seems to point to the quick formation of alkyl chloride and water from the small amount of ROH_2Cl formed in the initial stages of mixing, followed by the quick hydrolysis of the non-metal halide to give more hydrogen chloride. The main result may be expressed by the following overall equation. 1-Phenylethanol is also a very reactive alcohol and gives

$$SiCl_4 + 4ROH \longrightarrow SiO_2,H_2O + 4RCl$$

alkyl chloride during the mixing of the reagents (Gerrard *et al.*, 1951, 1952). There is a point in the relative rate of reaction between hydrogen chloride and the non-metal-oxygen–carbon system of bonds. Tetra-*n*-butoxysilane dissolves hydrogen chloride thus:

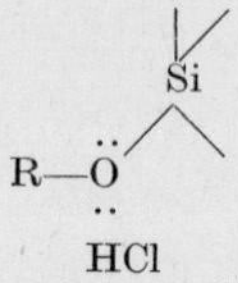

Further reaction to give RCl and HOSi$\leqslant$ is very slow at room temperature (R = *n*-Bu), and no evidence of the reaction which would give rise to ROH + Cl—Si$\leqslant$ has been mentioned. However, when R = *t*-Bu or 1-phenylethyl, alkyl chloride and hydrated silica are immediately formed when hydrogen chloride is passed into the alkoxysilane.

The essential function of a tertiary base, typified by pyridine, in inorganic non-metal halide systems, is to co-ordinate with the alcohol, ROH,NC_5H_5, and prevent the formation of hydrogen chloride, rather than to react with it as it is formed.

CHAPTER II

ESTERS OF BORIC ACID

I. Preparation

USUALLY alkyl and aryl borates are prepared by the azeotropic removal of water from an alcohol (or phenol)–boric acid system.

$$3ROH + H_3BO_3 \longrightarrow (RO)_3B + 3H_2O$$

$$3ROH + B_2O_3 \longrightarrow (RO)_3B + H_2O$$

There are examples for which this procedure is inadequate, and an alternative method has to be chosen. Boric oxide is a better reagent than the acid, because less water has to be removed. The hygroscopic oxide can be bought cheaply, or it may be prepared before use by heating boric acid at about 200°C at low pressure for several hours; it then is a light powder containing 29.5% of boron.

Individual operators have introduced modifications, such as the formation of a ternary azeotrope by the addition of benzene, toluene or carbon tetrachloride, and the situation is best shown as follows.

Modifications of the H_3BO_3—ROH System

Note	Reference
Addition of HCl, or H_2SO_4 (Me, Et, n-Pr, i-Bu)	Cohn (1911)
Water azeotroped with excess ROH	Bannister (1928) Ballard (1948) Haider *et al.* (1954) Wuyts *et al.* (1939) Scattergood *et al.* (1945)
Similarly for phenyl borate	Colclough *et al.* (1955)
Not suitable for t-alkyl	Scattergood *et al.* (1945) Haider *et al.* (1954) Ahmad *et al.* (1954)
Water removed as ternary azeotrope	Thomas (1946)
Suitable for triaryl borates	Prescott *et al.* (1941, 1942)
Suitable for t-alkyl borates	Lippincott (1953)

Modifications of the B_2O_3—ROH System

Note	Reference
B_2O_3 first used in 1867. (Me, Et, n-pentyl prepared in digestor)	Schiff (1867)
n-Pr, i-Pr, i-Bu, allyl borates (prepared in auto-clave at 110–170°C)	Councler (1871)
Toluene in Soxhlet extractor (containing an-hydrous copper sulphate)	Dupire (1936) Thomas (1946)
Not suitable for t-alkyl	Kahovec (1939) Thomas (1946)

Methyl borate has been particularly difficult to prepare, because the borate–methanol azeotrope is the constituent with the lowest boiling point; and as it happens this borate has been most prominent in the development of certain boron compounds, as will be shown. Calcium chloride has been used to remove methanol from the borate (70%)–methanol (30%) azeotrope obtained by reaction in a pressure bottle, shaken to avoid caking of boron trioxide (Etridge et al., 1928), and a device for the gradual addition of the trioxide to methanol has been described by Webster et al. (1933). A large-scale production due to Schlesinger et al. (1953) entails the use of the oxide and methanol in the molecular ratio 1 : 4 to get the borate–methanol azeotrope, from which the methanol is extracted by zinc chloride, lithium chloride, concentrated sulphuric acid (ligroin suspension), or by carbon disulphide azeotrope with methanol. In another procedure due to Schechter (1955), the trioxide (2 moles) was added to methanol (3 moles) at 35–60°C, the mixture then being allowed to digest for 1 hour at 25°C.

$$2B_2O_3 + 3CH_3OH \longrightarrow 3HBO_2 + B(OCH_3)_3$$

For small amounts of borate the practice (Schlesinger et al., 1940) has been to heat methanol (4 moles) with boron trioxide (1 mole) under reflux for 3 hours, the methanol–borate azeotrope, b.p. 59–61°C, being then distilled off and treated twice with anhydrous lithium chloride (50 g. for 250 c.c of azeotrope) in a separating funnel. The borate layer is distilled through an 18-in. spiral column, the fraction b.p. 68.5–69°C (30% yield) having boron 10.38%, calc. 10.41%. This ester is now commercially available.

There are a number of patents on the preparation and purification of trimethyl borate (Bragdon, 1957; May, 1957, 1958; and Ton, 1957), and the formation of this borate has been used by Porter (1957) and by Ehrlich et al. (1959) to determine boron in glasses. The viscosity and thermal stability have been investigated by Makishma et al. (1957), and Servoss et al. (1957) have reported on the vibrational spectra of normal and [10]B-labelled methyl borate. Methyl borate, prepared *in situ* (Strizhevskii *et al.*, 1956), has been suggested as a liquid flux for use with acetylene.

Alkoxyl exchange (see Schiff, 1867) has been used to prepare methyl borate by heating tri-*n*-butyl borate with methanol under reflux, the ester being separated from methanol by concentrated sulphuric acid, as in Vaughn's patent (1937). Further interest has been taken in this procedure by Wuyts *et al.* (1939), and although tri-*t*-butyl borate

$$(RO)_3B + 3R'OH \longrightarrow (R'O)_3B + 3ROH$$

could not be obtained from the alcohol and triphenyl borate (Colclough *et al.*, 1955), it was obtained from triethyl borate and a trace of sodium, as shown by Bannister (1928).

It is natural that attempts should be made to use borax and mineral acid in place of boric acid. This was tried by Rose in 1856, and by Frankland in 1862. Hydrogen chloride gas was used by Khotinskii *et al.* (1929), and concentrated sulphuric acid by Schlesinger *et al.* (1953). It was natural to try silver borate and an alkyl iodide, but early unsuccessful attempts by Nason (1857) and Schiff (1867) do not appear to have been followed up.

There is a patent (Wilson, 1959) relating to the preparation of lower alkyl borates by interaction of a high molecular weight poly-alcohol with borax or boron-containing ores, and converting the ester into a low molecular weight ester by transesterification with a lower alcohol.

Papers and patent specifications continue to appear on the preparation of boric esters with reference to some use or other, as the following list shows.

PATENTS

Levens *et al.* (1959)	Tris(di-*iso*butylcarbinyl) borate
Barnes *et al.* (1956)	Boric acid esters, for use as scintillation counters ([10]B enriched)
Stafiej (1959)	Esters of an oxyacid of boron
Tyson (1959)	Borate esters
Trautman (1957)	Preparation of trialkyl borates in connection with hydrocarbon oil additives
Rick (1957)	Triethyl borate used to produce inorganic oxide coatings
Lawrence *et al.* (1955)	Organo-boron compounds for fuels and lubricants, probably relates to boric esters

PATENTS (*cont.*)

Kirkpatrick (1956)	Crude borates relating to polyoxyalkylene
Hunter *et al.* (1957)	Tri*cyclo*hexyl borates Substituted tri*cyclo*hexyl borates Branched chain alkyl borates
Stoll (1957)	Complex alcohols purified as borate
Brown (1958)	An artificial resin from furfuryl alcohol and boric acid
Gates (1956)	Hot extruded metal is protected against oxidation by blowing a trialkyl borate against the hot metal
Johnson (1936)	Boric esters of polyhydric alcohols as textile agents

PAPERS

Baltimore Paint & Varnish Production Club (1955)	Organic borates in fire retardant paints
Owen *et al.* (1956)	Boric ester method for isolation of alcohols. Transesterification with tributyl borate, e.g. for trilinalyl borate
Kuskov *et al.* (1956, 1957)	Alkylation of aromatic hydrocarbons by esters of boric acid Synthesis of a number of alkyl borates
Zwahlen *et al.* (1957)	Boric esters and enol lactones
Leets *et al.* (1957)	Purification of alcohols by conversion into borates
Seidel *et al.* (1957)	Isolation of alcohols by conversion into borates
Stoll *et al.* (1956)	Ethyl borate is used to prepare more complex borates for isomerization when hydrolysed
Gel' Perin *et al.* (1956)	Continuous process for the hydrolysis of alkyl borates of higher aliphatic alcohols
Bashkirov (1956)	Alkyl borates formed in the oxidation of paraffin hydrocarbons by catalytic influence of boric acid and boric oxide

Early in the history of esters of boric acid, it was expected (Schiff, 1867) that esters of metaboric acid could be prepared, and although ethyl metaborate was then obtained as a syrupy hygroscopic liquid, metaborates have received but occasional mention, for example by Etridge *et al.* (1928) and Schlesinger *et al.* (1953), until the preparation of trimethoxyboroxole (trimethoxyboroxine) by heating boron trioxide and trimethyl borate in a sealed tube was described by Goubeau *et al.*

(1951, 1953), and its trimeric or boroxole structure was indicated by molecular weight and Raman spectra. There is much evidence to show that the —B=O structure is of low probability of occurrence, and boron avoids this by forming the boroxole six-membered ring.

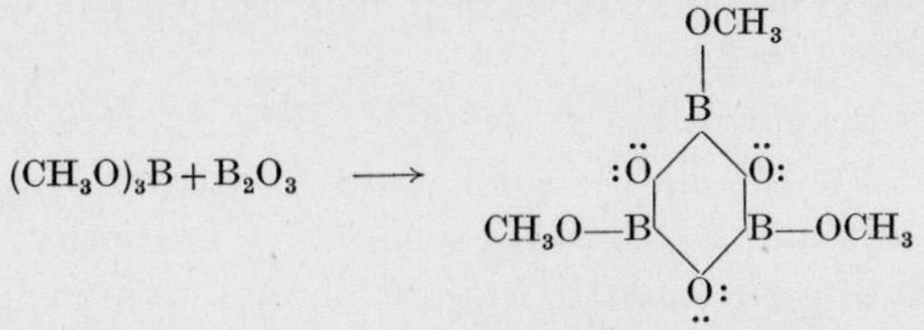

Trimethoxyboroxole
(Trimethoxyboroxine)

Trimethoxyboroxine freezes at about $10°C$; but commercial material contains traces of methyl borate, boric acid and methanol, and freezes about $-30°C$. It decomposes at temperatures above $150°C$, giving boric oxide and trimethyl borate (i); it burns in air to boric oxide, carbon dioxide and steam (ii); and with water it is immediately hydrolysed (iii).

(i) $(MeO)_3B_3O_3 \longrightarrow (MeO)_3B + B_2O_3$

(ii) $2(MeO)_3B_3O_3 + 9O_2 \longrightarrow 3B_2O_3 + 6CO_2 + 9H_2O$

(iii) $(MeO)_3B_3O_3 + 6H_2O \longrightarrow 3H_3BO_3 + 3MeOH$

Marked interest is being taken in actual and potential industrial applications of trimethoxyboroxine. Because of the high temperatures involved, certain metal fires (sodium, lithium, potassium, magnesium, zirconium and titanium) are extremely difficult to deal with by more conventional methods. Trimethoxyboroxine appears promising, as stated by Schechter (1957) and Commerford (1957); when sprayed on-to the fire it burns and forms a vitreous film of boric oxide which covers the metal, excludes air, and extinguishes the fire.

In the examples of fires with highly reactive metals, water, chlorinated hydrocarbons, and carbon dioxide all react with the reactive metals at the flame temperature, producing thereby gaseous inflammable or poisonous by-products. A solid blanketing material is necessary, and the boric oxide produced by the decomposition of the boroxole appears to be better than sand or sodium carbonate, which might be damp and difficult to apply. The process can be operated with other trialkoxyboroxines which are liquid at room temperature, triethoxy- and tri-n-propoxyboroxine being specifically mentioned in the patent specification. It should be understood that alkoxylboroxoles are very quickly hydrolysed, and storage of such material over long periods will not be free from difficulties.

1-Menthyloxyboroxine, m.p. 113–115°C, has been prepared by the azeotropic distillation of water with toluene, the equation being given as $ROH + B(OH)_3 \rightarrow ROBO + 2H_2O$, although infra-red studies showed the absence of B=O, and hence the presence of a trimeric form. Similarly, *cyclo*hexyloxyboroxine, m.p. 165–167°C, has been described by O'Connor et al. (1955).

Information on acid esters such as $ROB(OH)_2$, $(RO)_2BOH$, is very vague; but the hydrolysis of *l*-menthyloxyboroxine appears to have afforded the compound $ROB(OH)_2$, where $R = l$-menthyl. Olefins were obtained by heating the menthyl or *cyclo*hexyloxyboroxine to 270°C.

*Cyclo*hexylmetaborate, prepared by means of metaboric acid or boric acid (Clark, 1956), is suggested for use in ointment compositions. Some recent work on esters of metaboric acid has been described by Lappert (1958).

II. Properties of Alkyl and Aryl Borates

A. Physical Properties

The alkyl esters, $(RO)_3B$, are usually liquids with a quickly increasing boiling point as R increases in size. Thus the b.ps. are: CH_3, 68.7°C; C_2H_5, 117.3°C; n-C_3H_7, 177°C; i-C_3H_7, 140°C; n-C_4H_9, 115°C/15 mm. Tricyclohexylborate has b.p. 203°C/17 mm, m.p. 54°C, and tris-2,2,2-trichloroethyl borate has b.p. 112°C/0.05 mm, m.p. 97–98°C. The triaryl esters are usually solids or very high boiling-point liquids, such as phenyl, m.p. 92–93°C, α-naphthyl, m.p. 108°C, o-MeC_6H_4, b.p. 198°C/2.5 mm.

From studies on the usual physical properties (see list) of the simple alkyl borates, viscosity, Trouton constant, parachor, Raman, infra-red and X-ray spectra, and electron diffraction, the general conclusions are that the molecule of a borate such as trimethyl borate is planar, with the three B—O bonds at angles of 120°. The B—O bonds have some double-bond character due to back co-ordination, and this gives rise to restricted rotation, although valency force constants support single-bond character. Infra-red spectra show a strong characteristic absorption band at 1340 ± 10 cm^{-1}, attributable to the asymmetric stretching frequency of the B—O bond system. From the heats of hydrolysis the mean bond dissociation energies of the B—O bonds are 110 ± 5 kcal/mole.

Note	Reference
Molecular weight	French et al. (1938)
Parachor	Etridge et al. (1928)
	Laubengayer et al. (1941)
	Jones et al. (1946)
	Arbuzov et al. (1947)

Note	Reference
Viscosities	Jones *et al.* (1946)
	Haider *et al.* (1954)
Trouton constant	Wiberg *et al.* (1931)
Dipole moments	Cowley *et al.* (1935)
	Otto (1935)
	Mitra (1938)
	Lewis *et al.* (1940)
Polarizability	Ploquin (1958)
Raman spectra	Joglekar *et al.* (1936)
	Ananthakrishnan (1937)
	Milone (1938)
	Mitra (1938)
	Kahovec (1938, 1939)
	Becher (1952)
Infra-red spectra	Werner *et al.* (1955)
	George *et al.* (1955)
	Duncanson *et al.* (1958)
	Gerrard *et al.* (1959, 1960)
	Blau *et al.* (1960)
X-ray spectra	Katz *et al.* (1928)
Electron diffraction	Bauer *et al.* (1941)
Heats of hydrolysis	Charnley *et al.* (1952)

B. Chemical Properties

Alkyl and aryl borates have such remarkable thermal stability that some application for this property has naturally been sought. Their use as heat exchange media appeared at first sight promising; but their sensitiveness to moisture, and to air at elevated temperatures, seems to prohibit their general application for this purpose. As a general rule they are extremely sensitive to water, and are readily hydrolysed; intermediates such as $HOB(OR)_2$, or $ROB(OH)_2$ have not been isolated, except with rather special structures.

Wherever one turns in the organic chemistry of boron, propensity to ready hydrolysis has to be contended with, and this is a common frustration to the application of organic boron compounds. Boron is strongly electrophilic by virtue of its tendency to fill the vacant orbital and complete the octet. Consequently, hydrolysis of an alkyl or aryl borate has been deemed to involve prior co-ordination of water or hydroxyl to the boron atom as shown. It is probable that the breakdown of the complex would have to involve the steps shown in brackets.

Suggested Mechanism of Hydrolysis

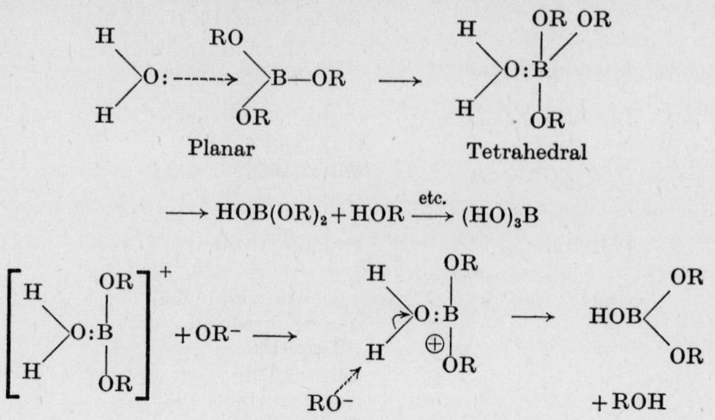

Planar   Tetrahedral

$$\longrightarrow HOB(OR)_2 + HOR \xrightarrow{etc.} (HO)_3B$$

Now the physical evidence mentioned points to back-co-ordination from oxygen to boron in alkyl borates, and this is indicated by the low acceptor power of boron in these compounds; for although tri-methyl borate will form co-ordination compounds, $(MeO)_3B,N\lessgtr$, with ammonia and amines, e.g. $MeNH_2$, Me_2NH, Me_3N, other alkyl borates do not do so, and even trimethyl borate will not co-ordinate with pyridine (Venkataramaraj Urs *et al.*, 1952). Triphenyl borate does complex with ammonia and amine (Colclough *et al.*, 1955, 1956), pre-sumably because the back-co-ordination to boron is offset by meso-meric interaction between the lone-pair electrons on oxygen, and the π-electrons of the aromatic ring.

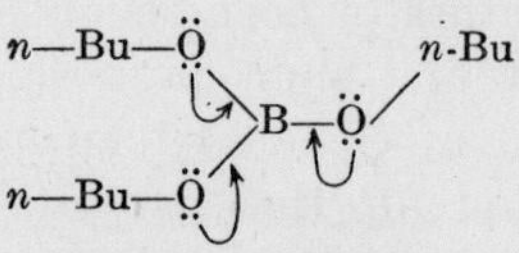

Back-co-ordination 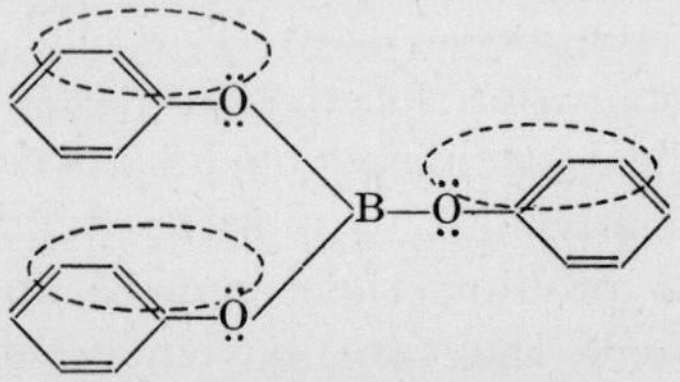   Back-co-ordination reduced

Factors influencing the co-ordinating power of boron attached to oxygen have been concisely summarized by Abel *et al.* (1958). In the aryl borates substituents in the ring influence the co-ordinative power of boron in two ways: by an inductive effect, and by a steric hindrance effect; and, of course, the donating power of nitrogen will be influenced by the same factors.

Thus although triphenyl borate will form a complex with ethylamine and pyridine, the co-ordination is less definite with di- or triethylamine; and 2,4,6-trichlorophenyl and 2,6-dimethylphenyl borates will not give a definite complex with pyridine. There are several interesting examples of this interplay of electronic and steric factors in the donor and acceptor systems involving aryl borates.

It might be expected that alkyl borates which do not co-ordinate with nitrogen bases are not likely to be easily hydrolysed by water if prior co-ordination by water is a requirement. On the other hand, aryl borates should be more susceptible. In ordinary circumstances there is little sign of any material difference. Alkyl borates are readily hydrolysed, and attempts to reduce hydrolytic propensity by internal or external co-ordination are often not very effective. There are examples of reduced ease of hydrolysis caused presumably by steric hindrance, as in the examples of the following borates: di-*iso*propyl-carbinyl, neopentyl (Scattergood *et al.*, 1945; *cyclo*hexyl, Cook *et al.*, 1950); and 1,1,1,3,3,3-hexachloro*iso*propyl (Gerrard *et al.*, 1955). Qualitatively, *s*-alkyl borates hydrolyse less readily than straight-chain ones (Scattergood *et al.*, 1945), and as the rate of hydrolysis is not reduced by 1N hydrochloric acid, it has been suggested that the hydroxyl ion addition is not the rate step.

It seems fairly certain that hydrolysis entails B—O bond fission, because there is no loss in rotatory power nor change in configuration when esters of optically active alkyl groups, such as 1-methylheptyl and 1-phenylethyl borates, are engaged (Gerrard *et al.*, 1951). Further-more, neopentyl alcohol is obtained from neopentyl borate.

As the B—O—C linkage often appears in organic boron compounds other than the ortho esters of boric acid, some general views on its properties are of considerable importance. In a trialkyl borate such as tri-*n*-butyl borate which contains an alkyl group of ordinary reactivity (Chapter I), electron density on oxygen is low, and that on boron comparatively high, because of back-co-ordination. This comparatively low electron density on oxygen is shown in an elegant way by the low solubility of hydrogen chloride in the ester (Gerrard *et al.*, 1959, 1960); for the solubility of hydrogen chloride in organic compounds containing oxygen is clearly related to electron density on oxygen. The rate of further reaction of hydrogen halides is comparatively slow, but increases in the order HCl < HBr < HI in the usual way. If the alkyl group is of considerable reactivity such as 1-phenyl-ethyl, *t*-butyl, then formation of alkyl halide is quick. The borates with an alkyl of ordinary reactivity, such as *n*-butyl, are remarkably resistant to reaction with phosphorus trichloride, thionyl chloride, and silicon tetrachloride at the temperature of reflux; but in the presence of a trace of ferric chloride, hydrogen chloride, butene, *n*-butyl chloride and degeneration products containing the non-metals were obtained, as shown by Chainani *et al.* (1960). Lack of reaction in absence of catalyst is in accordance with the apparent requirements for mutual replacement of alkoxyl and halogen in other systems such as alkyl phosphite–boron trichloride (Gerrard *et al.*, 1960), aryl phosphite–boron trichloride (or tribromide) (Frazer *et al.*, 1959), alkoxysilane–boron trichloride (Gerrard *et al.*, 1958), systems which may be formulated in terms of a four-centre broadside mechanism as mentioned on page 3. In a trialkyl borate such as tri-*n*-butyl borate the low electron density on oxygen will reduce its nucleophilic power, and little help comes from the boron because of back-co-ordination.

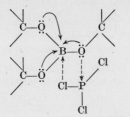

Inductive effect helps but
only weak for *n*-Bu

In the examples of very reactive alkyls, such as 1-phenylethyl, one might expect strong electron release to oxygen on demand, for its nucleophilic interaction with phosphorus here will give rise to alkyl chloride by a carbonium cation mechanism. This has not been tested yet.

Although phosphorus oxychloride forms a complex with boron trichloride, it does not do so with tri-*n*-butyl borate (Chainani *et al.*, 1960), in accordance with the principles mentioned. At 130°C a vigorous reaction occurred giving hydrogen chloride, butene, *n*-butyl chloride, and oxides of phosphorus and boron. Phosphorus pentachloride (3 moles) slowly disappeared from its mixture with tri-*n*-butyl borate (2 moles) at 70°C during 4 hours; butene, hydrogen chloride, *n*-butyl chloride, and phosphorus oxychloride were obtained, whereas with 5 moles of the borate and 3 of the pentachloride, the phosphorus oxychloride reacted on its own account when the temperature was raised sufficiently. The higher reflux temperature attained in the oxychloride system probably explains the occurrence of a reaction in that example, although the distribution of electron density in the oxychloride is much more complex than for the other non-metal halides mentioned, except the pentachloride. Phosphorus pentachloride is always peculiar (Gerrard *et al.*, 1952), and in the present instance may owe its ready reactivity to the existence of $[PCl_4]^+ [PCl_6]^-$, the strongly electrophilic $[PCl_4]^+$ competing with boron for the lone-pair electrons on oxygen. The catalytic effect of ferric chloride might be related to a similar Lewis acid effect. Attention has been given to this topic here because of the significance of removing OX from boron, and replacing it by Cl,

$$B—OX \longrightarrow B—Cl$$

Acetyl chloride has no action on tri-n-butyl borate unless ferric chloride or a similar catalyst is present; then n-butyl chloride, n-butyl acetate are obtained.

A number of reactions with oxy-acids have been recorded (Schiff, 1867; Cherbuliez et al., 1953). Thus sulphuric acid gave ethyl hydrogen sulphate, ethylene, metaboric acid and water when heated at 140–150°C with tri-ethyl borate, as might have been expected; and nitric acid afforded ethyl nitrate and boric acid. Ethyl esters of carboxylic acids are produced when ethyl borate is heated with the acid (acetic, benzoic, succinic, oxalic). A procedure for the preparation of acetates involving this reaction has been described by Hirao et al. (1953); the previously formed, but unisolated, borate is heated with glacial acetic

$$B(OH)_3 + 3ROH \longrightarrow (RO)_3B \longrightarrow 3CH_3COOR + B(OH)_3$$

acid. Methyl salicylate was obtained from methyl borate and salicylic acid (Cohn, 1911).

The kinetics of the solvolysis of alkyl borates have been investigated by Perkins et al. (1956) and by Denson et al. (1957), who have considered the catalytic function of amines and phenols. The constitution of alkyl borates in aqueous solutions has been examined by Kehiaian (1957).

In connection with the discussion on borohydride compounds, attention is drawn to the preparation of a number of metal-alkoxyboron compounds of the type $M^+[B(OR)_4]^-$. An early observation (Copaux, 1898) relates to the formation of sodium tetraethoxyboron from tri-ethyl borate and sodium ethoxide.

$$(EtO)_3B + EtONa \longrightarrow Na^+[B(OEt)_4]^-$$

The following list is of other examples similarly prepared (Cambi, 1914; Meerwein et al., 1929; Mead, 1955).

Na B(OMe)$_4$	Tl B(OEt)$_4$
Na B(OPr^n)$_4$	Tl B(OPr^n)$_4$
K B(OMe)$_4$	Zn [B(OMe)$_4$]$_2$
K B(OEt)$_4$	Na B(O pentyl^i)$_4$
K B(OPr^n)$_4$	K B(OCH$_2$Ph)$_4$
Li B(OMe)$_4$	Na B(OPh)$_4$
Ca]B(OMe)$_4$]$_2$	Na B(OPr^i)$_4$

Mead (1955) reported that the t-butoxy-compound could not be thus made. With respect to their reactivity with alkoxides, Meerwein (1948) placed borates in the order:

$$CCl_3CH_2-> EtOCH_2CH_2-> MeOCH_2CH_2-> PhCH_2-> CH_2{=}CHCH_2-$$
$$> Me > Et > i\text{-}Pr.$$

More recently, Wiberg *et al.* (1955) have prepared such compounds by adding the metal to a mixture of the alcohol and borate.

$$M + 2ROH + 2B(OR)_3 \longrightarrow M[B(OR)_4]_2 + H_2$$
$$\text{(Where R = Me or Et and M = Mg, Ca, Sr or Ba)}$$

The corresponding beryllium compound could not be isolated, although several methods were tried.

There is a link with the borohydride chemistry to be discussed later. Thus the sodium and lithium methoxyboron compounds have been obtained.

$$NaBH_4 + 4MeOH \longrightarrow Na^+[B(OMe)_4]^- + 4H_2$$

The metal tetra-alkoxyboron compounds are crystalline solids that are soluble in alcohols and tetrahydrofuran. They have a certain thermal stability; but at elevated temperatures they decompose to the metal alkoxide and alkyl borate.

III. Dihydric Alcohol Systems

Early observations on the exaltation of the optical activity of malic and tartaric acids on the addition of boric acid, and the increase in acid strength of boric acid in the presence of glycerol, observed by Thomson (1893), were properly correlated by later systematic work on various diols (Böeseken, 1913, and subsequent papers; Meulenhoff, 1925; Maan, 1929). Measurements on the conductivity of polyol–boric acid solutions led to the view that a boron–oxygen ring system was present, and with such diols the acetone condensation formula would prevail.

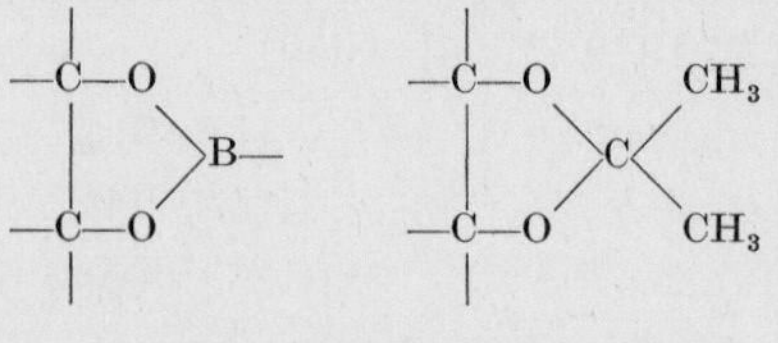

Certain such compounds were isolated (Hermans, 1925; Maan, 1929). For instance, 1-phenylcyclopentan-*cis* 1:2-diol and aqueous boric acid were mixed, and on cooling, an unhydrated hydrogen borate crystallized out. The same diol by a slightly different procedure gave a type formulated as shown. Indeed, the optical isomers of the boro-disalicylic acid complex were isolated by means of strychnine (Meulenhoff, 1925).

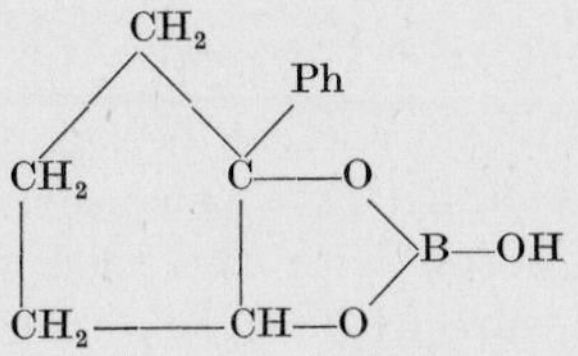

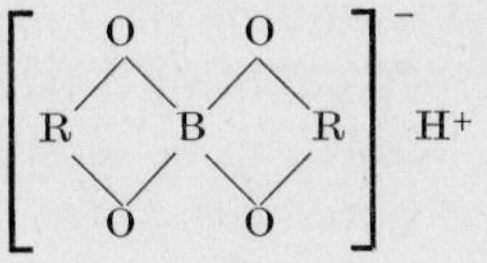

Polyol-boric acid interactions have been formulated by Hermans (1938).

It has been stated by Isbell *et al.* (1948) that compounds of type BD_2^- should be formed in concentrated solutions of carbohydrates containing a small amount of borate, whereas type BD^- should predominate in dilute solutions of carbohydrate with large amounts of borate, and type A will be in dilute solutions. An increase in pH should favour BD_2^- or BD^-. The existence of the borate ion $[B(OH)_4]^-$ has been confirmed (Edwards *et al.*, 1955; Fornaseri, 1949, 1950) by the studies on certain inorganic borates, and the tetrahedral disposition of the four groups is indicated. The B—O bond length in solid boric acid is stated to be 1.35 Å, the angle between the B—O bonds,

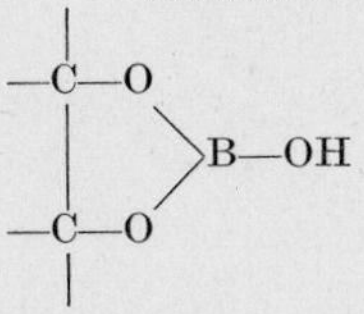

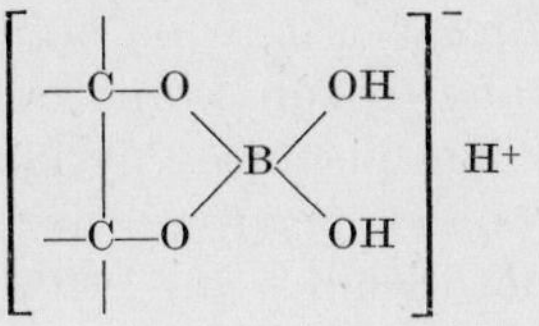

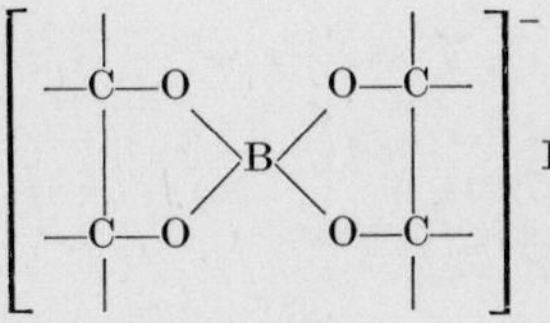

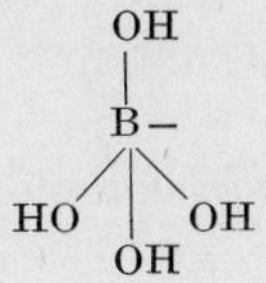

120°, and the inter-oxygen distance 2.32 Å (Zachariasen, 1934, 1954). In the tetrahedral boron system, the last number is less than 2.42 Å (Hermans, 1938).

In consideration of the following data for diols, it is likely that all reported compounds of 1,3-diols and boric acid belong to type A, whereas 1,2-diols will be of type BD^- and BD_2^-; because 1,2-diols can give ring closure with tetravalent boron with the least strain, and 1,3-diols are stated to give better closure with trivalent boron.

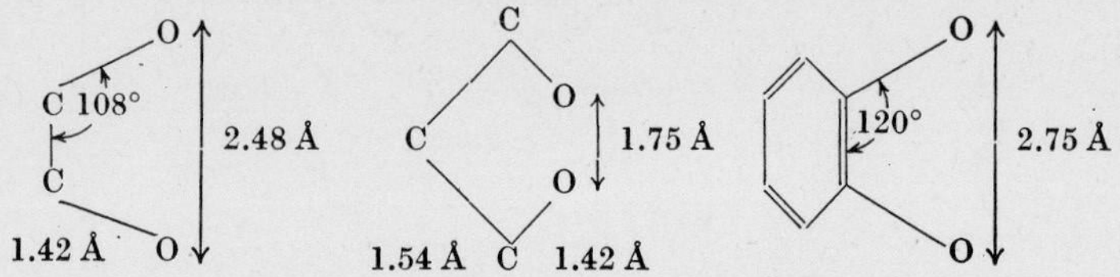

In a review (Zittle, 1951) on substances of biological interest it is agreed that it will be *cis*-hydroxyl groups on adjacent carbon atoms that will be involved in bonding to boron, and such reactions should be diagnostic of structure. For instance, pyridoxine (vitamin B_6) reacted with borate, but compounds resulting from substitution at the 4-hydroxymethyl group did not (Scudi *et al.*, 1940).

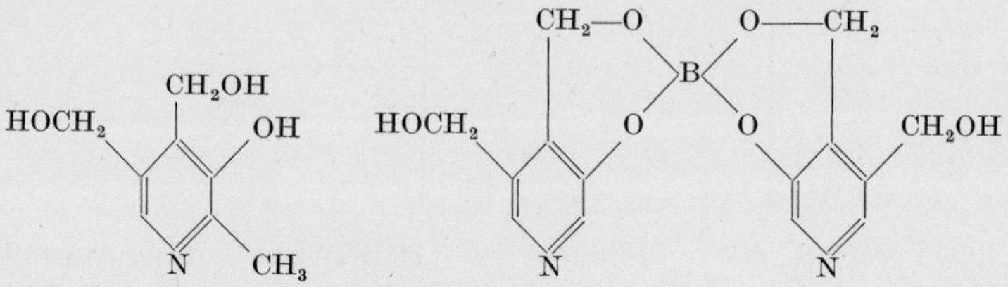

Inositols have little action with a borate; the adjacent *cis*-hydroxyl groups are presumably distorted from a single plane (Krantz *et al.*, 1938). Cyclitols having *cis*-hydroxyl groups in the 1,3,5-positions give complexes with sodium borate in aqueous solution. As compound formation reduces pH, such changes in pH are diagnostic of the formation of such compounds. (See Böeseken *et al.*, 1926; Karrer, 1926; Angyal *et al.*, 1952, 1956.)

Polyhydroxy compounds show ionophoretic mobility in aqueous sodium borate (Consden *et al.*, 1952; Foster *et al.*, 1953, 1955; Angyal *et al.*, 1957; and Weissbach, 1958), and from such studies a considerable fund of information has been gathered relating to polyols. The references cited should serve to complete this discussion for the purpose in hand.

IV. Polyhydroxy–Boric Acid Polymers

By the interaction of dihydric or polyhydric alcohols with boric acid one might expect to get poly-esters, and there are indeed a number of scattered references (see Gerrard, 1957, 1959) to the production of systems of rather indefinite constitution which were supposed to contain such polymers. The viscous material obtained by heating ethylene glycol and boric acid to 140–160°C was referred to as "aquaresin" and was stated to have applicability as a flexibilizer or plasticizer with glues, gelatin, and gum arabic (Bennett, 1934).

$$
\begin{array}{cccccccc}
& | & | & & & | & | & \\
-\text{O}-\text{B}-\text{O}-\text{C}----\text{C}-\text{O}-\text{B}-\text{O}-\text{C}----\text{C}- \\
& | & | & | & | & | & | & | \\
& \text{O} & & & \text{O} & & & \\
& | & & & | & & & \\
& -\text{C}- & & & -\text{C}- & & & \\
& | & & & | & & & \\
& -\text{C}- & & & -\text{C}- & & & \\
& | & & & | & & & \\
& \text{O} & & & \text{O} & & & \\
& | & & & | & & & \\
& -\text{B}-\text{O} & & & \text{O}-\text{B}- & & &
\end{array}
$$

Indefinite products of high viscosity were obtained by the interaction of an ester of boric acid with certain polymerized materials containing free hydroxyl groups (I. G. Farbenind, 1933). Boronated resins are referred to by Irany (1946), and it was stated that boric acid tends to form complex products with polyhydroxy compounds such as polyvinyl acetates, cellulose esters and ethers; and uses were suggested. Glycerol, phthalic anhydride, boric acid and a phenol-aldehyde resin are associated with the production of a clear condensate, soluble in toluene (Rosenblum, 1934), and there is another patent (Carpmael, 1930) dealing with boric acid and alkyd resins. Glycol borates are mentioned in connection with the chemistry of synthetic resins (Ellis, 1935), and the subject is also referred to in connection with protective and decorative coatings (Mattiello, 1946). There are references to boron-modified polyhydroxy polymers (Marvel et al., 1938; Hyman et al., 1948); "polyvinyl alcohol borate" is stated to be insoluble in water due to random cross-links.

Out of the quick and effective technical developments in the production and application of silicone polymers there emerged in the United States of America a belief that even more interesting materials might come from the organic chemistry of boron. Several sponsored research projects were formulated. It was hoped that the substitution of other high-melting elements such as boron might even enhance the thermal stability shown by the silicone compounds. Hence the expression "boron polymers" appeared with increasing frequency in the American technical press. In one of the first reports (Stout *et al.*, 1952) it was stated that condensation polymers were formed from boric acid and various glycols, but these had a low melting point and were water-sensitive. The same remarks apply to polymers formed from various mixtures of glycerol, boric acid and phthalic anhydride (or maleic anhydride). No polymers, which could be used for the purpose, came from these researches.

V. Trends

Interest in the various aspects of the organic boron chemistry related to di- or polyhydroxy-compounds is obviously being maintained. Recently reported investigations are listed as follows:

Cyclic esters of boric acid (Garner, 1955).

Analysis of flavones (Hörhammer *et al.*, 1955).

Formation and study of boric acid–mannitol complexes (Antikainen, 1955).

Polymer–boric anhydride reaction products (Edmonds, 1956).

Boric esters in motor fuels (Lyons, 1939; Hughes, 1955; Schechter, 1957).

Boric ester of 2-methyl-2,4-pentanediol for motor fuels (Darling *et al.*, 1956).

Separation of isomers as boric esters (Schmidt, 1955).

Boric acid esters of propane diols (Watt, 1956).

Interaction between borate and dextran (Zakrzewski *et al.*, 1956).

Glycerol boric acids, studied by potentiometric methods (Antikainen, 1956).

Imparting of fire resistance to water-base paints by poly(vinyl acetate) aqueous dispersions containing borates (Orth, 1957).

Triethanolamine borate as epoxy resin catalyst (Langer *et al.*, 1957).

Boric acid complexes of glycols and polyhydroxybenzenes as lubricant compounds (Thomas *et al.*, 1957).

Behaviour of boric acid with polyhydric alcohols (Kinoshita *et al.*, 1957).

Thermodynamic study of anionic chelate formation, with reference to mannitoboric acids (Antikainen, 1957).

Chelation of boric acid with hexoses (Antikainen, 1958).

Tetracycline–borate complex (Sekiguchi, 1958).

Borate polyester plastic with reference to slow neutron scintillation counters (Sun, 1956).

A solubility study of the boric acid–water-sorbitol system (Sciarra *et al.*, 1958).

Ion-exchange resins with boric acid groups (Solms *et al.*, 1957).

VI. Analysis

Determination of boron and boric acid is a subject of continuous investigation. Boric acid has been determined by barium borosaccharate (Ishibashi *et al.*, 1955), by a pH measurement of mannitol-boric acid (Csapo *et al.*, 1956), and a semimicro method has been described by Kuck *et al.* (1959). A polarographic determination using mannitol has been described by Lewis (1956).

Chapter III

BORON TRICHLORIDE—ALCOHOL, PHENOL AND ETHER SYSTEMS

I. Boron Trichloride—Alcohol Systems

In a large textbook on organic chemistry published in 1952, organic boron compounds are dismissed in six lines which are used to declare that "boric esters" are prepared from boron trichloride and alcohol by *warming them together*, and that these esters combine with alcohol to form a complex acidic substance $R'OB(OR)_3{}^-H^+$. From the beginning of the study of organic chemistry as such, there have been scattered references to the interaction of boron trichloride and alcohols; an ether-like odour was noticed by Berzelius (1824) when these reagents were mixed, and the definite isolation of products, e.g. trialkyl borates $(R=Me, Et, n\text{-}C_5H_{11})$, from sealed tube technique (Ebelmen *et al.*, 1846) constituted the first preparation of organic compounds of boron. In these conditions it was found (Councler, 1871, 1878) that allyl alcohol gave triallyl borate, but benzyl alcohol gave benzyl chloride.

The first systematic study on the interaction of alcohols and boron trichloride was not reported until 1931. This study, undertaken by Wiberg *et al.* (1931), was confined to methanol and ethanol; experimentation entailed high-vacuum line technique, and none of the products was removed from the vacuum system. The reagents were mixed at $-60°$ to $-80°C$, and the reactions were found to be quick and quantitative. Stepwise replacements of the chlorine atoms were controlled by mixing the reagents in the proper proportions.

$$BCl_3 + EtOH \longrightarrow EtOBCl_2 + HCl$$
Ethyl dichloroborinate

$$EtOBCl_2 + EtOH \longrightarrow (EtO)_2BCl + HCl$$
Diethyl chloroboronate

$$(EtO)_2BCl + EtOH \longrightarrow (EtO)_3B + HCl$$
Triethyl borate

Two observations, one by Kinney *et al.* (1935), that boron trichloride (1 mole) and isoamyl alcohol (2 moles) in chloroform at $-20°C$ gave the chloroboronate in only 15% yield, and the other by Martin *et al.* (1951), that 2-chloroethanol (1 mole) and the trichloride (1 mole) gave small yields of the chloroboronate and dichloroborinate are rather fortuitous, and bear little relation to the systematic reaction pattern and experimental technique.

Merely as part of a comprehensive investigation into the organic chemistry of inorganic non-metal halides such as phosphorus, thionyl, and silicon halides (Gerrard *et al.*, 1951), a systematic investigation of the boron trichloride system was undertaken. As in the previous work on other inorganic halides, ordinary bench techniques were used and, whenever possible, products were isolated for storage and further use. As reactions in the boron trichloride are usually quick, advantage was taken of the conveniently low temperature afforded by solid carbon dioxide, and the usual practice was to mix the reagents at $-80°C$, immediate products, and where inevitable derived products, being isolated at as low a temperature as possible by the use of low pressure technique. In this way the intermediate chloro-esters, alkyl dichloroborinates, $ROBCl_2$, and dialkyl chloroboronates, $(RO)_2BCl$, were isolated in a number of examples.

As in other inorganic non-metal halide systems, not only must the degree of reactivity of the central carbon atom (the C—O carbon) be carefully considered, but also the susceptibility of the chloro-esters to decomposition. Furthermore, mutual replacement of alkoxyl and chlorine might take place during subsequent operations involved in the isolation of primary products. The intervention of the hydrogen chloride generated in the system must also be taken into account (see Chapter I).

Outstanding features in these systems appear to be electron density on the oxygen atom, and the electrophilic function of the non-metal, in the present instance, boron. The primary clash of boron trichloride and an alcohol of ordinary reactivity, such as 1-butanol, may be depicted in terms of a three-centre end-on replacement of boron, followed by a three-centre end-on replacement on hydrogen, it being accepted that a proton cannot be ejected as a unimolecular process, because of the extremely high energy entailed.

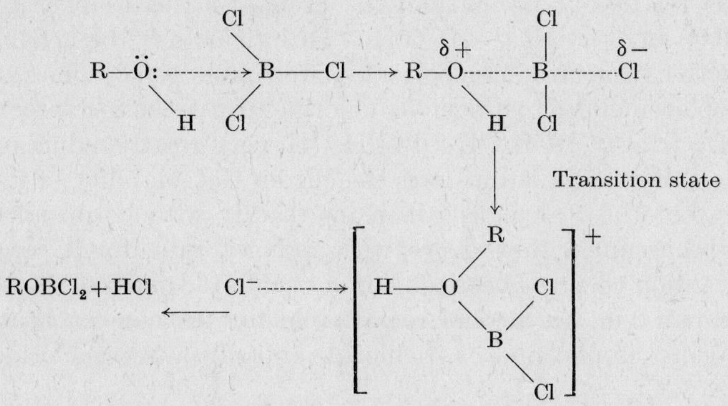

Transition state

$ROBCl_2 + HCl$

Thus there are two transition states, because replacement on hydrogen will involve one, and furthermore the second one will have some reversibility, judging from recent work on the solubility of hydrogen chloride in organic compounds containing oxygen, carried out by Gerrard *et al.* (1959, 1960). The reversibility, however, may not be a simple one, because the constitution of the $ROBCl_2$, HCl entity may involve an ion pair, or mere hydrogen bonding.

$$ROBCl_2 + HCl \rightleftharpoons \begin{cases} ROBCl_2 \\ HCl \end{cases}$$

It is argued that as boron has only six electrons in the tervalent state, addition of alcohol (or water) will occur first. This statement is a much oversimplified description of the situation, as will be shown in later discussion. If complex formation does occur, however, an irreversible decomposition of the complex will involve a second transition state, and one must consider precisely how hydrogen chloride is extruded.

$$ROH + BCl_3 \longrightarrow R\overset{\displaystyle ..}{-O}: \rightarrow BCl_3 \longrightarrow ROBCl_2 + HCl$$
$$\underset{H}{|}$$

An alternative formulation of mechanism which has been useful involves a four-centre, broadside approach of reagents, there being only one transition state.

$$\begin{array}{ccc} Cl & Cl \dashrightarrow H \\ \diagdown \diagup & \diagup \\ B \dashleftarrow O & \longrightarrow ROBCl_2 + HCl \\ | & \diagdown \\ Cl & R \end{array}$$

Although there is no direct evidence in this particular system, and indeed all three steps to trialkyl borate are absolutely quick, there are indirect indications that the second step, $ROBCl_2 + ROH \longrightarrow (RO)_2BCl + HCl$, is relatively slower than the first, and this quicker than the third, $(RO)_2BCl + ROH \longrightarrow (RO)_3B + HCl$. This may be attributed to the reduction in electrophilic power of boron as oxygen becomes attached to it. Replacement of hydrogen by chlorine to give the 2-chloroethanols, $ClCH_2CH_2OH$, Cl_2CHCH_2OH, Cl_3CCH_2OH, very greatly reduces electron density on the oxygen atom, as is clear from the solubility of hydrogen chloride; but these alcohols still react readily with boron trichloride. It may be assumed they do so with reduced rate, for a perceptibly slower reaction occurred between boron trichloride and 1,1,1,3,3,3-hexachloropropan-2-ol, an alcohol remarkable for its non-reactivity with such reagents as phosphorus trichloride and silicon tetrachloride under ordinary reaction conditions (Gerrard, *et al.*, 1955). The borate produced is slow to hydrolyse.

$$Cl_3C—CH—CCl_2$$

$$O \quad Cl \quad + BCl_3 \xrightarrow{\text{More slowly}} (RO)_3B \quad \text{m.p. } 313°C$$

$$H$$

Although chloretone, $CCl_3C(CH_3)_2OH$, is a tertiary alcohol, it has an extremely small electron density on oxygen, and a central atom of low reactivity, as is shown by its behaviour with phosphorus trichloride Gerrard *et al.* (1949). It would be an interesting alcohol for study in the boron trichloride systems.

Neopentyl alcohol, $(CH_3)_3CCH_2OH$, shows almost the same reactivity as ethanol, so far as the breaking of the hydrogen–oxygen bond is concerned, and reacts readily with boron trichloride. There is, however, a strong hindrance to the SN2 fission of the C—O bond.

When the alcohol is of considerable reactivity due to high electron density on oxygen (very reactive central carbon atom) such as in *t*-butanol or 1-phenylethanol, the effect mainly or entirely observed is the production of alkyl chloride and boric acid.

$$3Bu^t OH + BCl_3 \longrightarrow 3Bu^t Cl + (HO)_3B$$

$$3Ph(Me)CHOH + BCl_3 \longrightarrow 3Ph(Me)CHCl + (HO)_3B$$

$$3PhCH_2OH + BCl_3 \longrightarrow 3PhCH_2Cl + (HO)_3B$$

At this stage the influence of hydrogen chloride must be considered. With alcohols of ordinary reactivity, or with those of comparatively much lower reactivity, there is no noticeable intervention of hydrogen chloride under the experimental conditions which need to prevail; but with very reactive alcohols, the situation is very different; for these alcohols react quickly with hydrogen chloride, giving alkyl chloride and water. Therefore the ready production of alkyl chloride in the boron trichloride–reactive alcohol system can be due to one or both of two sequences of reactions. The general impression is that the normal

$$Bu^t OH + BCl_3 \longrightarrow Bu^t OBCl_2 \xrightarrow{HCl} Bu^t Cl$$

$$\Big\downarrow \text{Or directly} \Big\uparrow$$

$$\text{or } Bu^t OH + HCl \longrightarrow Bu^t Cl + H_2O \quad H_2O + BCl_3 \longrightarrow (HO)_3B + 3HCl$$

replacement reaction first to the alkyl dichloroborinate tends to occur, but the hydrogen chloride reacts quickly enough with the alcohol to give alkyl chloride and water, which can produce more hydrogen chloride, the main reactions being hydrolysis of the boron trichloride with water to give hydrogen chloride, and the reaction of this with alcohol.

It has been the intervention, in other ways, of hydrogen chloride in inorganic non-metal halide systems that has led to the use of a tertiary base such as pyridine. It is wrong to state that the base reacts with the hydrogen chloride formed; rather does it prevent the formation by first giving a very effective hydrogen bond, ROH,NC_5H_5. In the presence of pyridine, the borates of even the very reactive alcohols may be prepared, the reactions being quick even at $-80°C$.

$$
\begin{array}{c}
HNC_5H_5 \\
\nearrow \uparrow \\
R—O \quad Cl \\
\downarrow \diagdown \diagup \\
Cl—B \\
| \\
Cl
\end{array}
\quad \longrightarrow \quad
\begin{array}{c}
ROBCl_2 \\
+ C_5H_5NHCl
\end{array}
\xrightarrow{\text{Etc.}} (RO)_3B
$$

From time to time chemists have made casual statements to the effect that the function of pyridine is to activate the inorganic halide by forming a complex with it. In every case where the formation of such a complex can be demonstrated with any degree of reliance, the complex is always very much less reactive than the inorganic halide itself. The boron trichloride system provides absolutely indisputable evidence of this tendency; for when pyridine and boron trichloride are united, pyridine–boron trichloride is very quickly formed as a white crystalline substance, which has rather low reactivity with alcohols, as was shown by Gerrard et al. (1952). Consequently it can hardly be an intermediate in the reaction sequence that very rapidly gives a trialkyl borate when an alcohol in the presence of pyridine reacts with boron trichloride.

It was suggested that pyridine–boron trichloride might be a convenient way of transporting or storing boron trichloride which is to be used for certain reactions. It is a white crystalline solid, m.p. 115°C, b.p. 175°C/1 mm. It is insoluble in n-pentane, petroleum, ether, and cold water, but is soluble in chloroform, benzene, toluene, acetone, methanol, ethanol, and is hydrolysed by hot water. The preparation of a number of borates (including t-butyl) was described in the same paper; the complex (1 mole) was heated in chloroform with the alcohol (3 moles) and pyridine (2 moles).

$$C_5H_5NBCl_3 + 3Bu^tOH + 2C_5H_5N \longrightarrow 3C_5H_5NHCl + (Bu^tO)_3B$$

A. Alkyl Dichloroborinates

Methyl dichloroborinate, $MeOBCl_2$, was first prepared by the interaction of boron trichloride and dimethyl ether in a vacuum apparatus, and ethyl dichloroborinate was similarly prepared (Wiberg et al., 1931; Ramser et al., 1930). These chloro-esters were also prepared from the proper proportion of the corresponding alcohol.

$$Et_2O + BCl_3 \xrightarrow[-80°C]{} Et_2OBCl_3 \xrightarrow{56°C} EtOBCl_2 + EtCl$$

$$EtOH + BCl_3 \longrightarrow EtOBCl_2 + HCl$$

In more recent work (Gerrard *et al.*, 1951, and subsequent papers), *n*-butanol (1 mole) was added to boron trichloride (1 mole) at −80°C. After hydrogen chloride had been removed at low pressure, eventually at 15°C, *n*-butyl dichloroborinate remained in 100% yield. It has b.p. 42°C/20 mm. It is quickly hydrolysed to boric acid, *n*-butanol and hydrogen chloride. At 15°C slow decomposition into *n*-butyl chloride, boron trioxide and boron trichloride occurs; but at 100° the decomposition is 81% advanced in about 15 hours.

$$Bu^nOH + BCl_3 \xrightarrow{-80°C} Bu^nOBCl_2$$

$$3Bu^nOBCl_2 \xrightarrow{heat} 3n\text{-}BuCl + [3BOCl] \longrightarrow B_2O_3 + BCl_3$$

By mixing (+)-2-octanol (1 mole) with boron trichloride (1 mole), (−)-2-chlorooctane was obtained; but the rotatory power showed that considerable loss in optical activity had occurred. It is assumed that in this example, too, the alkyl chloride was formed by decomposition of the dichloroborinate, which is very unstable as is the case for ordinary secondary alkyl groups in general. The conclusion is that not every molecule of alkyl chloride could have been formed by an SN2 mechanism. Therefore the choice lies with the following mechanisms: not SN2, nor SNi alone, could be SN2 + SN1, or SN2 + SNi, or SN1 + SNi, or SN1 alone.

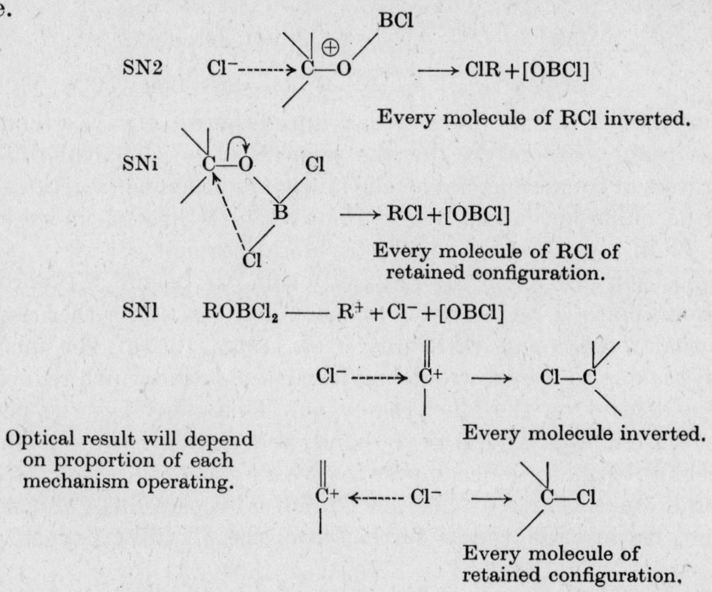

SN2 Cl^- ------> $\overset{\oplus}{C}$—O $\longrightarrow$ ClR + [OBCl]

Every molecule of RCl inverted.

SNi $\longrightarrow$ RCl + [OBCl]

Every molecule of RCl of retained configuration.

SN1 $ROBCl_2$ —— $R^+ + Cl^- + [OBCl]$

Cl^- ------> $\overset{\parallel}{C^+}$ $\longrightarrow$ Cl—C

Every molecule inverted.

Optical result will depend on proportion of each mechanism operating.

$\overset{\parallel}{C^+}$ <------ Cl^- $\longrightarrow$ C—Cl

Every molecule of retained configuration.

*iso*Butyl dichloroborinate, b.p. 34°C/20 mm, as is usual with the *iso*butyl group, is more stable than *n*-butyl dichloroborinate, and on decomposition gives a mixture of *iso*butyl and *t*-butyl chlorides, to be expected from the operation of an SN1 (carbonium ion) mechanism, which could give rise to a Wagner–Meerwein rearrangement.

$$\mathrm{Me{\underset{Me}{\overset{H\ \ \ H}{>}}\!C\!\!-\!\!C\!\!-\!\!OBCl_2} \longrightarrow Me{\underset{Me}{\overset{Me}{>}}}C^+ + Cl^- + [OBCl] \longrightarrow t\text{-}BuCl.}$$

Evidence of the formation of *s*-butyl dichloroborinate at $-80°C$ was observed; but decomposition into *s*-butyl chloride occurred quickly even at that temperature.

In a further detailed study of alkyl dichloroborinates it has been shown that they are obtainable also by quick mutual replacement between boron trichloride and a trialkylborate, or a chloroboronate.

$$BCl_3 + B(OR)_3 \longrightarrow 3ROBCl_2$$
$$BCl_3 + (RO)_2BCl \longrightarrow 2ROBCl_2$$

Two modes of decomposition, and a related one of mutual replacement, were discerned, the proportion of each in the total change depending upon the electronic nature of R, its structure, and upon experimental conditions. The elimination reaction appears to be of significance only when longer alkyl groups, such as *n*-octyl, are involved. When the

$$3ROBCl_2 \xrightarrow{\text{``Decomposition''}} B_2O_3 + BCl_3 + 3RCl$$
$$3ROBCl_2 \xrightarrow{\text{``Elimination''}} B_2O_3 + BCl_3 + 3HCl + 3 \text{ Olefin}$$
$$2ROBCl_2 \underset{\phantom{\text{``Mutual replacement''}}}{\overset{\text{``Mutual replacement''}}{\rightleftharpoons}} BCl_3 + (RO)_2BCl$$

pressure above a dichloroborinate is reduced there is always a tendency, even at room temperature, for the removal of boron trichloride, and this process of course encourages the mutual replacement reaction which gives the chloroboronate. When the system is heated under strong reflux, or in a sealed tube, boron trichloride cannot get away, and so decomposition into alkyl chloride and olefin can occur. 2-Chloroethyl dichloroborinate is rather stable to decomposition, and rather responds to mutual replacement (Edwards *et al.*, 1955, 1957). The increased stability is due to the electronic influence of chlorine which reduces the electron density on the alkoxyl oxygen. This effect has the peculiar result of strengthening the C—O bond, and reducing the tendency of the dichloroborinate to decompose to give alkyl chloride. 4-Chlorobutyl dichloroborinate shows a tendency to mutual replacement and decomposition, because chlorine is further from the alkyloxy oxygen atom.

The following Wagner–Meerwein rearrangements were shown for the dichloroborinates:

$$Bu^tCH_2 \longrightarrow CMe_2Et; \ CHMePr^i \longrightarrow CMe_2Et; \ CHMeBu^t \longrightarrow CMe_2Pr^i.$$

In recording the stability of dichloroborinates, caution must be exercised, because they are susceptible to the influence of traces of Lewis acids, and ferric chloride, and aluminium chloride are particularly effective. Thus when prepared from rigorously purified trichloride, n-butyl dichloroborinate remained unchanged after being held at 100°C for 9 hours; whereas 1% by weight of ferric chloride or aluminium chloride was sufficient to effect complete decomposition within 30 seconds at room temperature.

The steric course of the decomposition, however, remained the same and so the mechanisms may be formulated thus. A may be a second

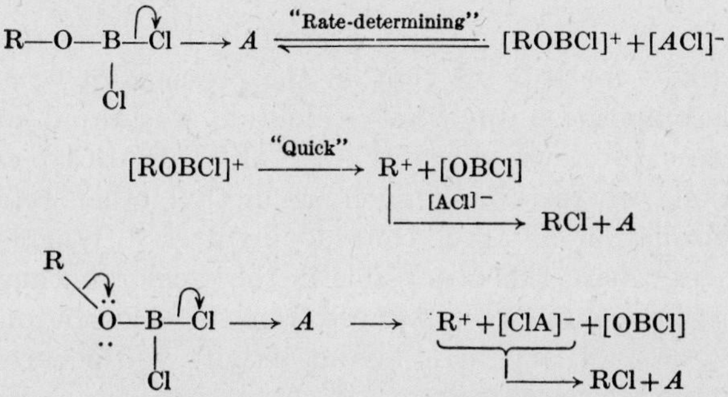

$$R—O—B—Cl \longrightarrow A \ \underset{\text{"Rate-determining"}}{\overset{}{\rightleftharpoons}} \ [ROBCl]^+ + [ACl]^-$$

$$[ROBCl]^+ \xrightarrow{\text{"Quick"}} R^+ + [OBCl]$$
$$\xrightarrow{[ACl]^-} RCl + A$$

$$O—B—Cl \longrightarrow A \longrightarrow R^+ + [ClA]^- + [OBCl]$$
$$\longrightarrow RCl + A$$

molecule of the chloroester, the boron atom of which will be electrophilic. However, ferric chloride and aluminium chloride are much more effective. Pyridine hydrochloride does not appear to have any catalytic influence on the rate of decomposition; but from work on entirely different systems, it might catalyse the mutual replacement reaction (Gerrard *et al.*, 1952; Currell *et al.*, 1960).

Apart from the inductive influence of chlorine in increasing stability of dichloroborinates, there is a possibility of chelation in the examples of chloroethyl and chloropropyl, although 4-chlorobutyl rather loses this additional stabilizing influence.

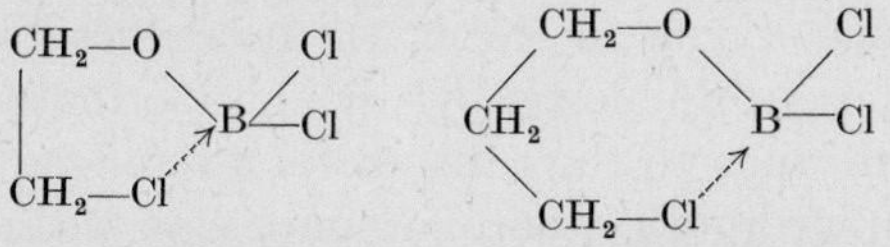

Stability of chloro-substituted alkyl dichloroborinates are in the order: $Cl(CH_2)_2 > Cl(CH_2)_3 \gg Cl(CH_2)_5 > Cl(CH_2)_4$; $Cl(CH_2)_2 \gg C_2H_5$; $Cl(CH_2)_3 \gg n\text{-}C_3H_7$; $n\text{-}C_4H_9 > Cl(CH_2)_4$; $n\text{-}C_5H_{11} > Cl(CH_2)_5$.

B. Chloroboronates

Dialkyl chloroboronates are prepared by the interaction of the proper proportion of alcohol and boron trichloride, by mutual replacement between a trialkyl borate and boron trichloride, or between two molecules of a dichloroborinate.

$$2(RO)_3B + BCl_3 \longrightarrow 3(RO)_2BCl$$

$$ROBCl_2 + (RO)_3B \longrightarrow 2(RO)_2BCl$$

Chloroboronates may decompose as shown in the equation, and concurrently by a reaction which produces an olefin. They will undergo

$$3(RO)_2BCl \longrightarrow 3RCl + [(ROBO)_3 \longrightarrow B(OR)_3 + B_2O_3]$$

mutual replacement, encouraged by low pressure removal of boron trichloride.

$$(RO)_2BCl \rightleftharpoons (RO)_3B + [ROBCl_2 \longrightarrow BCl_3 + (RO)_2BCl]$$

Di-n-butyl chloroboronate was essentially unchanged after being heated at 100°C for 3 hours, but in the presence of 0.05% of ferric chloride, decomposition to n-butyl chloride was rapid at 20°C. In general, traces of Lewis acids ($FeCl_3$, $AlCl_3$, $ROBCl_2$) catalyse the decomposition; whereas bases such as diethyl ether retard decomposition. Formation of alkyl chloride involves a Wagner–Meerwein rearrangement where this is possible in the dichloroborinate systems; but $l(-)$-2-chlorooctane was produced from the chloroboronate derived from $d(+)$-2-octanol, and little loss in activity had occurred, a result suggesting an SN2 mechanism.

As far as details have been discerned, substitution of chlorine in the 2 or 3 positions, as in 2-chloroethyl, or 3-chloropropyl, have little influence in increasing stability of the chloroboronates, and 4-chlorobutyl and 5-chloropentyl are even less stable than their unsubstituted analogues. There is a subtle point here. An alkyl dichloroborinate is considered as a catalyst in the decomposition of itself, by virtue of the electrophilic nature of the boron atom. Considering the electrophilic power of boron in ethyl dichloroborinate, substitution to give the 2-chloroethyl group will reduce the electron density on oxygen, and thereby increase the electrophilic power of boron, because the back-co-ordination from oxygen to boron will be reduced. We are reduced to a situation wherein 2-chloroethyl dichloroborinate should be more stable in itself, but open to the catalysis of a stronger Lewis acid. It is probable that an ordinary chloroboronate has weak catalytic effect owing to strong back-co-ordination from oxygen to boron. However, substitution by chlorine in the 2-position of ethyl, or the 3-position of propyl, should materially increase self catalysis, but reduce the uncatalysed rate of decomposition; the overall result could be but a slight increase in rate. Furthermore, some mutual replacement reaction would give the dichloroborinate and confuse the catalytic action.

C. Alkenyl and Alkynyl Compounds

A few unsaturated hydrocarbon groups have been studied with regard to their influence on boron trichloride–alcohol systems (Gerrard et al., 1956, 1957). Simple mixing of boron trichloride (1 mole) and the alcohol (3 moles) (allyl, prop-2-yn-1-ol, but-3-en-1-ol, 2-methylallyl) in n-pentane or methylene dichloride at − 80°C, followed by removal of hydrogen chloride at low pressure, led to excellent yields of the corresponding trialkynyl or trialkenyl borate; hydrogen chloride did not react with the olefinic or acetylenic bond. 1-Methylallyl alcohol gave only about 57% yield of borate, because of the concurrent formation of a mixture of 1- and 3-methylallyl chloride.

3-Methylallyl alcohol gave exclusively boric acid and a mixture of 1- and 3-methylallyl chlorides. Formation of alkenyl chloride in the examples cited is due to the increased reactivity of the alcoholic carbon atom by electron release from the double bond. It is evident that the 3-methylallyl system is more effective in this process.

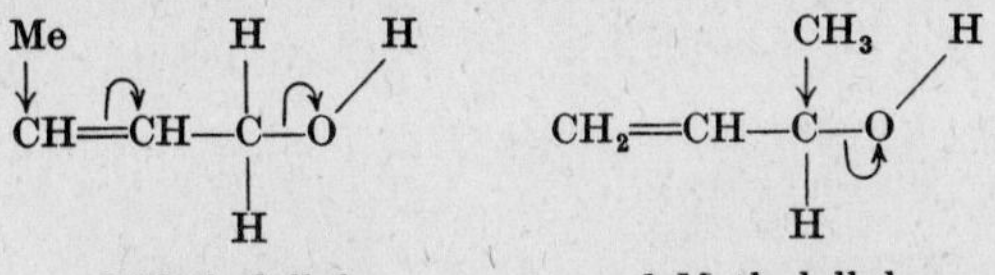

3-Methylallyl   1-Methylallyl

It is to be expected that the chloro-esters would be prone to quick decomposition (see Table II).

TABLE II

Decomposition of Chloro-esters

R =	Time—hours	Decomposition % at 20°C
Chloroboronates $(RO)_2BCl$		
Allyl	50	55
But-2-en-1-yl	22	5
1-Methylallyl	0.5	95
2-Methylallyl	27	31
3-Methylallyl	0.5	95
Dichloroborinates $ROBCl_2$		
Allyl	1	97
But-3-en-1-yl	21	93
2-Methylallyl	0.5	91

Hydrolysis of tri-1-methylallyl and tri-3-methylallyl borates gave the corresponding alcohol only; thus showing the boron–oxygen fission already observed for alkyl borates.

D. Interaction of Boron Trichloride and Ethylene Glycol

The first detailed studies on boron compounds derived from ethylene glycol have recently been recorded by Blau et al. (1957, 1960). In this work several compounds were isolated, and a foundation for detailed work on other glycols and polyhydroxy-compounds is thus established.

Addition of ethylene glycol (1 mole) to boron trichloride (1 mole) at $-80°C$ followed by warming to $20°C$ led to the isolation of ethylene chloroborinate, b.p. $40–44°C/0.2$ mm in 72% yield, and a residue of diethylene ethylene diborate, m.p. $162–164°C$. When 2 moles of the glycol were added at $-80°C$, ethylene 2-hydroxyethyl borate was obtained as a white solid in 95% yield, and when heated to $100°C/0.4$ mm, it gave diethylene ethylene diborate.

$$\begin{array}{l} CH_2OH \\ | \quad\quad + BCl_3 \longrightarrow \\ CH_2OH \end{array} \quad \begin{array}{l} CH_2-O \\ \quad\quad\quad\Large\rangle B-Cl + 2HCl \\ CH_2-O \end{array}$$

$$\begin{array}{l} CH_2OH \\ | \quad + 2 \\ CH_2OH \end{array} \quad \begin{array}{l} CH_2O \\ \quad\quad\Large\rangle BCl \longrightarrow \\ CH_2O \end{array} \quad \begin{array}{l} CH_2-O \\ \quad\quad\quad\Large\rangle B-OCH_2CH_2O-B\Large\langle \\ CH_2-O \end{array} \quad \begin{array}{l} O-CH_2 \\ \quad\quad | \\ O-CH_2 \end{array}$$

$$2 \begin{array}{l} CH_2OH \\ | \quad\quad + BCl_3 \longrightarrow \\ CH_2OH \end{array} \quad \begin{array}{l} CH_2O \\ \quad\quad\Large\rangle B-OCH_2CH_2OH + 3HCl \\ CH_2O \end{array}$$

The 2-hydroxyethyl compound did not react with acetyl chloride, and may have the tetravalent boron structure.

$$\left[\begin{array}{l} CH_2-O \\ \quad\quad\quad\Large\rangle B\Large\langle \\ CH_2-O \end{array} \quad \begin{array}{l} O-CH_2 \\ \quad\quad | \\ O-CH_2 \end{array} \right]^{-} H^{+}$$

Alkyl dichloroborinates (Pr^n, Bu^n) and ethylene glycol gave alkyl ethylene borates, whereas dialkyl chloroboronates (Bu^n, Bu^i, n-C_5H_{11}) gave tetra-alkyl ethylene diborates.

$$\begin{array}{l} CH_2OH \quad Cl \\ | \quad\quad\quad\quad\Large\rangle BOR \longrightarrow \\ CH_2OH \quad Cl \end{array} \quad \begin{array}{l} CH_2O \\ \quad\quad\Large\rangle B-OR \\ CH_2O \end{array}$$

$$\begin{array}{l} CH_2OH \\ | \quad\quad + 2(RO)_2BCl \longrightarrow \\ CH_2OH \end{array} \quad \begin{array}{l} RO \\ \quad\Large\rangle B-O-CH_2CH_2-O-B\Large\langle \\ RO \end{array} \quad \begin{array}{l} OR \\ \quad\quad \\ OR \end{array}$$

Ethylene chloroboronate is a colourless viscous liquid; it has considerable thermal stability, and even ferric chloride had little effect. The chlorine atom has the usual chemical reactivity and is readily replaced by hydroxyl, alkoxyl, thioalkyl, diethylamino, carboxyl, and $(CH_2)_2O_2BO$—. Very reactive alcohols such as tertiary butyl alcohol of course give the alkyl chloride unless pyridine is present.

$$(CH_2O)_2BCl \xrightarrow{H_2O} (CH_2O)_2BOH + HCl \xrightarrow{H_2O} (CH_2OH)_2 + (HO)_3B$$

$$(CH_2O)_2BCl + ROH \longrightarrow (CH_2O)_2B-OR + HCl$$

$$(CH_2O)_2BCl + Bu^tOH \longrightarrow (CH_2O)_2BOH + Bu^tCl$$

$$(CH_2O)_2BCl + Bu^tOH + C_5H_5N \longrightarrow (CH_2O)_2BOBu^t + C_5H_5NHCl$$

$$(CH_2O)_2BCl + Bu^nSH \xrightarrow{Slow} (CH_2O)_2BSR + HCl$$

$$(CH_2O)_2BCl + 2Et_2NH \longrightarrow (CH_2O)_2BNEt_2 + Et_2NH_2Cl$$

$$(CH_2O)_2BCl + CH_3COOH \longrightarrow (CH_2O)_2BOCOCH_3 + HCl$$

$$(CH_2O)_2BCl + (CH_2O)_2BOH \longrightarrow (CH_2O)_2B-O-B(OCH_2)_2 + HCl$$

The B—O bonds are susceptible to mutual replacement with B—Cl bonds, but in general the cyclic part is more stable and much less reactive. There is no example of a dialkyl hydrogen borate analogous to ethylene boric acid $(CH_2O)_2BOH$, and there is only one example of a mixed trialkyl borate, $(RO)_2BOR'$.

By adding ethylene glycol (1 mole) to boron trichloride (2 moles) at $-80°C$, ethylene bis-dichloroborinate was obtained as a liquid, stable at $20°C$, but unstable to distillation. On being heated it undergoes mutual replacement to ethylene chloroboronate and boron trichloride.

$$\begin{array}{c} CH_2OH \\ | \\ CH_2OH \end{array} + 2BCl_3 \longrightarrow \overset{Cl}{\underset{Cl}{>}}B-OCH_2CH_2O-B\overset{Cl}{\underset{Cl}{<}}$$

$$\overset{Cl}{\underset{Cl}{>}}B-OCH_2CH_2O-B\overset{Cl}{\underset{Cl}{<}} \xrightarrow{Heat} \begin{array}{c} CH_2-O \\ | \\ CH_2-O \end{array}{>}BCl + BCl_3$$

The chlorine atoms have the usual reactivity, being readily replaced by OH and OR; n-butanol (4 moles) gave tetra-n-butyl diethylene borate, and ethylene glycol (2 moles) afforded triethylene diborate. Diethylamine afforded the tetra-amine compound.

$$\begin{array}{c} CH_2OBCl_2 \\ | \\ CH_2OBCl_2 \end{array} + 8Et_2NH \longrightarrow \overset{Et_2N}{\underset{Et_2N}{>}}BOCH_2CH_2OB\overset{NEt_2}{\underset{NEt_2}{<}} + 4Et_2NH_2Cl$$

II. Boron Trichloride—Phenol Systems

A. Monohydric Phenols

The preparation of triaryl borates (Ar = Ph, m-MeC$_6$H$_4$, α—C$_{10}$H$_7$) was first recorded in 1901 by Michaelis *et al.*; the reagents were heated together in a sealed tube; no yields were stated, and chlorinated by-products were mentioned. Some very detailed work has recently been described by Colclough *et al.* (1955, 1956). Phenol and boron trichloride interact at −80°C to give triphenyl borate, diphenyl chloroboronate and phenyl dichloroborinate according to proportions of reagents. There is ready mutual replacement between boron trichloride and triphenyl borate to give either the chloroboronate or the dichloroborinate, neither of which can be distilled owing to mutual replacement within each.

$$PhOH + BCl_3 \longrightarrow PhOBCl_2 + HCl$$

$$2PhOH + BCl_3 \longrightarrow (PhO)_2BCl + 2HCl$$

$$3PhOH + BCl_3 \longrightarrow (PhO)_3B + 3HCl$$
$$\text{b.p. } 177\text{–}178°C/0.5 \text{ mm}$$
$$\text{m.p. } 92\text{–}93°C$$

$$2(PhO)_3B + BCl_3 \rightleftharpoons 3(PhO)_2BCl$$

$$(PhO)_3B + 2BCl_3 \rightleftharpoons 3PhOBCl_2$$

Because the benzene nucleus is not normally open to nucleophilic attack, there is no mechanism available in this system for the loss of easily hydrolysed chlorine as phenyl chloride. Consequently the chlorine remains in the system in an easily hydrolysed state. If the more volatile member, usually boron trichloride, of a mutual replacement system is withdrawn then triphenyl borate will remain as the only residue.

The following phenols, shown in Table III, have been studied in the boron trichloride system.

TABLE III

Melting Points of Triphenylborates (RO)$_3$B

R =	m.p.°C	R =	m.p.°C
p-ClC$_6$H$_4$	55	o-NO$_2$C$_6$H$_4$	108
2,5-Cl$_2$C$_6$H$_3$	106	2,6-(MeO)$_2$C$_6$H$_3$	134
2,4,6-Cl$_3$C$_6$H$_2$	173	α-C$_{10}$H$_7$	108
2,6-Me$_2$C$_6$H$_3$	135	β-C$_{10}$H$_7$	120
4-n-octylC$_6$H$_4$	82	o-I C$_6$H$_4$	90

p-Nitrophenyl, o-chlorophenyl, and o-tolyl dichloroborinates readily undergo mutual replacement at room temperature and reduce pressure; but o-nitrophenyl dichloroborinate had marked stability probably due to chelation. Interaction of the chloroboronate with ethanol affords one of the rare examples of a mixed borate.

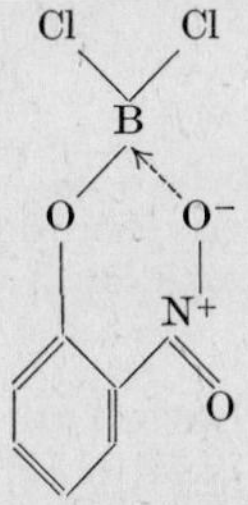

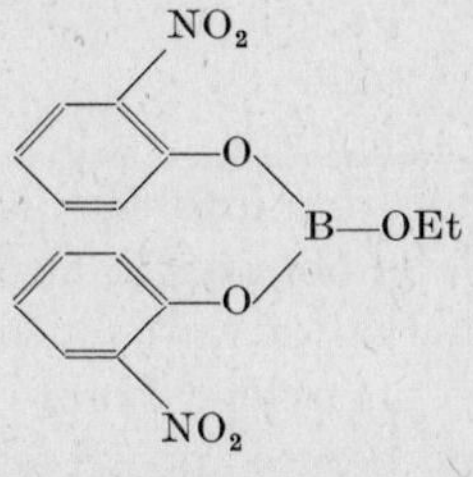

Aryl borates are referred to in connection with the preparation of esters and hydroxyketones (Kuskov *et al.*, 1956), and the complex formation of boron with aromatic hydroxy-compounds has been discussed by Khan *et al.* (1959). It is interesting to notice a patent specification by Mitchell (1958) on "borated thermosetting phenolic resins".

B. Dihydric and Polyhydric Phenols

The first report on a systematic and detailed study of the organic boron chemistry of di- and polyhydric phenols has recently been given by Gerrard *et al.* (1959).

Catechol, and boron trichloride in more than one molecular proportion, afforded the stable *o*-phenylene chloroboronate, and not the bis-dichloroborinate. This and catechol gave tri-*o*-phenylene bisborate.

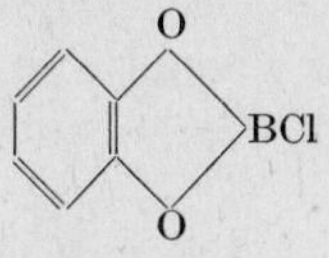

b.p. 64°C/10 mm   Could *not* be prepared
m.p. 57°C

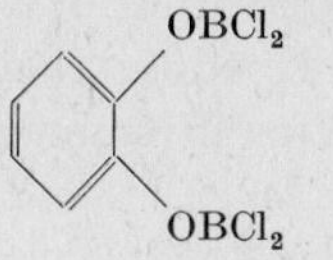

The chloroboronate had to be heated to 220–230°C for many hours to demonstrate mutual replacement to boron trichloride and the bisborate.

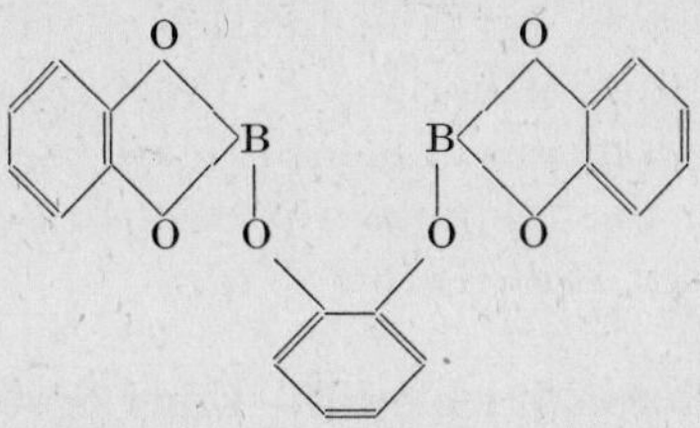

The chlorine has normal reactivity, being replaceable by OH, OR, SR, or NR$_2$. *o*-Phenylene hydrogen borate is one of the rare examples of an acid ester of boric acid, and on being heated it gave the anhydride formulated. The anhydride can be converted into the chloroboronate by boron trichloride, and into the alkyl phenylene borate with alcohol, and the anhydride is obtainable also from catechol and boric acid by the azeotropic removal of water in benzene.

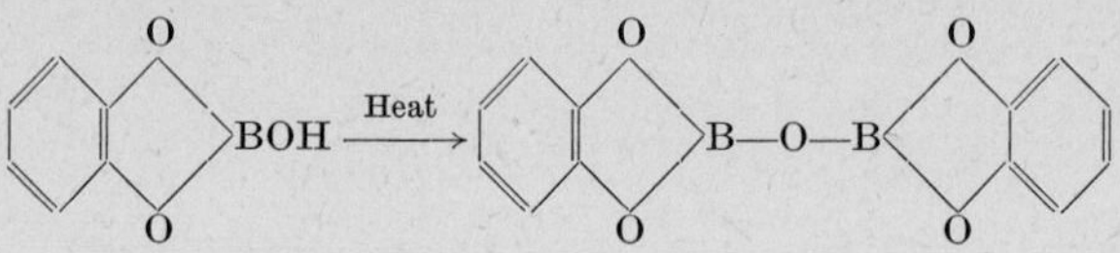

When methyl o-phenylene borate was heated at 240°C for about 16 hours, trimethyl borate and tri-o-phenylene bisborate were isolated in about 90% yield, showing a slow mutual replacement of RO groups. However, n-octyl o-phenylene borate was recovered after it had been heated at 350°C for 12 hours. At 330°C (12 hours) the s-butyl ester gave but-2-ene in 90% yield.

The general thermal stability of o-phenylene borate compounds may be due in part to 6 π-electron resonance.

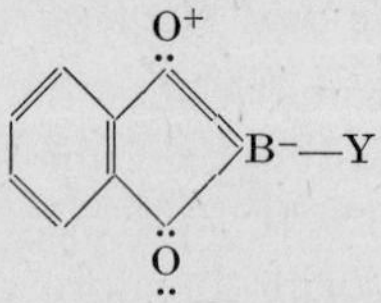

With resorcinol, quinol and pyrogallol, no volatile compounds or bis-dichloroborinates could be obtained. Polymeric materials remained, and in particular the one from pyrogallol was very tough and hard. They were easily hydrolysed to their parent phenol and boric acid. This work has significance in connection with the so-called boron polymers.

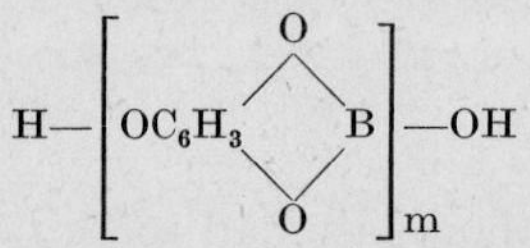

The structure and composition of catechol–boric acid–pyridine complexes are referred to by Kuemmel *et al.* (1956), and there is a patent dealing with esters of boric acid, Prochazka (1955), and formation of a boric ester as a means of identification is evidently a process in increasing use, and in this instance was used to detect 1,8-dihydroxy-naphthol groups (Haber, *et al.* 1956).

III. Boron Trichloride—Ether Systems

About early experiments (Ebelmen *et al.*, 1846) with ethers and boron trichloride it was reported that hydrogen chloride was evolved from diethyl ether, and ethyl protoborate formed. Methyl and n-pentyl protoborates were also mentioned. Gattermann (1889) referred to a co-ordination compound, boron trichloride–diethyl ether; but more extended observations and systematic investigations were reported very much later by Ramser *et al.* (1930) and Wiberg *et al.* (1931). By an enclosed low-pressure apparatus, similar to that used in the corresponding experiments with methanol and ethanol, the complex, Et_2OBCl_3, m.p. 56°C, was prepared, and decomposed by melting to give ethyl dichloroborinate and ethyl chloride. Furthermore the dichloroborinate also formed a complex Et_2OBCl_2OR, which underwent replacement as shown. The chloroboronate would not complex with ether, and it is

$$Et_2OBCl_3 \longrightarrow EtOBCl_2 + EtCl$$
$$\downarrow Et_2O$$
$$\begin{matrix} (EtO)_2BCl \\ + Et_2OBCl_3 \end{matrix} \Big\{ \longleftarrow Et_2OBCl_2OEt$$

evident that the electrophilic function of boron has been considerably reduced by back-co-ordination from the two atoms of oxygen. Since the complex, Et_2OBCl_3, gave diethyl ether with water it may be concluded that the ethyl–oxygen bonds were both intact.

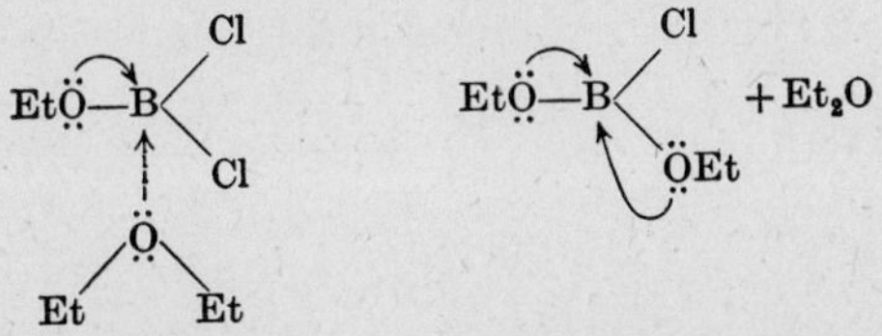

Will form complex.
Back-co-ordination
is insufficient to
prevent formation.

Does not form complex.
Back-co-ordination is
too strong.

In a systematic study a number of mixed ethers were allowed to react with boron trichloride (Gerrard *et al.*, 1951, 1952). The ethers were mixed with boron trichloride at $-80°C$, and the system was allowed to warm to room temperature. There was no isolation of a complex; instead the ethers had undergone fission by the time attempts were made to analyse the system. The mechanism suggested is formulated thus:

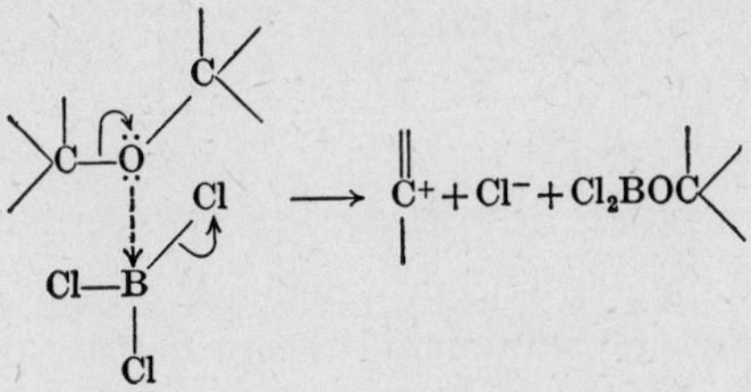

The carbonium ion was produced from the more reactive of the two carbon atoms in the ether, and of course the formation of the alkyl chloride from the ion was accompanied by the usual effects according to the groups attached to the carbon atom. And of course the other part of the cleaved molecule, the alkyl dichloroborinate, behaved according to its nature, as determined by the reactivity of the alkyl group. In the example of ethyl (+)-1-methylheptyl ether, the primarily formed 2-chlorooctane showed preponderant inversion with much loss in rotatory power, in accordance with the usually accepted result from an intermediate carbonium cation. Similarly, ethyl (+)-1-phenylethyl ether afforded completely racemized 1-chloro-1-phenylethane, a result not unexpected from a very reactive alkyl group. In alkyl group structures in which a Wagner–Meerwein rearrangement can occur, it does so, again in accordance with a mechanism involving a carbonium cation. Thus *iso*butyl *n*-butyl ether affords *n*-butyl dichloroborinate and a mixture of *iso*butyl and *t*-butyl chlorides.

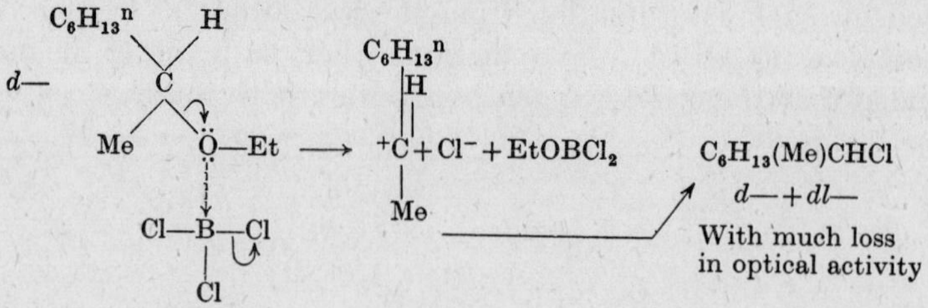

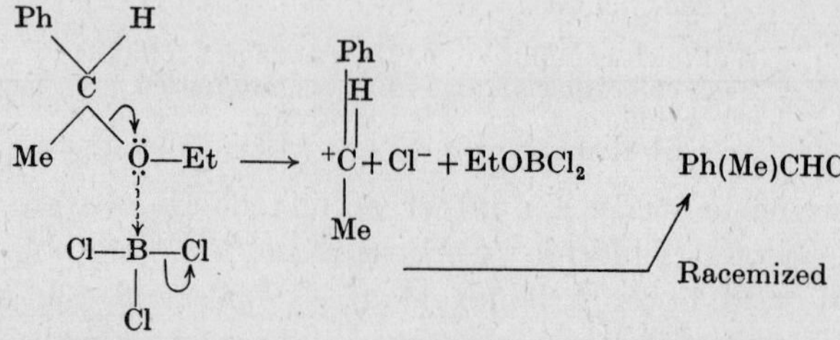

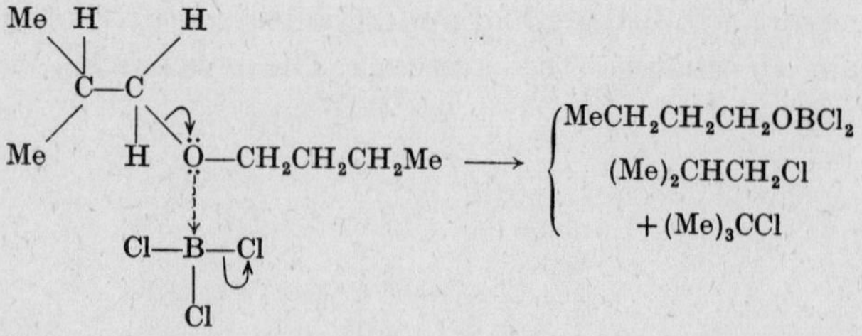

$$
\begin{array}{l}
d-\quad
\underset{\substack{\displaystyle | \\ Me \quad \overset{\cdot\cdot}{O}-Et \\ | \\ Cl-B-Cl \\ | \\ Cl}}{\overset{C_6H_{13}{}^n \quad H}{\diagdown\diagup}{C}}
\longrightarrow
\;\overset{C_6H_{13}{}^n}{\underset{Me}{\overset{|H}{|}}}{}^+C + Cl^- + EtOBCl_2
\qquad
\begin{array}{l} C_6H_{13}(Me)CHCl \\ d- + dl- \\ \text{With much loss} \\ \text{in optical activity} \end{array}
\end{array}
$$

$$
\underset{\substack{\displaystyle | \\ Me \quad \overset{\cdot\cdot}{O}-Et \\ | \\ Cl-B-Cl \\ | \\ Cl}}{\overset{Ph \quad H}{\diagdown\diagup}{C}}
\longrightarrow
\;\overset{Ph}{\underset{Me}{\overset{|H}{|}}}{}^+C + Cl^- + EtOBCl_2
\qquad
\begin{array}{l} Ph(Me)CHCl \\ \text{Racemized} \end{array}
$$

$$
\underset{\substack{\displaystyle | \\ Me \quad H \quad \overset{\cdot\cdot}{O}-CH_2CH_2CH_2Me \\ | \\ Cl-B-Cl \\ | \\ Cl}}{\overset{Me \ H \quad H}{\diagdown | \diagup}{C-C}}
\longrightarrow
\left\{ \begin{array}{l} MeCH_2CH_2CH_2OBCl_2 \\ (Me)_2CHCH_2Cl \\ + (Me)_3CCl \end{array} \right.
$$

The fate of the dichloroborinate depends on the nature of the alkyl group, and the factors influencing the properties of dichloroborinates apply here, and can of course give rise to experimental difficulties in the analysis of these systems.

Boron trichloride has a zero dipole moment in benzene as is to be expected from its planar symmetry; but when ether was added to the system, a value of 5.89 D was obtained, indicative of a tetrahedral structure for the 1 : 1 complex, Et_2O,BCl_3 (Ulich *et al.*, 1931, and Nespital, 1932).

Having regard to the use of *diglyme*, diethylene glycol dimethyl ether, in the recent startling development of sodium borohydride reactions, its reaction with boron trichloride is pertinent; it gave methyl chloride and presumably the dichloroborinate, $MeOCH_2CH_2OCH_2CH_2OBCl_2$, and the chloroboronate, $(MeOCH_2CH_2OCH_2CH_2O)_2BCl$, according to Brown *et al.* (1958).

Anisole and phenetole formed solid complexes at $-80°C$ (Gerrard *et al.*, 1952); but these decomposed on melting when warmed to 20°C, and gave phenyl dichloroborinate and the alkyl chloride. However, the complexes returned the ether when treated with water, indicating their simple structure as

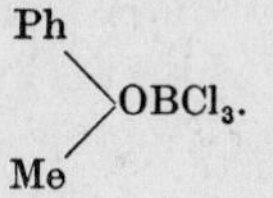

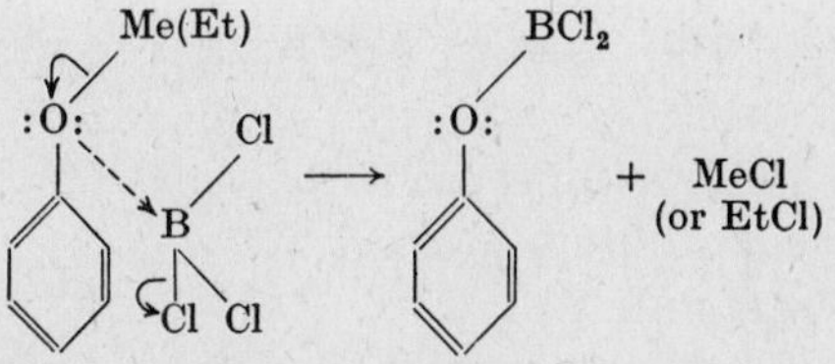

$$\begin{array}{c} Ph \\ \diagdown \\ \diagup OBCl_3. \\ Me \end{array}$$

$$+ \quad MeCl \atop (or\ EtCl)$$

Although one might expect electron release from phenyl to help the formation of the complex, mesomeric interaction between the oxygen lone pairs and the π-electron system of the ring can reduce the electron density on oxygen, and hence its co-ordinating power. The comparatively low electron density on oxygen is evident in the low solubility of hydrogen chloride in these ethers (Gerrard *et al.*, 1956, 1959, 1960). Accepting di-*n*-butyl ether as a convenient standard, the solubility of hydrogen chloride in this ether being 1.061 mole per mole of ether at 0°C, the corresponding value for anisole, about 0.21, and for phenetole, 0.22, show a decided reduction in electron density. The fission into methyl (or ethyl) chloride and primarily phenyldichloroborinate is in accordance with the difficulty in breaking the bond between the oxygen atom and the benzene ring; for this would require a nucleophilic substitution reaction for which the aromatic ring is not normally fitted. It is for this reason that phenyldichloroborinate undergoes mutual replacement in this system, giving eventually boron trichloride and triphenyl borate. Not without acute interest is the correlation between the lower solubility of hydrogen chloride (0.12)

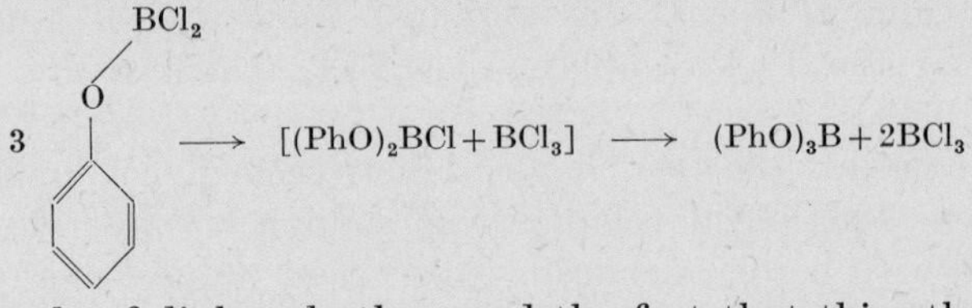

$$3 \quad \text{[PhO-BCl}_2\text{]} \longrightarrow [(PhO)_2BCl + BCl_3] \longrightarrow (PhO)_3B + 2BCl_3$$

mole per mole of diphenyl ether, and the fact that this ether will not even form a complex with boron trichloride. A limiting value of the electron density has evidently been reached in the diphenyl ether.

A. Cyclic Ethers

Cyclic ethers afford a variety of results, and have been investigated by Edwards et al. (1955, 1957) and Grimley et al. (1954). Ethylene oxide reacts at $-80°C$ to give 2-chloroethyl dichloroborinate, and as the latter has a propensity to undergo mutual replacement, rather than irreversible decomposition, 2-chloro-ethyl borate is an eventual product. Tetrahydrofuran does form an isolable complex, stable at 20°C, giving,

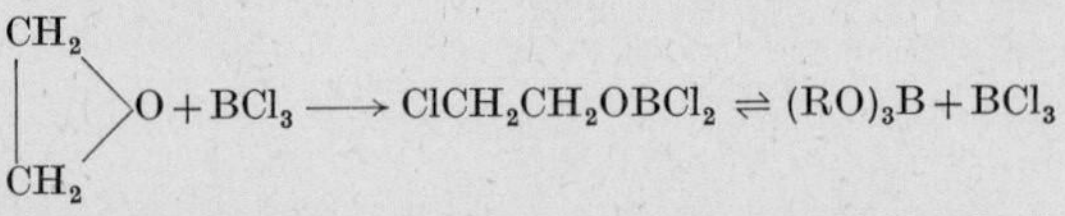

$$\text{(CH}_2\text{-CH}_2\text{)O} + BCl_3 \longrightarrow ClCH_2CH_2OBCl_2 \rightleftharpoons (RO)_3B + BCl_3$$

on being heated, 4-chlorobutyl dichloroborinate, which decomposed to 1,4-dichlorobutane.

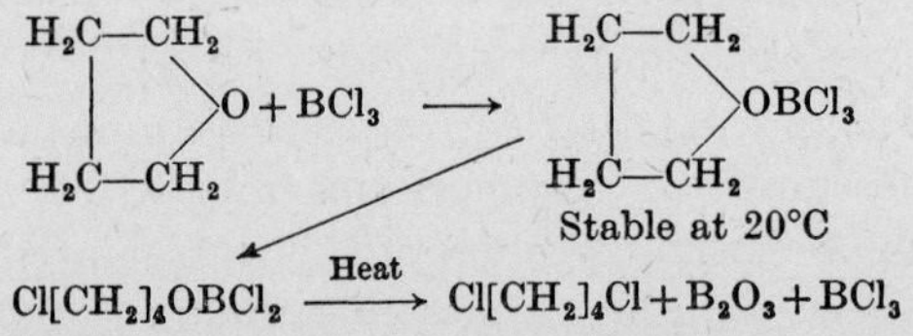

$$\text{(H}_2C\text{-CH}_2\text{, H}_2C\text{-CH}_2\text{)O} + BCl_3 \longrightarrow \text{(H}_2C\text{-CH}_2\text{, H}_2C\text{-CH}_2\text{)OBCl}_3$$

Stable at 20°C

$$Cl[CH_2]_4OBCl_2 \xrightarrow{\text{Heat}} Cl[CH_2]_4Cl + B_2O_3 + BCl_3$$

Remarkable effects were observed in experiments with propylene oxide, epichlorohydrin, trimethylene oxide and tetrahydropyran. Tetrahydropyran formed an isolable complex, stable at 20°C; but trimethylene oxide did not, fission having occurred before analysis could be conducted. Nor was there evidence of complex formation with epichlorohydrin and propylene oxide.

When the reactants are mixed in equimolecular proportions, ω-chloroalkyl dichloroboronites are formed, but other reactions occur concurrently and consecutively, and give rise to dimeric and polymeric esters.

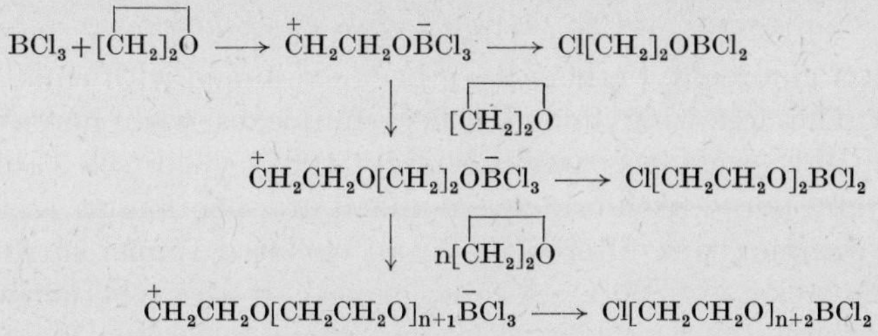

$$BCl_3 + [CH_2]_2O \longrightarrow \overset{+}{C}H_2CH_2O\overset{-}{B}Cl_3 \longrightarrow Cl[CH_2]_2OBCl_2$$

$$\downarrow \quad [CH_2]_2O$$

$$\overset{+}{C}H_2CH_2O[CH_2]_2O\overset{-}{B}Cl_3 \longrightarrow Cl[CH_2CH_2O]_2BCl_2$$

$$\downarrow \quad n[CH_2]_2O$$

$$\overset{+}{C}H_2CH_2O[CH_2CH_2O]_{n+1}\overset{-}{B}Cl_3 \longrightarrow Cl[CH_2CH_2O]_{n+2}BCl_2$$

A similar scheme serves to account for results with three- and four-membered ring systems. With propylene oxide and epichlorohydrin two isomeric dichloroborinates may form, according to mode of fission.

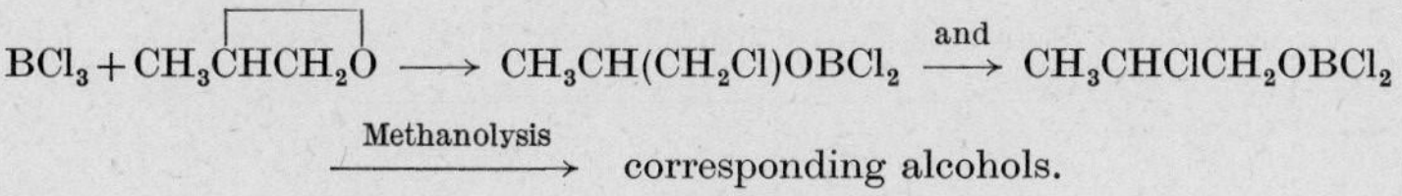

$$BCl_3 + CH_3\overset{\frown}{CHCH_2}O \longrightarrow CH_3CH(CH_2Cl)OBCl_2 \xrightarrow{\text{and}} CH_3CHClCH_2OBCl_2$$

$$\xrightarrow{\text{Methanolysis}} \text{corresponding alcohols.}$$

With five- and six-membered rings, an extension of the mechanism suggested for ethylene oxide is evident.

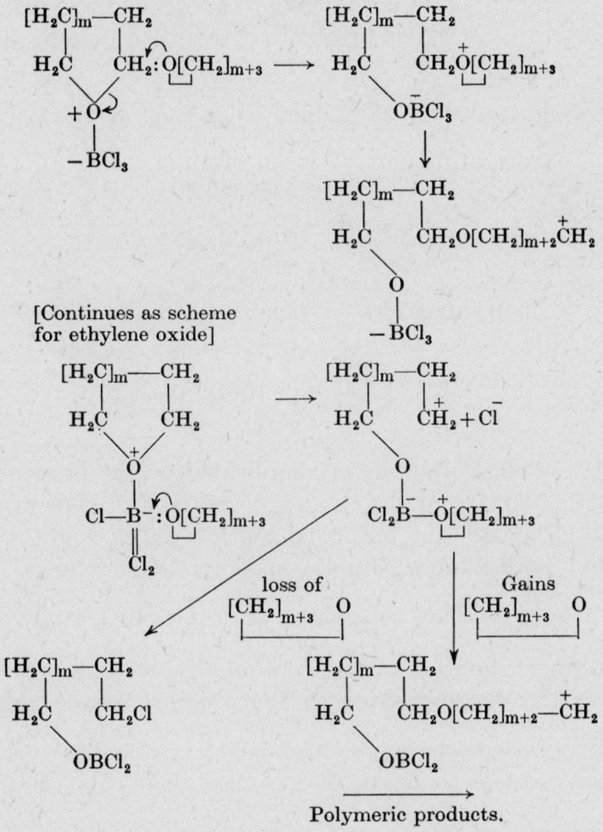

Polymeric products.

B. Alkyl Chloroalkyl Ethers

Alkyl chloroalkyl ethers have been examined with equal discrimination. The following mixed ether complexes with boron trichloride, ROR',BCl_3, were prepared in nearly 100% yield (see Table IV). The ether and boron trichloride were mixed in n-pentane at $-80°C$, and the solid complex was filtered off in an enclosed funnel surrounded by a cooling jacket at $-66°C$. A complex with di-n-butyl ether was isolated, but it had a m.p. below 20°C, and there was some indication of the formation of a complex of n-butyl *iso*butyl ether at $-80°C$.

TABLE IV

Melting Points of Boron Trichloride–Ether Complexes

Complex	m.p.°C
$MeOCH_2Cl . BCl_3$	20
$MeOCH_2CH_2Cl . BCl_3$	35–36
$EtOCH_2CH_2Cl . BCl_3$	20
$(ClCH_2CH_2)_2O . BCl_3$	42–49

The following results were obtained when each complex was heated.

$$ClCH_2CH_2OMe,BCl_3 \longrightarrow ClCH_2CH_2OBCl_2 + MeCl$$
$$100\% \text{ yield}$$

$$ClCH_2CH_2OEt,BCl_3 \longrightarrow ClCH_2CH_2OBCl_2 + EtCl$$
$$92\%$$

$$(ClCH_2CH_2)_2O,BCl_3 \longrightarrow ClCH_2CH_2OBCl_2 + ClCH_2CH_2Cl$$
$$25\%$$

Concurrent dissociation
$$\longrightarrow (ClCH_2CH_2)_2O + BCl_3$$
$$56\%$$

Investigation of these systems is complicated by the propensity of the chloroethyl dichloroborinate to undergo mutual replacement giving the tri-chloroalkyl borate and boron trichloride (Abel *et al.*, 1957). The chloromethyl methyl ether complex decomposed as follows.

$$6ClCH_2OMeBCl_3 \longrightarrow 6MeCl + 3(ClCH_2)_2O + 4BCl_3 + B_2O_3$$

This interesting formation of bis-chloromethyl ether was extended by carrying out the decomposition in the presence of the chloromethyl methyl ether.

$$2ClCH_2OMe,BCl_3 + 4ClCH_2OMe \longrightarrow 6MeCl + 3(ClCH_2)_2O + B_2O_3$$

Bischloromethyl ether did not even form a complex with boron trichloride, nor was there any other reaction. This is in accordance with the low electron density on the oxygen atom, due to the inductive effect of two adjacent chlorine atoms.

H   H

| \

H   C—H   H   C→Cl

Cl←C←O H   Cl←C←O H

H   H

Can complex with BCl_3 Cannot complex

This correlates well with the solubility of hydrogen chloride, which is only 0.072 mole in 1 mole of bis-chloromethyl ether at 0°C.

Diphenyl ether does not react with boron trichloride (Colclough *et al.*, 1955), presumably because of the mesomeric interaction of the oxygen lone-pair electrons with the benzene π-electrons.

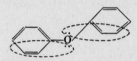

With reference to correlation of donor strength and stability of the complex formed, there is some confusion to which attention may now be directed, although this is not the place for an adequate discussion in detail. The mixed unsubstituted dialkyl ethers mentioned above either give a very elusive complex with boron trichloride, or none at all. The argument must be that the donor power of these ethers is so great by comparison that the electron density of the ether is disturbed so much as to give an irreversible decomposition primarily to an alkyl chloride and a dichloroborinate. When the electron density is not so great, i.e. the donor power is not so great, a complex of some stability is formed. The more disturbance to the electron density distribution in the ether the donation gives, the easier will irreversible decomposition occur. On the other hand the bis-2-chloroethyl ether complex, though stable enough, undergoes more reversible than irreversible decomposition, and, as mentioned, bis-chloromethyl ether does not react at all.

That boron trichloride can be removed from the established complexes just mentioned, either by pyridine or 1-butanol, points to the formula, $RR'OBCl_3$, and not to $R'^+[ROBCl_3]^-$.

$$ROR'BCl_3 + C_5H_5N \longrightarrow C_5H_5NBCl_3 + ROR'$$

$$ROR'BCl_3 + 3Bu^nOH \longrightarrow (Bu^nO)_3B + ROR' + 3HCl$$

C. Methanolysis

It will be clear that in such systems as the cyclic ethers identification of the constituents of mixtures will be a protracted process. A process referred to as *methanolysis* is a useful one in such circumstances. This comprises the addition of methanol in large excess, so that all the boron esters are converted into methyl borate and hydrogen chloride (from chloro-esters), whereupon fractional distillation gives the methanol–methyl borate azeotrope (b.p. 56°C), then methanol (b.p. 65°C) and finally the alcohols, ROH. In this procedure the undesirable use of water is avoided.

The following are illustrations of the application of the procedure.

$$BCl_3 + CH_3.\overline{CH.CH_2.O} \longrightarrow \begin{cases} CH_3.CH(CH_2Cl).OBCl_2 \\ CH_3.CHCl.CH_2.OBCl_2 \end{cases}$$

$$\xrightarrow{\text{Methanolysis}} \quad \text{corresponding alcohols}$$

$$BCl_3 + (3n+3)[\overline{CH_2]_m.O} \longrightarrow (Cl.[CH_2]_m.\{O.[CH_2]_m\}_n.O)_3B$$

$$\xrightarrow{\text{Methanolysis}} 3Cl.\{[CH_2]_m.O\}_{n+1}.H$$

$$(3n+2)[\overline{CH_2]_m-O} + [\overline{CH_2]_m-O},BCl_3 \longrightarrow (Cl-[CH_2]_m-\{O-[CH_2]_m\}_n-O)_3B$$

$$\xrightarrow{\text{Methanolysis}} 3Cl-\{[CH_2]_m-O\}_{n+1}-H$$

D. Unsaturated Ethers

A number of ethers containing alkenyl groups have recently been investigated by Gerrard *et al.* (1956, 1957).

Boron trichloride formed a white solid when mixed with diallyl ether in equimolecular proportions in *n*-pentane at −80°C; but if it were the complex, it was too unstable to characterize, for it decomposed quickly into the primary products of fission, allyl chloride and allyl dichloroborinate, which then behaved in accordance with their individual properties.

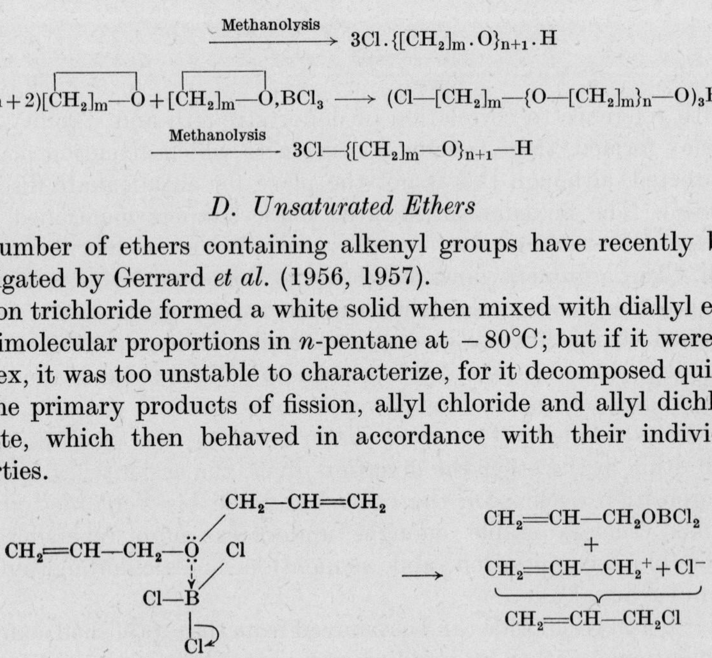

$$ROR + BCl_3 \longrightarrow [1:1 \text{ Complex}] \longrightarrow$$

$$RCl + ROBCl_2 \longrightarrow 2RCl + [BOCl \longrightarrow BCl_3 + B_2O_3]$$

$$3ROR + 2BCl_3 \longrightarrow 6RCl + B_2O_3$$

When the reagents were mixed in the ratio of 3 moles of ether to 2 moles of boron trichloride, the latter produced by the decomposition of the dichloroborinate $3BOCl \longrightarrow B_2O_3 + BCl_3$ was able to effect ether fission on its own account.

Di-2-methylallyl ether behaved similarly to the diallyl ether. The allyl 2-methylallyl ether evidently was cleaved at the methylallyl—O bond, as might have been expected from the greater reactivity of the "central" carbon atom,

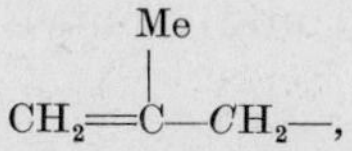

$$\overset{\displaystyle Me}{\underset{\displaystyle |}{CH_2{=}C{-}CH_2{-}}},$$

in the methylallyl group due to the inductive effect of the methyl group.

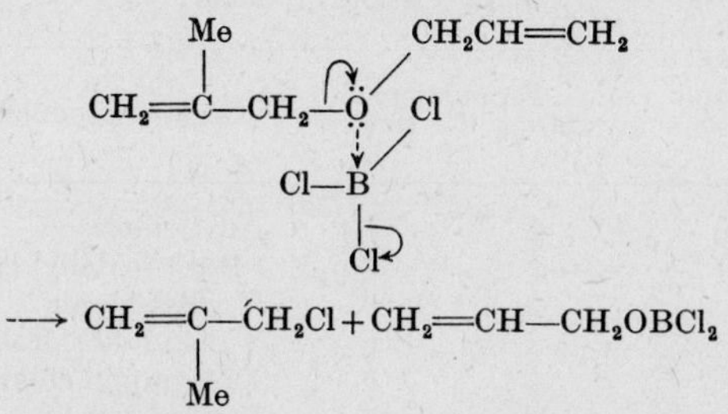

$$\overset{Me}{\underset{|}{CH_2{=}C{-}CH_2{-}}} \overset{CH_2CH{=}CH_2}{\underset{O}{\diagup}} Cl$$

$$Cl{-}B \diagup$$

$$\underset{Cl}{|}$$

$$\longrightarrow \ \underset{\displaystyle Me}{\underset{\displaystyle |}{CH_2{=}C{-}CH_2Cl}} + CH_2{=}CH{-}CH_2OBCl_2$$

In order to discern which way the primary fission has occurred in examples where both dichloroborinates and chloroboronates are too unstable to separate from the alkenyl chloride, advantage is taken of the further fission of the ether by the dichloroborinate, which is assumed to occur by the same mechanism as that by boron trichloride itself. This second fission results in the formation of a chloroboronate, which gives the trialkenyl borate as one of the products of decomposition. This is why experiments were conducted with the proportion of the ether to boron trichloride as $2:1$.

$$3ROR' + 3BCl_3 \longrightarrow ROR', BCl_3 \longrightarrow 3ROBCl_2 + 3R'Cl$$

$$3ROBCl_2 + 3ROR' \longrightarrow 3(RO)_2BCl + 3R'Cl$$

$$3(RO)_2BCl \longrightarrow 3RCl + (RO)_3B + B_2O_3$$

$$6ROR' + 3BCl_3 \longrightarrow 3RCl + (RO)_3B + 6R'Cl + B_2O_3$$

For example, allyl t-butyl ether was cleaved at the t-Bu—O bond; for although the system affords a mixture of t-butyl chloride and allyl chloride, isolation of triallyl borate pointed to the source of the allyl chloride as being the chloroboronate, formed from the dichloroborinate during the fission of a second molecule of ether. The assumption is of course that the fission with the dichloroborinate as reagent followed the same course as with boron trichloride. The fission of the allyl s-butyl ether occurred at the s-Bu—O bond, showing greater electron release from the s-Bu group than from the allyl group, evident from the isolation of triallyl borate. Allyl n-butyl ether, however, afforded n-butyl dichloroborinate (which is stable enough to be clearly isolated) and thus underwent allyl-oxygen fission, in accordance with the greater electron release in the allyl group. From such results it has been concluded that the reactivity of the *central* carbon in the allyl group lies between that in n-butyl and s-butyl. A summary of results is shown in Table V.

TABLE V

Ether Fission Products

Unsaturated ether	Proportions		Organic products isolated
	Ether	BCl$_3$	
Diallyl	3	2	Allyl chloride
Allyl Bu^t	2	1	t-BuCl + triallyl borate
Allyl B^s	2	1	Triallyl borate
Allyl Bu^n	1	1	n-Butyl dichloroborinate
Allyl Pr^n	2	1	Di-n-propyl chloroboronate
Allyl Pr^i	2	1	Triallyl borate
2-Methylallyl			
Allyl	2	1	Triallyl borate
Di-2-methylallyl	3	2	2-methylallyl chloride
Di-3-methylallyl	3	2	Butenyl chloride
Furan	1	1	A polymer
Vinyl ether	3	1	A polymer

The expression "butenyl chloride" in Table V calls for explanation. Owing to the nature of the mesomeric ion produced from both 1-methylallyl and 3-methylallyl groups,

$$CH_3—CH=CH—\overset{+}{C}H_2 \longleftrightarrow CH_3—\overset{+}{C}H—CH=CH_2,$$

the chloride formed will be a mixture of 1-methylallyl chloride and 3-methylallyl chloride, even though only the 3-methylallyl ether was used. The proportion of 1-methylallyl to 3-methylallyl chloride was 34 : 66.

Although small amounts of boron trichloride did not react with 3-methylallyl chloride, especially at the lower temperatures used in the ether fission experiments, polymerization of the organic part slowly occurred at room temperature in the presence of larger amounts of boron trichloride, which was completely recovered after the reactions.

The following reaction scheme was suggested.

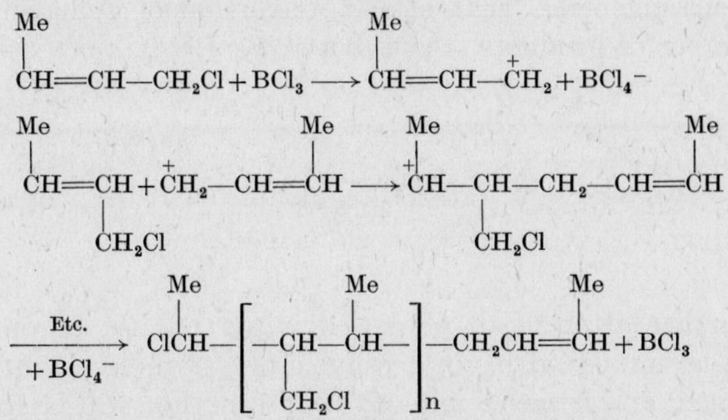

$$CH=CH-CH_2Cl + BCl_3 \longrightarrow \overset{Me}{CH}=CH-\overset{+}{CH_2} + BCl_4^-$$

$$\overset{Me}{CH}=\overset{+}{CH} + \overset{Me}{CH_2}-\overset{Me}{CH}=CH \longrightarrow \overset{Me}{CH}-CH-CH_2-\overset{Me}{CH}=CH$$

$$\xrightarrow[+BCl_4^-]{Etc.} \quad ClCH-\left[-\overset{Me}{CH}-CH-\right]-CH_2CH=\overset{Me}{CH} + BCl_3$$

This points to a possibility of using boron trichloride as a cationic catalyst.

Allyl phenyl ether calls for special mention because it undergoes the ortho-Claisen rearrangement on fission with boron trichloride. The product isolated from the allyl phenyl ether (3 moles) and boron trichloride (1 mole), mixed at − 80°C, was tri-o-allylphenyl borate, b.p. 188°C/0.05 mm, and this on methanolysis gave o-allylphenol.

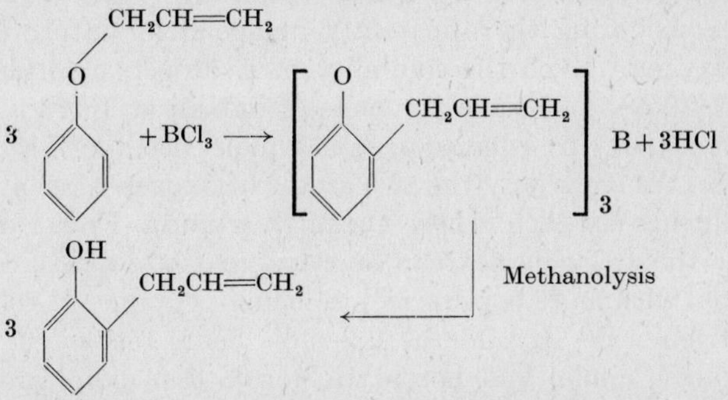

An example of one of the modifications of the Claisen rearrangement is the conversion of allyl phenyl ether into o-allylphenol by heating the ether at 200°C for some 6 hours (see Gerrard *et al.*, 1957 for other refs.). Therefore the boron trichloride procedure seems to offer an attractive alternative. The mechanism suggested is shown in the diagram.

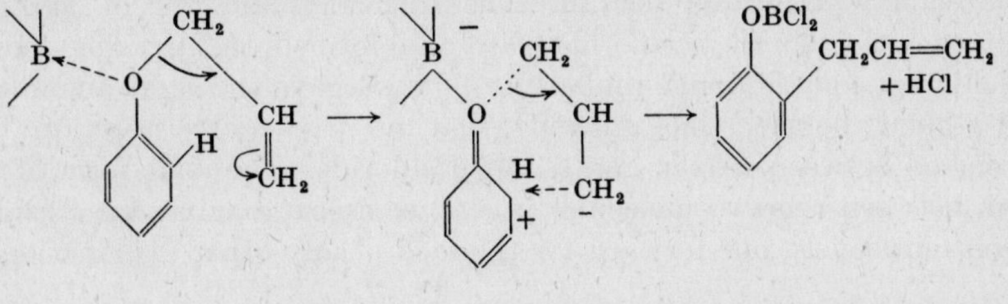

There appears to be only one other mention of a Lewis acid catalysed Claisen rearrangement, that of the conversion of guaiacol allyl ether into a number of products (Bryusova *et al.*, 1941).

In a further attempt to prepare trivinyl borate, boron trichloride (1 mole) was mixed with vinyl ethyl ether (3 moles); but a polymer was obtained, and it seems unlikely that a method will be found. From Table V it will be noticed that furan also formed a polymer.

E. Dioxygen Cyclic Ethers

Very little indeed has been published on the interaction of boron halides and dioxygen cyclic ethers. 1,4-Dioxan is naturally one which has been selected for study. At first it appeared to form not more than a 1 : 1 complex with boron trichloride (Holliday *et al.*, 1952), which is surprising considering the apparently independent nature of the two atoms of oxygen. Even the solubility of hydrogen chloride (Gerrard *et al.*, 1956, 1959), 1.05 mole per mole of dioxan at 10°C, is not much greater than that in *n*-butanol, 0.964 mole per mole at the same temperature. In some way the two atoms of oxygen have a significant mutual influence on each other. On further study, Frazer *et al.* (1958) found that the 1 : 1 complex can be converted into a 3 : 2 complex on being treated with more boron trichloride in *n*-pentane at − 80°C, excess of boron trichloride beyond the 0.5 mole being recovered. Similarly when dioxan (1 mole) and boron trichloride (1.5 mole) are mixed in *n*-pentane at − 80°C, the 3 : 2 complex is precipitated, and any boron trichloride added in excess of this amount is recovered. Furthermore the 3 : 2 complex is readily converted into the 1 : 1 complex by the addition of the right amount of dioxan. Both complexes are white solids which fume in air and are readily hydrolysed by water to boric acid, hydrochloric acid and dioxan. The calculated amount of pyridine removed the boron trichloride part and formed the pyridine boron trichloride complex, and similarly *n*-butanol gave the right amount of tri-*n*-butyl borate, thus showing that no irreversible reaction had occurred between dioxan and boron trichloride. The ionic form of the link between the two molecules of 1 : 1 complex to give the 3 : 2 has been tentatively put forward for the lack of any other explanation.

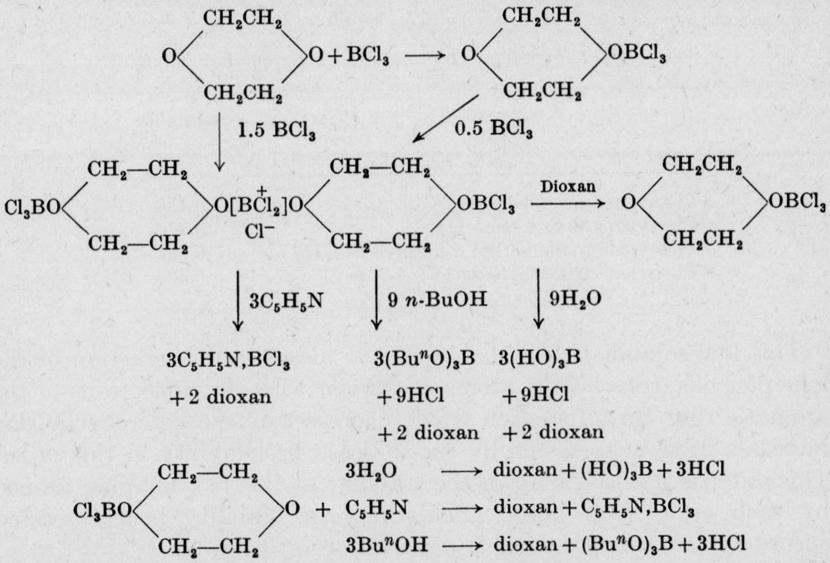

$$
\underset{\substack{\text{CH}_2\text{CH}_2}}{\overset{\text{CH}_2\text{CH}_2}{O}}\!\!\diagdown O + BCl_3 \longrightarrow \underset{\substack{\text{CH}_2\text{CH}_2}}{\overset{\text{CH}_2\text{CH}_2}{O}}\!\!\diagdown OBCl_3
$$

$$\big\downarrow 1.5\ BCl_3 \qquad\qquad 0.5\ BCl_3$$

$$
Cl_3BO\!\!\diagdown \underset{\text{CH}_2-\text{CH}_2}{\overset{\text{CH}_2-\text{CH}_2}{}}\!\!\diagup O[\overset{+}{B}Cl_2]O \underset{Cl^-}{} \diagdown \underset{\text{CH}_2-\text{CH}_2}{\overset{\text{CH}_2-\text{CH}_2}{}}\!\!\diagup OBCl_3 \xrightarrow{\text{Dioxan}} \underset{\text{CH}_2\text{CH}_2}{\overset{\text{CH}_2\text{CH}_2}{O}}\!\!\diagdown OBCl_3
$$

$$\big\downarrow 3C_5H_5N \qquad \big\downarrow 9\ n\text{-BuOH} \qquad \big\downarrow 9H_2O$$

$$
\begin{array}{ccc}
3C_5H_5N,BCl_3 & 3(Bu^nO)_3B & 3(HO)_3B \\
+2\ \text{dioxan} & +9HCl & +9HCl \\
 & +2\ \text{dioxan} & +2\ \text{dioxan}
\end{array}
$$

$$
Cl_3BO\!\!\diagdown \underset{\text{CH}_2-\text{CH}_2}{\overset{\text{CH}_2-\text{CH}_2}{}}\!\!\diagup O \quad + \quad
\begin{array}{l}
3H_2O \longrightarrow \text{dioxan} + (HO)_3B + 3HCl \\
C_5H_5N \longrightarrow \text{dioxan} + C_5H_5N,BCl_3 \\
3Bu^nOH \longrightarrow \text{dioxan} + (Bu^nO)_3B + 3HCl
\end{array}
$$

Although the systems involving boron tribromide will be discussed separately, it is convenient to mention now that a 1:1 complex between dioxan and the tribromide was obtained as a white solid (Frazer *et al.*, 1958), which with water, pyridine and *n*-butanol behaved analogously to the boron trichloride complex. The 1:1 complex would not react further with boron tribromide, but there were indications that two molecules of the complex annex 1 mole of boron trichloride.

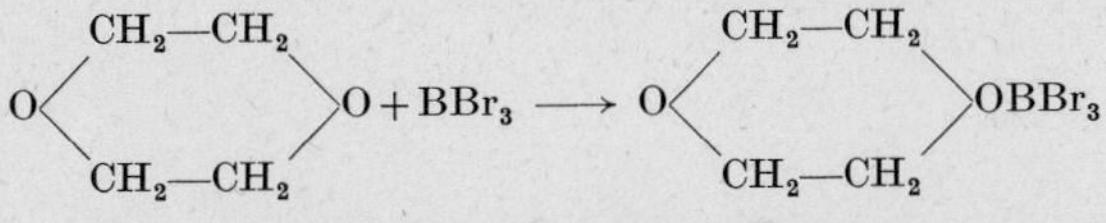

$$
\underset{\text{CH}_2-\text{CH}_2}{\overset{\text{CH}_2-\text{CH}_2}{O}}\!\!\diagdown O + BBr_3 \longrightarrow \underset{\text{CH}_2-\text{CH}_2}{\overset{\text{CH}_2-\text{CH}_2}{O}}\!\!\diagdown OBBr_3
$$

Other systems under investigation (Cooper *et al.*, 1960) are the dioxygen ethers derived from catechol.

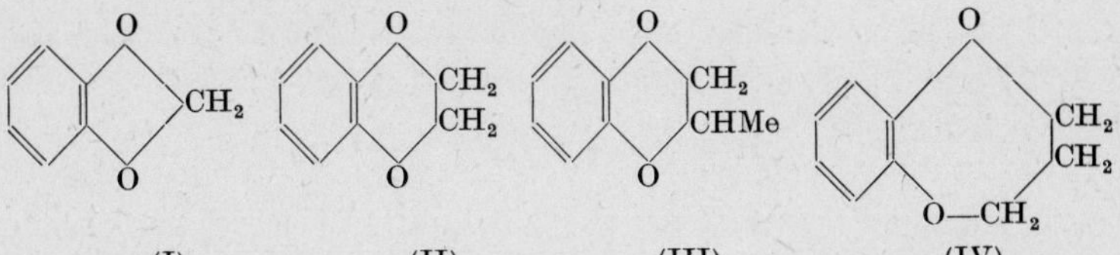

(I) (II) (III) (IV)

The electron density on oxygen is remarkably small, as indicated by the solubility of hydrogen chloride as shown in Table VI (Gerrard *et al.*, 1960).

TABLE VI

Solubility of Hydrogen Chloride in *o*-Phenylene Ethers at 0°C

Ether	Solubility moles per mole
o-Phenylene methylene (I)	0.130
o-Phenylene dimethylene(II)	0.240
o-Phenylene monomethyl dimethylene (III)	0.255
o-Phenylene trimethylene (IV)	0.438

This low solubility is probably due to mesomeric interaction of the lone-pair electrons on the atoms of oxygen with the π-electrons of the aromatic ring, an interaction which decreases as the heterocyclic ring increases in size, as shown by the increase in solubility in this order. This order is precisely that of the stability of the 1 : 1 complex formed by each ether with boron trichloride, the stability being assessed according to ease of reversible dissociation on heating.

1,3-Dioxalane and its derivatives are being examined, and in these systems the products of irreversible decomposition are being examined in detail.

Chapter IV

THE BORON CHEMISTRY OF CARBOXYLIC ACIDS, ESTERS AND CARBONYL COMPOUNDS

I. Acids and Esters

EARLY experiments (Pictet *et al.*, 1903) on the interaction of boric acid and acetic anhydride led to the assignment of the formula $(CH_3COO)_3B$ for the "boron acetate" produced. This material has been used from time to time for the production of trialkyl borates (Cook *et al.*, 1950), and except for one investigator (Dimroth, 1925), the assigned formula was accepted until it was strongly challenged by Gerrard *et al.* (1954, 1956) in recent work.

Based on a titration method the ratio acetic acid : boric acid was found to be 2:1 and the "pyroacetate" formula for the so-called "boron acetate" was assigned (Dimroth, 1925):

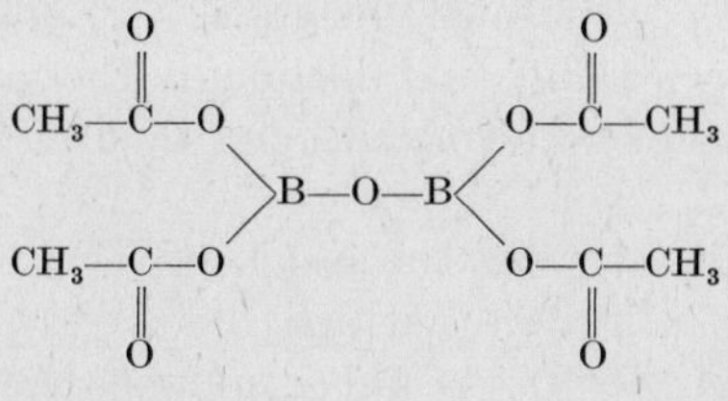

now called tetraacetyl diborate. This work was apparently ignored, for those using "boron acetate" as a reagent for the production of alkyl borates still gave the 3:1 formula $(CH_3CO_2)_3B$.

"Boron acetate" was prepared by Gerrard *et al.* (1954) a number of times by each of the following methods:

$$B(OH)_3 + 3.6(CH_3CO)_2O \longrightarrow \text{tetraacetyl diborate}$$
$$2B(OH)_3 + 5(CH_3CO)_2O \longrightarrow \text{tetraacetyl diborate}$$
$$2B(OH)_3 + 10(CH_3CO)_2O \longrightarrow \text{tetraacetyl diborate}$$
$$B(OH)_3 + CH_3COCl \longrightarrow \text{tetraacetyl diborate}$$
$$BCl_3 + 3CH_3COOH \longrightarrow \text{tetraacetyl diborate}$$

In every example quoted the product had acetic acid : boric acid ratio of 2:1, never 3:1, when determined by a specially selected titration method (Jones, 1898; Morgan *et al.*, 1924). In one example, boric acid (2 moles) and acetic anhydride (10 moles) gave the tetra-acetyl diborate (99.3% yield), the acetic anhydride added in excess of the stoicheiometric amount being recovered.

$$2B(OH)_3 + 5(CH_3CO)_2O = (CH_3CO_2)_2BOB(CH_3CO_2)_2 + 6CH_3COOH$$

Concurrently, however, it was stated (Ahmad *et al.*, 1954) that the product of the interaction of boric acid and acetic anhydride was the triacetate, $(CH_3CO_2)_3B$; although the method of titration was not described in sufficient detail to be followed by others. More recent detailed work by Hayter *et al.* (1957) confirms beyond doubt the elusiveness of the triacetate, and indeed this could not be isolated, even though the prescribed methods of Pictet *et al.* (1903) and Ahmad *et al.* (1954) were carefully followed. Boric acid (a little more than 2 moles) and acetic anhydride were heated together at 50–100°C with vigorous stirring. Colourless needles, recrystallized in suitable ways, had m.p. 147–148°C, fairly agreeing with the views of workers assigning the tetraacetyl formula, but in gross disagreement with the m.p. of about 120°C cited by others (Pictet *et al.*, 1903; Ahmad *et al.*, 1954; Cook *et al.*, 1950; and Meerwein *et al.*, 1932). The molecular weight by ebullioscopic procedure in chloroform, the titration procedure with a Beckmann pH meter, and X-ray powder diffraction patterns all point exclusively to the tetraacetyl formula. It had previously been mentioned (Kahovec, 1939) that although Raman frequencies for a number of boron esters conformed among themselves, "boron acetate" did not.

There is some relevance with the assigned "Tri" formula for "glyceryl borate" (Ahmad *et al.*, 1954), which has been strongly refuted by Gerrard *et al.* (1954, 1958), on the grounds that not only does the

"Tri" structure   "Diborate" structure

"tri" structure entail an unlikely distortion of valency angles (Fischer–Herschelder models), but experimental evidence, molecular weight by ebullioscopic procedure, infra-red spectrum, interaction with thionyl chloride and pyrolysis, point exclusively to the diborate formula.

In another series of experiments by Frazer *et al.* (1955) it was found that whereas an ethyl ester of a hydroxy-acid (glycollate, β-hydroxy-propionate, malate, and lactate) (3 moles) and boron trichloride (1 mole) readily afford the borate, e.g. $[CH_3CHO(CO_2Et)]_3B$, in high yield, attempts to prepare the dichloroborinate, or the chloroboronate, were unsuccessful, because of the fission of the carbethoxyl group which gave ethyl chloride. Ethyl acetate formed a solid complex stable at 20°C, there being no permanent severance of the B—Cl or a O—C bond at this stage; for when 1-octanol was added, hydrogen chloride, 3 moles, tri-1-octyl borate, and ethyl acetate were isolated. On being heated at 100°C the complex decomposed to acetyl chloride and presumably to ethyl dichloroborinate, judging from the final decomposition products, ethyl chloride, boron trichloride, and boric oxide.

In accordance with the properties of phenyl dichloroborinate, phenyl acetate gave triphenyl borate and acetyl chloride.

$$EtOAc + BCl_3 \longrightarrow EtOAcBCl_3 \longrightarrow$$
$$AcCl + [EtOBCl_2 \longrightarrow EtCl + BCl_3 + B_2O_3]$$
$$PhOAc + BCl_3 \longrightarrow PhOAcBCl_3 \longrightarrow$$
$$AcCl + [PhOBCl_2 \longrightarrow BCl_3 + B(OPh)_3]$$

This work was extended by Gerrard *et al.* (1956) mainly to acetates CH_3CO_2R' in which R' was varied. Three modes of decomposition were discerned when the complexes were heated:

(i) $RCO_2R'BCl_3 \xrightarrow{\text{Dissociation}} RCO_2R' + BCl_3$

(ii) $RCO_2R'BCl_3 \xrightarrow{\text{Fission A}} R.COCl + R'OBCl_2$

(iii) $2RCO_2R'BCl_3 \xrightarrow[\text{+ } RCO_2R']{\text{Fission B}} RCOOB—O—BOCOR + 3R'Cl + R.COCl$
$$\phantom{2RCO_2R'BCl_3 \xrightarrow{\text{Fission B}}} \underset{Cl}{|} \quad \underset{Cl}{|}$$

For simple primary alkyl groups, Et,*n*-Bu, fission occurs by mechanism (ii), the dichloroborinate decomposing in accordance with its own properties.

When R' was *iso*butyl or neopentyl, a tertiary alkyl chloride was formed, in accordance with the mode of decomposition of the dichloroborinate. When R (in RCO) was Pr or CMe_3, ethyl dichloroborinate was stabilized by the acyl chloride, PrCOCl, Me_3CCOCl, from which it was difficult to separate. Benzyl acetate gave a complex which decomposed vigorously at 0°C, and gave a hydrocarbon polymer, similar to, but not identical with, one formed by the interaction of boron trichloride and benzyl alcohol. Substitution of hydrogen by chlorine at the 2-carbon atom of R' did not alter the mode of fission, but did give a different reaction sequence for the reaction of the dichloroborinate, again in accordance with the properties of such examples.

$$2ClCH_2CH_2OBCl_2 \longrightarrow (ClCH_2CHO)_2BCl + BCl_3$$
$$3(ClCH_2CH_2O)_2BCl \longrightarrow 2(ClCH_2CH_2O)_3B + BCl_3$$

Substitution of chlorine at the 1-carbon atom of R in RCOOR′ caused dissociation (i) to accompany decomposition (ii). Whereas ethyl dichloroacetate was recovered in 75% yield, ethyl trichloroacetate did not even form a complex. Ethyl benzoate was also partly recovered, because of dissociation (i).

It is remarkable that when R′ is a secondary group (examined only for R=CH₃), the reaction sequence is shown in (iii), and when the dichlorodiacetyl diborate was heated, acetyl chloride and boric oxide were obtained; whereas with acetic acid, the tetraacetyl diborate was formed.

$$(AcOBCl)_2O \xrightarrow{\text{Heat}} 2AcCl + B_2O_3$$

$$(AcOBCl)_2O + 2AcOH \longrightarrow [(AcO)_2B]_2O + 2HCl$$

t-Butyl acetate in *n*-pentane gave an immediate precipitate of dichlorodiacetyl diborate when treated with boron trichloride.

Certain trifluoroacetoxyboron compounds of indefinite composition are mentioned by Muetterties (1957), and tris-trifluoroacetyl borate $(CF_3COO)_3B$ has been prepared by Gerrard *et al.* (1958) from boron trichloride and trifluoroacetic acid. It is a solid, m.p. 88°C with decomposition; and sublimation at $100°C/10^{-4}$ mm gives tetrakistrifluoroacetyl diborate. Further studies might lead to an explanation for the elusiveness of triacetyl borate.

Kuskov (1956) refers to the use of tetra-acetyl diborate in connection with the condensation of aldehydes with acetic anhydride; and Ploquin (1956) refers to boro-oxalates.

II. Carbonyl Compounds

A. Aldehydes

There appears to be only one report (Frazer *et al.*, 1957) so far on the systematic study of boron trichloride–aldehyde systems. Eleven aldehydes, each having different electronic and steric features, were examined.

Ten of the aldehydes reacted with vigour even at −80°C. Matter volatile at 20°C/15 mm was removed, and the weight of residue gave the molal proportion of reactants. For each of the majority of aldehydes there was but one reaction sequence:

$$6CH_3CHO + 2BCl_3 \longrightarrow 3(CH_3CHCl)_2O + B_2O_3$$

Acetaldehyde   Bis-1-chloroethyl ether.

Similarly:

n-Butyraldehyde ⟶ Bis-1-chlorobutyl ether.

isoButyraldehyde ⟶ Bis-1-chloro-2-methylpropyl ether.

Two mechanisms of the initial attack present themselves.

$$[A]$$

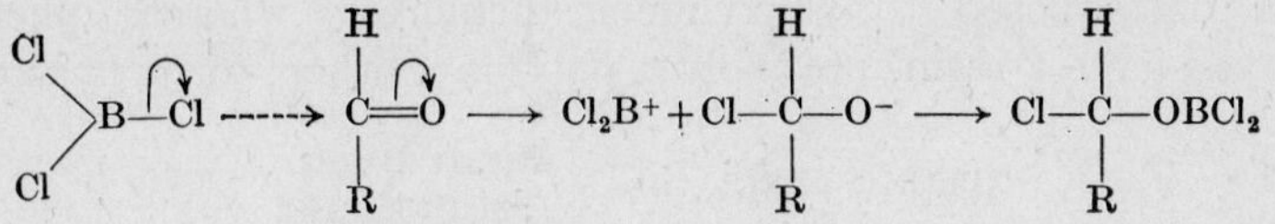

$$\underset{Cl}{\overset{Cl}{>}}B\!-\!Cl \dashrightarrow \underset{R}{\overset{H}{\underset{|}{C}}}\!=\!O \longrightarrow Cl_2B^+ + Cl\!-\!\underset{R}{\overset{H}{\underset{|}{C}}}\!-\!O^- \longrightarrow Cl\!-\!\underset{R}{\overset{H}{\underset{|}{C}}}\!-\!OBCl_2$$

$$[B]$$

$$R\!-\!\overset{H}{\underset{|}{C}}\!=\!O \longrightarrow R\!-\!\underset{Cl}{\overset{H}{\underset{|}{C}}}\!-\!OBCl_2$$

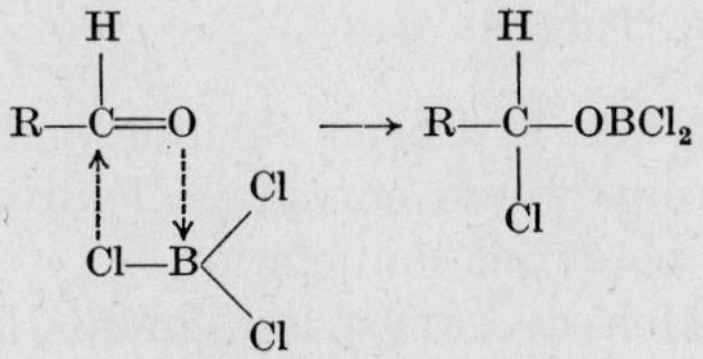

4-Centre broadside reaction.

Of the two, the 4-centre reaction [B] appears more attractive, as the first would be against the usual trend for boron to allow the loss of electrons so easily. Contrary to expectations, the formation of a co-ordination compound, $RCHOBCl_3$, does not appear to be an alternative, for bromal does not react with boron trichloride. This aldehyde is hindered by F-strain from reaction by the first two mechanisms, and this could account for non-reaction. On the other hand, there is no steric hindrance to formation of the complex.

Crotonaldehyde formed a white solid dichloroborinate, but there was some doubt about the choice of its being $CH_3CHClCH=CHOBCl_2$ or $CH_3CH=CHCHClOBCl_2$. In accordance with electron release effect from the vinyl group, the dichloroborinate decomposed on being heated to give $CH_3CHClCH=CHCl$.

Monochloroacetaldehyde gave bis-1,2-dichloroethyl ether and tris-1, 2-dichloroethyl borate, which on further heating gave the bis-ether.

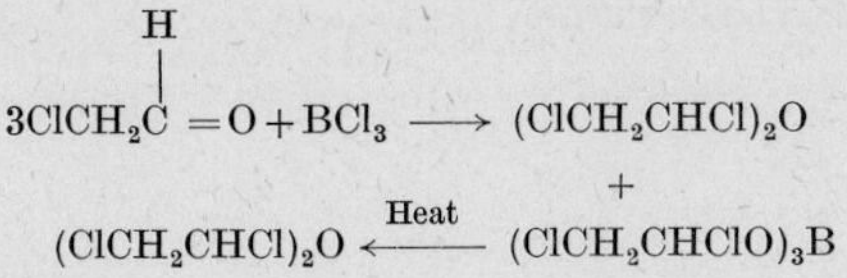

$$3ClCH_2\overset{H}{\underset{|}{C}}=O + BCl_3 \longrightarrow (ClCH_2CHCl)_2O$$
$$+$$
$$(ClCH_2CHCl)_2O \xleftarrow{\text{Heat}} (ClCH_2CHClO)_3B$$

This shows the beginning of observable reduction in reactivity of the central carbon atom

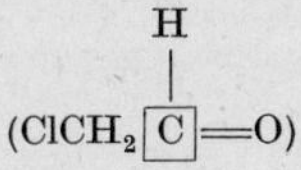

$$(ClCH_2\boxed{\overset{H}{\underset{|}{C}}}=O)$$

3

caused by the electron attracting chlorine atom. Consequently, dichloroacetaldehyde gave the borate $(Cl_2CHCHClO)_3B$, b.p. 116–118°C/ 11 mm, stable at 200°C, and trichloroacetaldehyde gave the borate, $(Cl_3CCHClO)_3B$, b.p. 98–105°C/0.5 mm, of similar stability.

Benzaldehyde gave a mixture of dichloroborinate and chloroboronate which decomposed to benzylidene dichloride, whereas phenylacetaldehyde gave a resin, probably a polyvinyl ether; for it is a tendency

$$3PhCHO + 2BCl_3 \longrightarrow \begin{cases} PhCHClOBCl_2 \\ (PhCHClO)_2BCl \end{cases}$$

$$B_2O_3 + PhCHCl_2 \longleftarrow \underline{\hspace{4cm}}$$

of α-chloro-ethers to lose hydrogen chloride and form vinylic ethers, and the β-phenyl group would encourage hydrogen chloride elimination, because of the potential conjugation.

β-Phenylpropionaldehyde gave tris-1-chloro-3-phenyl-propyl borate $(PhCH_2CH_2CHClO)_3B$, which on attempted distillation gave unidentified products.

Thus the play is on the reactivity of the *central* carbon atom, and the following general scheme may be formulated:

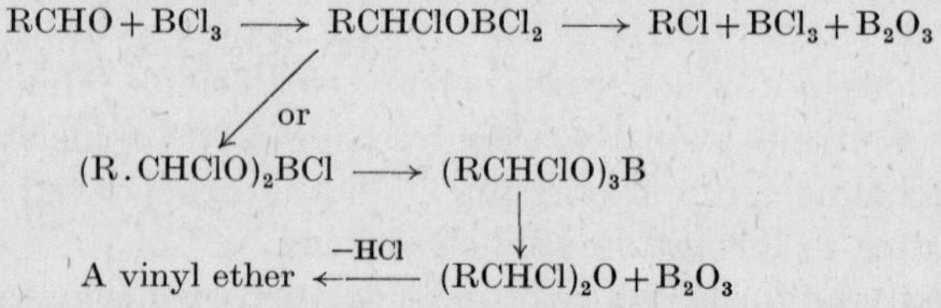

$$RCHO + BCl_3 \longrightarrow RCHClOBCl_2 \longrightarrow RCl + BCl_3 + B_2O_3$$

$$\swarrow \text{ or}$$

$$(R.CHClO)_2BCl \longrightarrow (RCHClO)_3B$$

$$\downarrow$$

$$\text{A vinyl ether} \overset{-HCl}{\longleftarrow} (RCHCl)_2O + B_2O_3$$

Benzaldehyde and crotonaldehyde have the greatest electron release to the α-carbon atom, and consequently the dichloroborinates gave the corresponding chloride, $PhCHCl_2$, etc. On the other hand, the dichloro- and trichloroacetaldehydes go no further than the borates at temperatures below 200°C.

A probable mechanism of conversion of the borate into the ether is formulated as an SNi reaction.

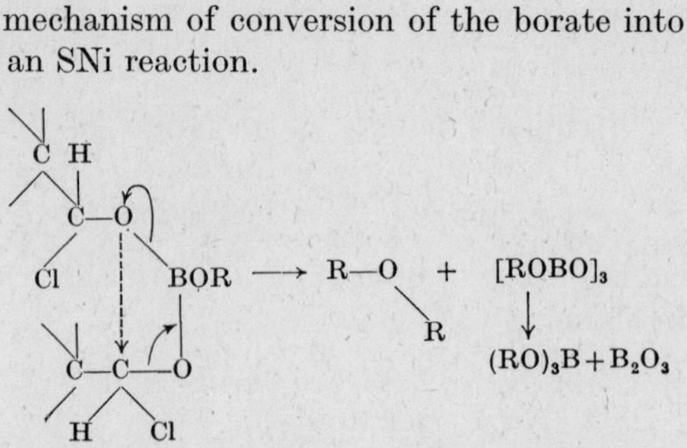

The α-chlorine atom must have a strong influence in reducing electron density on oxygen, and therefore in reducing back–co-ordination onto boron. Therefore an alternative mechanism based on a 4-centre broadside approach must not be disregarded.

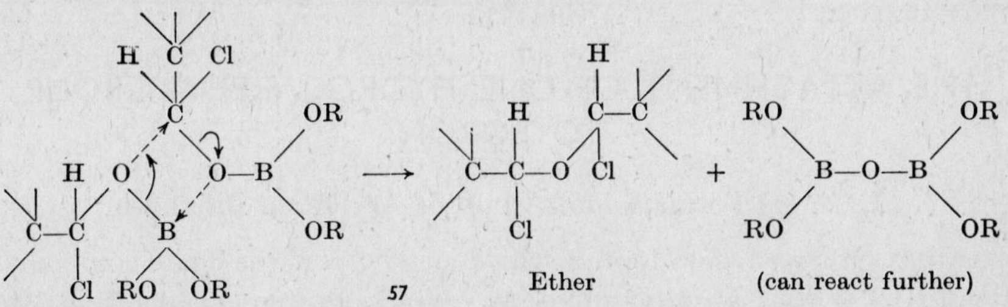

This formation of α-chloro-ethers is not unattractive in comparison with other methods of preparation, and the borates are esters of the elusive α-chloro-alcohols, RCHClOH.

B. Ketones

The interaction of boron trichloride with ketones is in the very early stages of investigation by Gerrard and his co-workers.

CHAPTER V

THE ATTACHMENT OF ONE HYDROCARBON GROUP TO BORON

I. Direct Formation of Alkyl- or Arylboron Dihalides

THE procedure to be adopted will depend upon the final compound required. If it is an advantage, as indeed it often is, to retain two readily functional chlorine atoms on boron, then a procedure for the replacement of one chlorine atom in boron trichloride to give an alkyl or arylboron dichloride, which can be immediately isolated as such, is the one to be sought. Otherwise it becomes necessary to approach the dichloride by a series of steps involving hydrolysis, dehydration, and attachment of chlorine, or involving attachment of three alkyl or aryl groups followed by stepwise replacement of these by chlorine.

Although dimethylzinc has been used to prepare methylboron dichloride, zinc alkyls are not in general attractive reagents.

$$ZnMe_2 + 2BCl_3 \longrightarrow ZnCl_2 + 2MeBCl_2$$

Phenylboron dichloride was prepared by heating diphenylmercury in a sealed tube for 1 hour at 180–120°C (Michaelis *et al.*, 1880, 1901), and phenylboron dibromide was similarly prepared.

$$HgPh_2 + 2BCl_3 \longrightarrow 2PhBCl_2 + HgCl_2$$
$$HgPh_2 + 2BBr_3 \longrightarrow 2PhBBr_2 + HgBr_2$$

The preparation of β-chlorovinylboron dichloride by a vapour phase reaction between acetylene and boron trichloride at 150–200°C in the presence of catalytic mercurous chloride, as shown by Arnold (1946), see also Ruigh *et al.* (1956), is reminiscent of the preparation of phenylboron dichloride from boron trichloride and benzene passed over catalytic palladium at 500–600°C, as described by Pace (1929).

$$C_2H_2 + BCl_3 \xrightarrow[\text{150-200°C}]{\text{Mercurous chloride}} Cl-\overset{\overset{\displaystyle H}{|}}{C}=\overset{\overset{\displaystyle H}{|}}{C}-BCl_2$$

b.p. 55°C/165 mm

$$C_6H_6 + BCl_3 \xrightarrow[\text{500-600°C}]{\text{Palladium}} C_6H_5BCl_2 + HCl$$

Researches into the formation of boron polymers (Ruigh *et al.*, 1956) led to a detailed investigation of the catalytic boronation of benzene

58

by the above palladium procedure, because this appeared to have the potentiality of technical development. The process proved fickle, and the final report offered little promise. Boron trichloride heated with an aromatic hydrocarbon in the presence of aluminium powder and a trace of aluminium chloride in an autoclave gives an arylboron dichloride (Muetterties, 1959), and a corresponding result is obtained with boron tribromide or triiodide. An attractive process would be the production of phenylboron dibromide in 50% yield by heating boron tribromide and benzene under reflux in the presence of aluminium bromide were it not for the cost of boron tribromide (Bujwid et al., 1959).

$$C_6H_6 + BBr_3 \xrightarrow{\text{AlBr}_3} C_6H_5BBr_2 + HBr$$
b.p. 100–101°C/19 mm,
m.p. 25–28°C

Phenoxyphenylboron dibromide, b.p. 124–126°C/0.1 mm, was similarly prepared without the aid of aluminium bromide.

$$C_6H_5OC_6H_5 + BBr_3 \xrightarrow[\text{reflux}]{170-180°C} C_6H_5OC_6H_4BBr_2 + HBr$$

Gerrard et al. (1960) have found that by far the most attractive direct method for the preparation of phenylboron dichloride is that using tetraphenyltin, which is now commercially available.

$$SnPh_4 + 4BCl_3 \longrightarrow 4PhBCl_2 + SnCl_4$$
b.p. 66°C/11 mm

This is based on the use of tetravinyltin for a similar formation of vinylboron dichloride reported by Brinckman et al. (1959). Much remains to be done in extending the method; for tetra-n-butyltin released only one n-butyl group.

$$n\text{-Bu}_4Sn + BCl_3 \longrightarrow n\text{-BuBCl}_2 + n\text{-Bu}_3SnCl$$
b.p. 66°C/98 mm

A. Properties of Alkyl- or Arylboron Dihalides

The properties of such alkyl- or arylboron dihalides may be illustrated best by referring mainly to phenylboron dichloride and to n-butylboron dichloride.

Whereas the two halogen atoms are vigorously removed by hydrolysis, the C—B link remains firm as long as oxidants are absent. The product is an aryl- or alkylboronic acid.

$$PhBCl_2 + H_2O \longrightarrow PhB(OH)_2 + 2HCl$$
Phenylboronic acid

$$n\text{-BuBCl}_2 + H_2O \longrightarrow n\text{-BuB(OH)}_2 + 2HCl$$
n-Butylboronic acid

The effect of alcoholysis depends upon the reactivity of the alcoholic carbon atom (p. 1). Thus, whereas alcohols of ordinary reactivity such as n-butanol afford the dialkyl ester, reactive alcohols such as t-butanol and 1-phenylethanol afford the alkyl chloride and boronic acid.

$$PhBCl_2 + 2 \; EtOH \longrightarrow PhB(OEt)_2 + 2HCl$$

$$PhBCl_2 + 2 \; n\text{-BuOH} \longrightarrow PhB(OBu^n)_2 + 2HCl$$

$$n\text{-BuBCl}_2 + 2 \; n\text{-BuOH} \longrightarrow n\text{-BuB}(OBu^n)_2 + 2HCl$$

$$PhBCl_2 + 2 \; t\text{-BuOH} \longrightarrow 2 \; t\text{-BuCl} + PhB(OH)_2$$

There is an almost universal production of esters when pyridine (p. 26), or some such reagent, is present to prevent the formation of hydrogen chloride (Michaelis et al., 1882; Brindley et al., 1955, 1956).

$$PhBCl_2 + 2 \; t\text{-BuOH} : NC_5H_5 \longrightarrow PhB(OBu^t)_2 + 2C_5H_5NHCl$$

$$\alpha\text{-}C_{10}H_7BCl_2 + 2 \; MeONa \longrightarrow \alpha\text{-}C_{10}H_7B(OMe)_2 + 2NaCl$$

Phenylboron dichloride undergoes a mutual replacement reaction with, e.g. di-n-butyl phenylboronate to afford n-butyl phenylchloroborinate, $PhBCl(OBu^n)$, and the corresponding butyl compound, $n\text{-BuBCl}(OBu^n)$, can be similarly made.

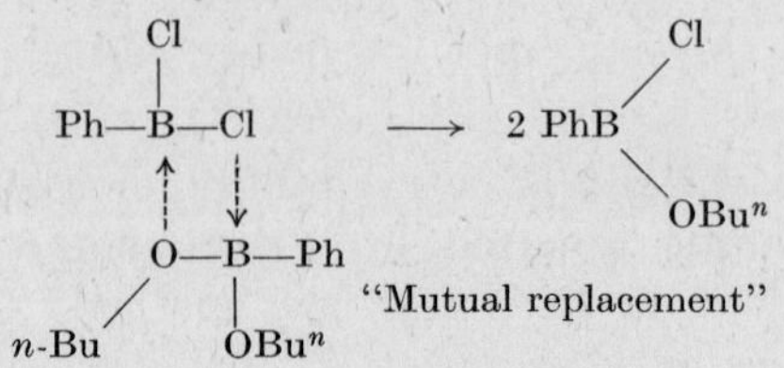

"Mutual replacement"

Such chloro-esters can also be prepared by the controlled interaction with an alcohol.

$$PhBCl_2 + ROH \longrightarrow PhBCl(OR) + HCl$$

Similar reactions have been described for phenylboron dibromide by Abel et al. (1957), for example:

$$PhBBr_2 + n\text{-BuOH} \longrightarrow PhBBr(OBu^n) + HBr$$

Perhaps of special interest is the direct conversion of phenylboron dihalide (Cl,Br) into the phenylboronic anhydride trimer by controlled addition of one mole of water, as shown by Abel et al. (1956, 1957).

$$3PhBCl_2 + 3H_2O \longrightarrow (PhBO)_3 + 6HCl$$

$$3PhBBr_2 + 3H_2O \longrightarrow (PhBO)_3 + 6HBr$$

The replacement of one atom of chlorine in boron trichloride by an alkyl or an aryl group decreases the power of the boron compound to split ethers, according to Dandegaonker *et al.* (1957). Thus phenylboron dichloride does not react with ethers at 20°C, but does so with certain exceptions at higher temperatures.

$$PhBCl_2 + ROR' \longrightarrow PhBCl(OR) + R'Cl$$

The suggested carbon cation mechanism is based on the fate of the more electron-releasing group, which appears in the alkyl chloride, R'Cl, the occurrence of a molecular rearrangement when R' = *iso*-butyl; and on the occurrence of preponderant inversion with much loss in optical activity when R' = 2-octyl.

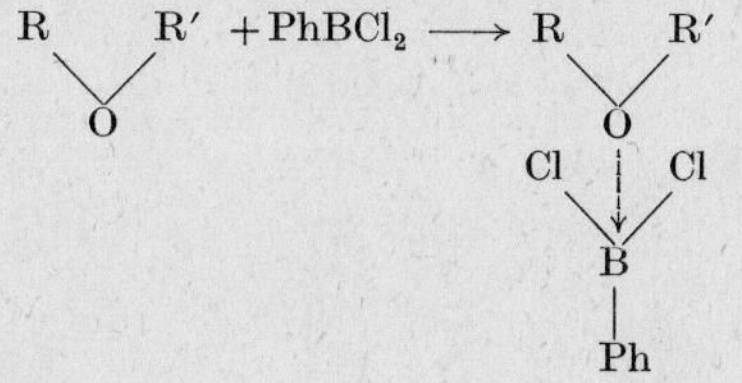

$$\longrightarrow R'^+[PhBCl_2OR]^- \longrightarrow R'^+ + Cl^- \longrightarrow R'Cl + PhBCl(OR)$$

TABLE VII

Alkyl Chlorides Formed from Ethers

Ether	Alkyl chloride	Ether	Alkyl chloride
Et_2O	EtCl	Bu^iOBu^t	Bu^tCl
$Bu^n{}_2O$	Bu^nCl	Bu^sOBu^t	Bu^tCl
Bu^nOBu^i	$\begin{cases} Bu^iCl \\ + Bu^tCl \end{cases}$	Bu^sOBu^i	Bu^sCl
Bu^nOBu^s	Bu^sCl	$\begin{cases} (-)\!-\!n\text{-}C_6H_{13}. \\ CHMe.OEt \end{cases}$	$\begin{cases} (+)\!-\text{ and }(\pm)\!- \\ n\text{-}C_6H_{13}.CHMeCl \end{cases}$
Bu^nOBu^t	Bu^tCl		
$Ph.CH_2.OMe$	$PhCH_2Cl$	PhOMe	MeCl
$Ph.CH_2.OEt$	$PhCH_2Cl$	PhOEt	EtCl
$Ph.CH_2.OBu^n$	$PhCH_2Cl$	$o\text{-}C_6H_4Me.OMe$	MeCl
		$o\text{-}C_6H_4Me.OEt$	EtCl

That the dichloride does not react with diphenyl ether or bis-2-chloroethyl ether is due to a considerable reduction of electron density on the oxygen atom due to electronic effects, and, to some extent, a reduction in availability of electron density due to steric hindrance, as discussed by Gerrard *et al.* (1959, 1960). Examples of results are given in the Table VII.

An indication that phenylboron dibromide is more reactive than the dichloride in this system is shown by the fact that it will cleave n-butyl t-butyl ether at room temperature.

$$PhBBr_2 + ROR' \longrightarrow PhBBr(OR) + R'Br$$

A hope to harness internal ether fission to the development of an applicable boron polymer was not realized because of the ready hydrolysis of the product. When p-methoxyphenylboron dichloride was heated, methyl chloride was quantitatively evolved, and a resin-like solid remained. This was easily hydrolysed, and from the mixture, p-hydroxyphenylboronic acid was isolated (Dandegaonker $et\ al.$, 1959).

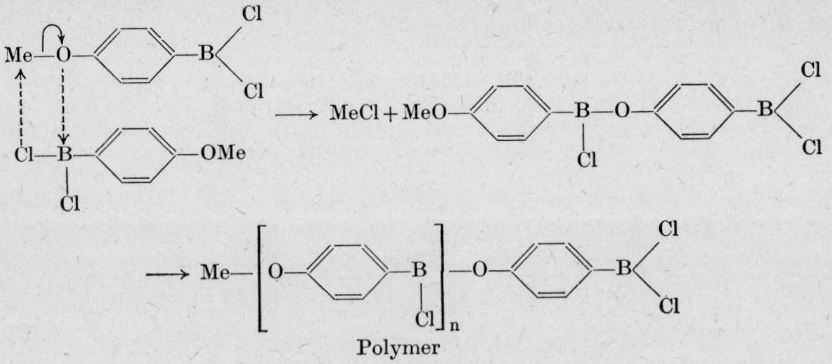

II. Attachment by Grignard Reagent

A very general procedure for the attachment of one alkyl or aryl group to boron is by interaction of a Grignard reagent, to supply the hydrocarbon group, and a boron trihalide (F,Cl,Br) or a boric ester to supply the boron. There has been a somewhat circumstantial choice of boron compound and experimental conditions, and many of the descriptions entail permissive rather than compelling details. Different workers have circumstantially changed too many factors independently, and it remains for someone to undertake an experimental survey into the factors determining the procedure to be adopted according to the group to be attached. In general, the isolable product is the boronic acid, $RB(OH)_2$, and not the intermediate dihalide or dialkyl ester.

In early years the Grignard reagent was used by Khotinsky $et\ al.$ (1909) to form a carbon to boron bond by interaction with an alkyl borate (Me,Et,n-Pr, i-Bu,i-Am) at 0°C.

$$B(OR)_3 \text{ (in ether) added to ArMgBr} \longrightarrow ArB(OR)_2 \xrightarrow{2H_2O} ArB(OH)_2$$

Phenyl- and m-tolylboronic acids, and a number of alkylboronic acids were thus prepared.

The majority of workers have used the Grignard technique, with variations in experimental detail and boron reagent. Boron trifluoride has been favoured, and boron trichloride or tribromide have not been without attention.

$$RMgBr + BF_3(Cl,Br) \longrightarrow RBF_2(Cl, Br) \xrightarrow{2H_2O} RB(OH)_2 + 2HF(Cl,Br)$$

Representative procedures and a useful apparatus design have recently been described (Brindley *et al.*, 1955). In a typical preparation an ethereal solution of *n*-butylmagnesium bromide (1 mole) is added to a vigorously stirred ethereal solution of tri-*n*-butyl borate (1 mole) at −70°C, the apparatus being filled with nitrogen. The mixture is then poured into ice-water, carefully neutralized with dilute hydrochloric acid, to obtain an ethereal solution of the ester. Although it is

$$n\text{-BuMgBr} + B(OBu^n)_3 \longrightarrow n\text{-BuB}(OBu)_2$$

possible to isolate the ester as such, the usual practice is to hydrolyse it to the boronic acid. When a boron trihalide is used, the initial aqueous treatment gives the boronic acid.

III. Attachment by Organo-lithium Reagents

Lithium alkyls and aryls can be used in place of the Grignard reagent, and offer advantages in certain circumstances.

$$R'Li + B(OR)_3 \longrightarrow R'B(OR)_2 + LiOR$$

A typical procedure (Brindley *et al.*, 1955) is to prepare *n*-butyllithium in ether by interaction of lithium and *n*-butyl bromide at −10°C in an atmosphere of nitrogen. The solution is filtered through a sinter, and a measured volume added dropwise, in an enclosed apparatus, to an ethereal solution of tri-*n*-butyl borate at −60°C. After treatment with water, the dried ethereal solution gives di-*n*-butyl *n*-butylboronate, $n\text{-BuB}(OBu^n)_2$, b.p. 108.5°C/13 mm in 60% yield. The magnesium method gave a 42% yield of the ester. The ester can be obtained directly in this way because of the greater ease of hydrolysis of the unchanged trialkyl borate, as compared with the boronate. By neither method was it found convenient to isolate the ester of phenylboronic acid, and so hydrolysis was deliberately performed to isolate the acid.

Potassium fluoroborate has been used by Sazonova *et al.* (1956) in recent work. ω-Bromostyrene, PhCH=CHBr in diethyl ether was added to a suspension of potassium fluoroborate, KBF_4, and magnesium in ether. The resulting system was treated with dilute acetic acid and ice, the organic layer washed with 2N-NaOH and the alkali extract washed with ether. Acidification of the alkali extract afforded styryl-boronic acid, $PhCH{=}CHB(OH)_2$, m.p. 129°C in 25% yield.

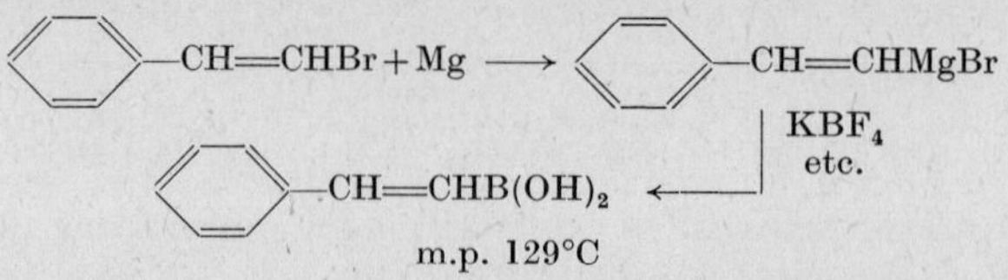

Other workers (Gerrard *et al.*, 1958) have experienced difficulty in the use of potassium fluoroborate. For discussion on *p*-vinylphenylboronic acid, see Cazes (1958) and p. 205. Lawesson (1957) has studied bromine substituted thiopheneboronic acids, and the crystal structure of *p*-bromophenylboronic acid has been discussed by Zvonkova (1958).

IV. Boronic Acids

Of special attraction is the production of diboronic acids, the essential requirement in the Grignard system being the formation of a bi-functional reagent (Nielsen *et al.*, 1957; Musgrave, 1957). This has been accomplished by using tetrahydrofuran (T.H.F.) as the solvent, in order to attain a higher temperature for completion of metallation.

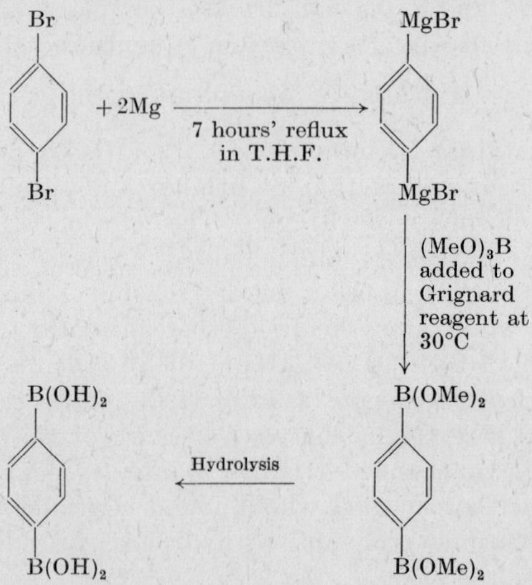

From 1,3-dibromobenzene, the 1,3-diboronic acid was similarly obtained. Lithiumaryl method was tried, but lower yields were obtained.

The diboronation of benzene was discovered independently by Musgrave (1957) during the purification of *p*-bromophenylboronic acid prepared by the interaction of p-bromophenylmagnesium bromide and tri-n-butyl borate in ether at $-70°C$. A small amount of p-phenylenediboronic acid was isolated, the *para* positions of the boron attachments being indicated by deboronation (or dephenylation) through the action of bromine in aqueous potassium bromide. It is known that a small amount of bifunctional Grignard compound is

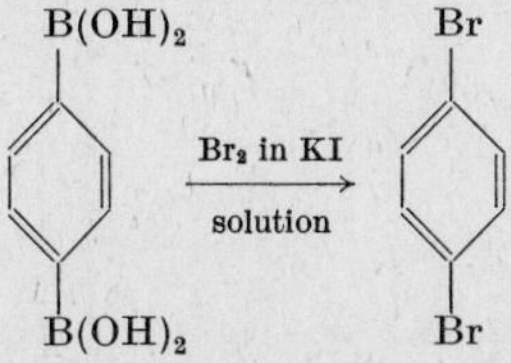

formed during the normal preparation of p-bromophenylmagnesium bromide from p-dibromobenzene (Quelet, 1927), and it was apparent that the diboronic acid had come from the bifunctional Grignard compound so formed. An ethereal solution of the bifunctional Grignard reagent was therefore deliberately prepared from the monofunctional compound by the use of ethyl bromide and magnesium (present in excess). The solution was added to methyl borate at 70°C, and the diboronic acid was isolated in 25% yield. Purification of the acid by recrystallization from hot water was followed by formation of a very interesting derivative, the neopentyl glycol ester, m.p. 233°C.

Modifications in procedure for the preparation of boronic acids continue to appear. A recent attractive one by Cowie *et al.* (1959) comprises the preparation of an arylboronic acid by heating a mixture of a trialkyl borate with magnesium and the aryl bromide. Thus a mixture of tri-n-butyl borate, bromobenzene, and magnesium began to react at about 150°C, and the magnesium had disappeared after heating under reflux at 230°C for 2–3 hours. After the usual procedure of hydrolysis, phenylboronic acid was isolated as the anhydride in 33–44% yield. Longer heating in the example of chlorobenzene gave similar results.

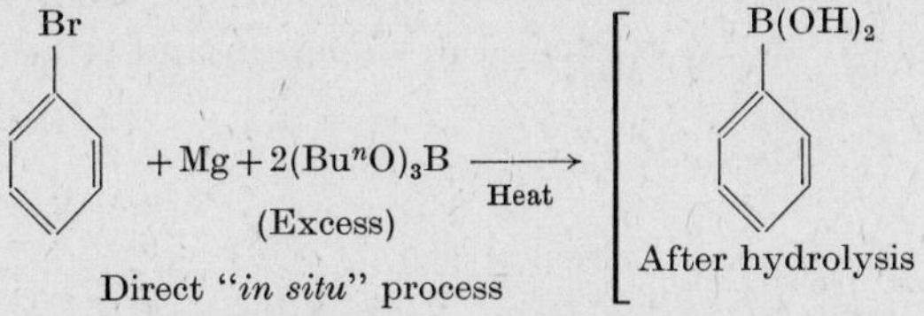

Essentially the same system is provided by lower boiling alkyl borates, $(Pr^nO)_3B$, b.p. 177°C (21–30% yield), $(Pr^iO)_3B$, b.p. 140°C (10% yield), $(EtO)_3B$, b.p. 117°C (12% yield), but owing to the lower reflux temperature, longer reaction times were required, and it is not surprising that trimethyl borate, b.p. 68°C, gave no measurable yield.

Even p-phenylenediboronic acid was obtained in 16% yield from tri-n-butyl borate with p-dibromobenzene and magnesium, present in sufficient amounts. p-Dichlorobenzene, however, responded poorly.

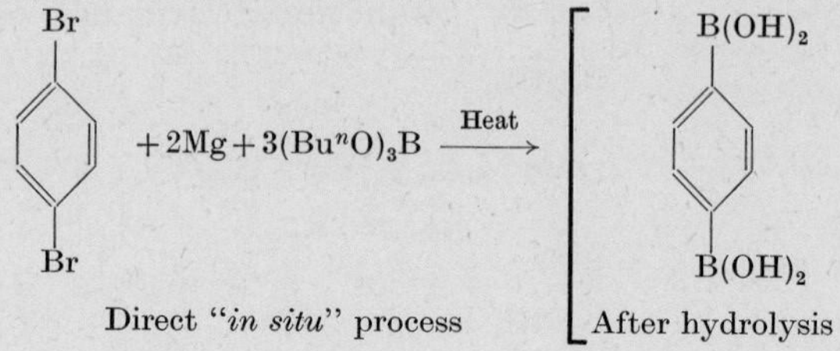

Direct *"in situ"* process   After hydrolysis

An attempt to prepare the o-phenylenediboronic acid by this process led to a peculiar result, for phenylboronic anhydride was obtained, together with a mixture of olefins. A possible mechanism is depicted in the diagram.

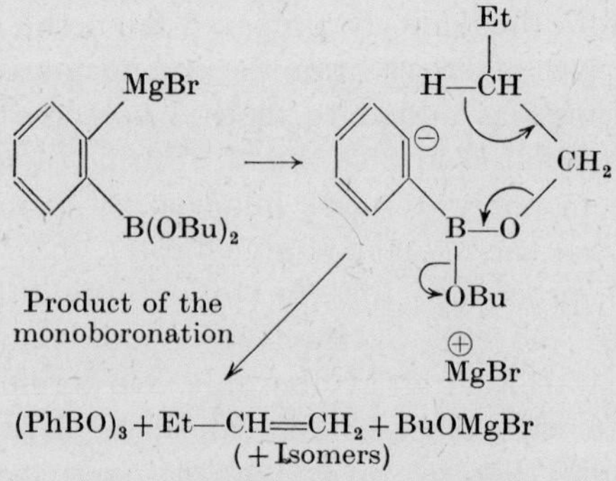

Product of the monoboronation

$(PhBO)_3 + Et—CH\!=\!\!=\!CH_2 + BuOMgBr$
(+ Isomers)

The boronic anhydride [PhBO] will appear as the trimer $[PhBO]_3$, see later; but this on hydrolysis will give phenylboronic acid.

In connection with investigations into the biological activity of boron compounds (p. 76), work reported by Gilman *et al.* (1957) and Santucci *et al.* (1958), sponsored by Division of Biology and Medicine of U.S. Atomic Energy Commission, has produced such boronic acids as the following, made by means of aryllithium and tributyl borate.

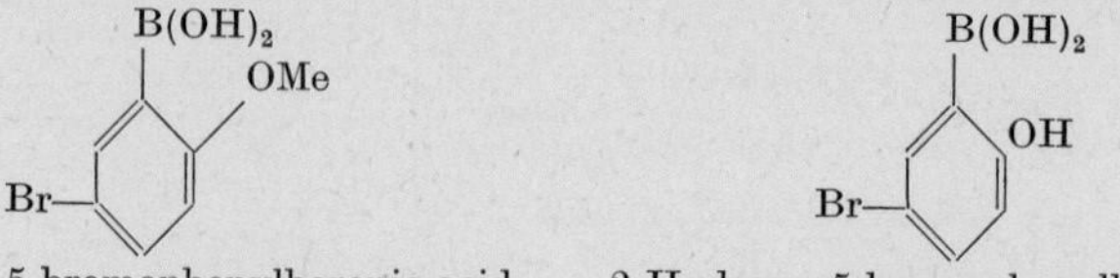

2-Methoxy-5-bromophenylboronic acid   2-Hydroxy-5-bromophenylboronic acid

Thielens (1958) refers to 9-phenanthrylboronic acid as a new luminescent organo-boron compound.

V. Properties of Boronic Acids

A. Acid Strength

Dissociation constants of 31 boronic acids $RB(OH)_2$ (mainly R = aryl) were determined by one school of research (Branch et al., 1934; Yabroff et al., 1933, 1934; Bettman et al., 1934; and Clear et al.. 1938).

Phenylboronic acid is three times as strong as boric acid, which of course is weak, and n-butylboronic acid is one-tenth as strong as phenylboronic acid. This is in accordance with the effect of substitution in acetic acid. Any effect which reduces electron density on the "proton" oxygen strengthens the acid; any effect which increases electron density thereon weakens the acid. More remote substitution by aryl has less effect; thus 2-phenylethylboronic acid is about as strong as n-butylboronic acid. Benzylboronic acid, however, is about as strong as phenylboronic acid. Substitution of an ortho hydrogen by alkyl or phenyl decreases the strength of phenylboronic acid; but meta or para substitution effects only a slight decrease in strength. However, the strength is considerably increased by electro-negative substituents ($NO_2 > F > Cl > Br > COOH$), the o-nitro group having the least effect.

B. Substitution in the Aromatic Nucleus

Ainley et al. (1930), König et al. (1930), Seaman et al. (1931), Bean et al. (1932), Bettman et al. (1934), Kuivila et al. (1952), and Gilman et al. (1957) have contributed examples of the following reactions which lead to the formation of a number of substituted aryl boronic acids.

Little has been written about the salts of boronic acids. Michaelis et al. (1882) mentioned a sodium and a calcium salt. n-Butylboronic acid gives a hydrated sodium salt when treated with concentrated solution of sodium hydroxide, and the structure of the barium salt of p-carboxyphenylboronic acid has been assigned the structure shown.

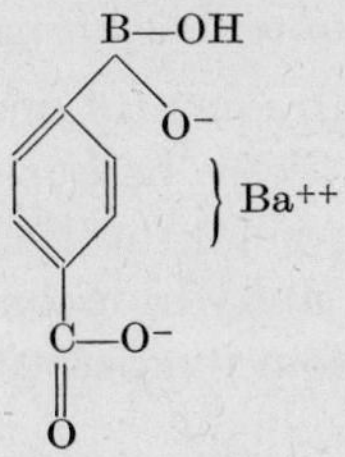

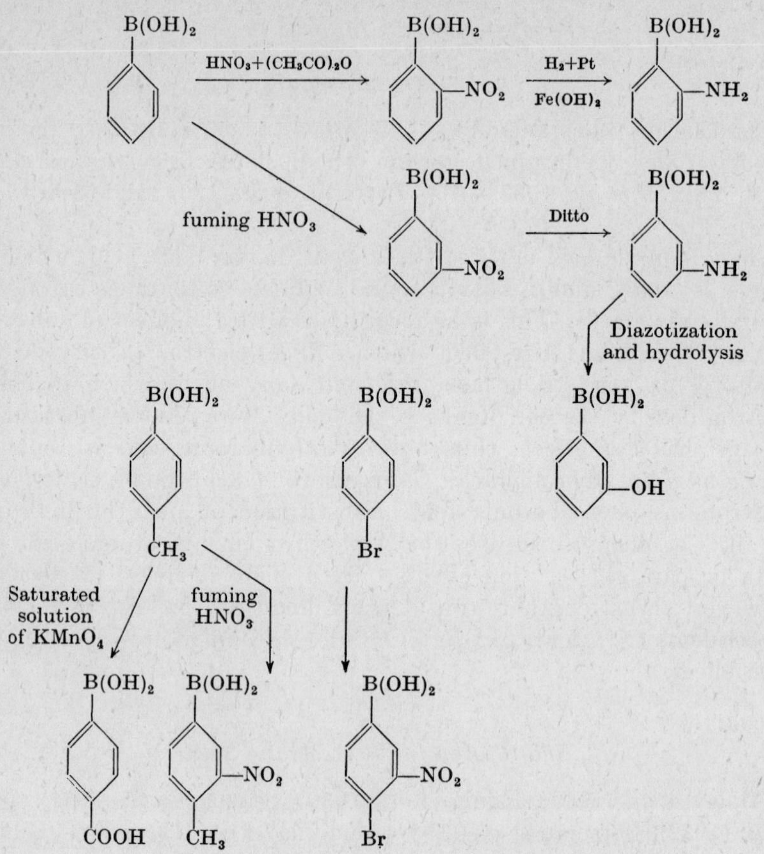

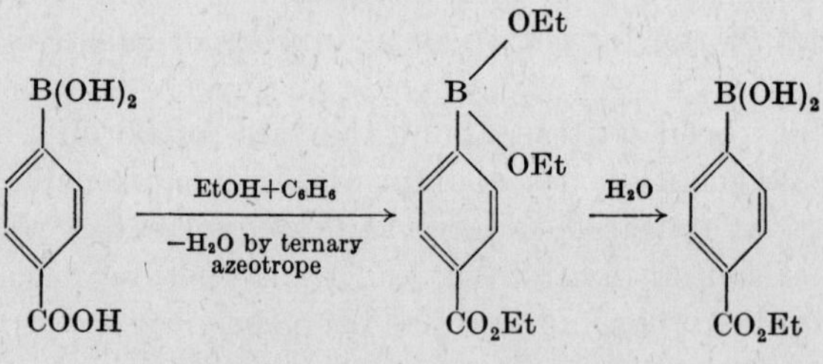

C. Anhydride Formation

The melting-point is not infrequently quoted differently for the same boronic acid. This is most likely because of the dehydration of the acid, and it is the melting-point of the anhydride which is more reliable. Formation of the anhydride has significance and may be achieved by desiccation at room temperature, or by heating, especially under reduced pressure.

Phenylboronic anhydride, m.p. 218°C, was first prepared in 1882 (Michaelis *et al.*). A standard procedure (Abel *et al.*, 1956) is to heat the acid at 110–115°C for 8 to 10 hours. Another example is p-methoxyphenylboronic anhydride, m.p. 206–207°C, prepared by Dandegaonker *et al.* (1959).

The existence of a B=O linkage must be a rarity if indeed it has been definitely established for any compound. The trimeric form of arylboronic anhydride was indicated in 1936 by Kinney *et al.*, and has since become firmly accepted for the boronic anhydrides.

The following investigators and co-workers have reported on this field of work: Krause (1921, 1931); König (1930); Seaman (1931); Bean (1932); Bettman (1934); Branch (1934); Yabroff (1934); Johnson (1938); Snyder (1938); Burg (1940); Bauer (1941); Grummitt (1942); Wiberg (1948); Goubeau (1951, 1953); Dworkin (1954); Kuivila (1954); McCusker (1955); Brindley (1956); Dandegaonker (1957); Mattraw (1956).

Accepted Structure   *Evidence*

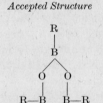

Molecular weight (cryoscopic).
Trouton constant.
Electron diffraction.
Bond distances and angles.
Raman spectra.

Electron diffraction for methylboronic anhydride pointed to a planar six-membered ring with alternating boron and oxygen atoms, each methyl group being bonded to a boron atom and in the plane of the ring. The heat of solution in water has been determined by Dworkin *et al.* (1954). The anhydrides are very easily hydrolysed, and readily form the ester of the corresponding acid by the usual azeotropic technique when heated with the alcohol or phenol.

$$(RBO)_3 + 6ROH \longrightarrow 3RB(OR)_2 + 3H_2O$$

$$\left(MeO\!\!\left\langle\!\!\!\bigcirc\!\!\!\right\rangle\!\!BO\right)_3 + 6Bu^nOH \longrightarrow 3MeO\!\!\left\langle\!\!\!\bigcirc\!\!\!\right\rangle\!\!B(OBu^n)_2 + 3H_2O$$

Conversion of an anhydride into the boron dichloride by means of boron trichloride seems attractive, for it constitutes a very convenient means of preparing the dichloride from the boronic acid, and the dichloride is the only liquid component.

$$
\begin{array}{c}
\text{Ph} \\
| \\
\text{B} \\
\diagup\diagdown \\
\text{O} \quad \text{O}
\end{array}
$$

$$
\text{Ph—B} \quad \text{B—Ph} + 3\text{BCl}_3 \xrightarrow[\text{at } -80°\text{C}]{\text{In inert solvent}} 3\text{PhBCl}_2 + \left\{ \begin{array}{l} 3\text{BOCl} \longrightarrow \\ \text{B}_2\text{O}_3 + \text{BCl}_3 \end{array} \right.
$$

$$
\begin{array}{c}
\text{C}_6\text{H}_4\text{OMe} \\
| \\
\text{B} \\
\diagup\diagdown \\
\text{O} \quad \text{O}
\end{array}
$$

$$
\text{MeOC}_6\text{H}_4\text{B} \quad \text{B—C}_6\text{H}_4\text{OMe} + 3\text{BCl}_3 \longrightarrow 3\text{MeO}\!\!\left\langle\!\!\bigcirc\!\!\right\rangle\!\!\text{BCl}_2 + \left\{ \begin{array}{l} 3\text{BOCl} \longrightarrow \\ \text{B}_2\text{O}_3 + \text{BCl}_3 \end{array} \right.
$$

Boron trifluoride, however, will not react with phenylboronic anhydride even at 20°C (Abel *et al.*, 1956). Aluminium trichloride and *n*-butylboronic anhydride was found to give a mixture of *n*-butylboron dichloride and di-*n*-butylboron chloride (McCusker *et al.*, 1955).

In organo-boron chemistry, complexing with nitrogen bases has attracted much attention. Methylboronic anhydride forms a 1 : 1 and a 1 : 2 complex with ammonia, $(\text{MeBO})_3\text{NH}_3$, $(\text{MeBO})_3 2\text{NH}_3$, as well as a 1 : 1 complex (m.p. 67°C) with trimethylamine.

A point of contrast discussed by Gerrard (1959) may be stressed between the result of dehydration of an alkyl- or arylboronic acid and the corresponding silanol. Whereas the latter gives rise to silicone polymers, the boronic acid does not appear to develop anything more than the trimer, and indeed the six-membered ring designated the boroxole, in this instance the triphenylboroxole, is a characteristic result of producing what would potentially be a boron–oxygen double bond.

$$
\text{R}_2\text{Si(OH)}_2 \longrightarrow
\begin{array}{ccccccc}
& \text{R} & & \text{R} & & \text{R} & \\
& | & & | & & | & \\
-\text{Si}&-&\text{O}-&\text{Si}&-\text{O}-&\text{Si}&-\text{O}- \\
& | & & | & & | & \\
& \text{R} & & \text{R} & & \text{R} &
\end{array}
$$

A siloxane polymer

$$
3\text{RB(OH)}_2 \longrightarrow
\begin{array}{c}
\text{Ph} \\
| \\
\text{B} \\
\diagup\diagdown \\
\text{O} \quad \text{O} \\
| \qquad | \\
\text{Ph—B} \quad \text{B—Ph} \\
\diagdown\diagup \\
\text{O}
\end{array}
\qquad \text{Boroxole}
$$

D. Formation and Properties of Esters

As in the example of boric acid, boronic acids may be converted directly into their esters by interaction with a hydroxy-compound, water being removed as an azeotrope (Torssell, 1954; Brindley et al., 1955).

$$RB(OH)_2 + 2R'OH \longrightarrow RB(OR')_2 + 2H_2O$$

Esters of alkylboronic acids are usually more slowly hydrolysed than esters of arylboronic acid, or of boric acid, and indeed di-n-butyl n-butylboronate could be isolated from the Grignard system because of this relative stability. Nevertheless, as a rule the esters are far from being hydrolytically stable, although they have good thermal stability. For aryl groups and alkyl groups of ordinary reactivity (p. 1) both esterification and hydrolysis probably involves B—O fission, and not C—O fission (Brindley et al., 1956), since (+)-2-methylheptanol is recovered unchanged in rotatory power after being converted into and obtained from the ester of phenylboronic acid. This result conformed with the optical result when the ester was prepared from the (+)-alcohol and phenylboron dichloride.

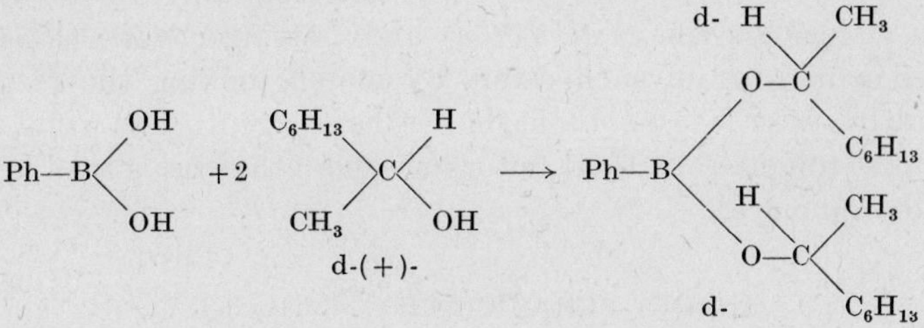

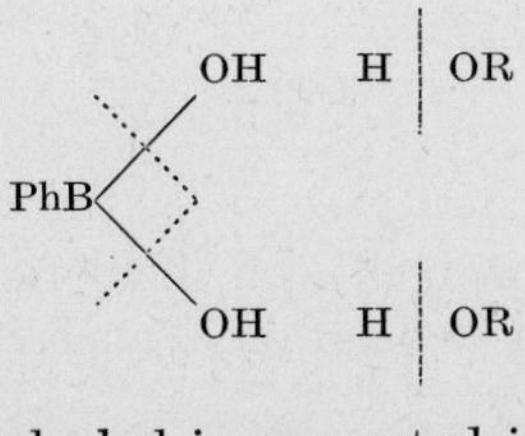

Furthermore, neopentyl alcohol is converted into, and obtained from the ester of phenylboronic acid, thus showing the absence of a mechanism which would lead to rearrangement.

The diethanolamine esters are of special interest; for, because of their relative hydrolytic stability, they can be used for the purification of arylboronic acids (Letsinger et al., 1955; Musgrave et al., 1955). 2-Aminoethanol and the arylboronic acid are heated in toluene, and on cooling crystals of the boronate separate. Infra-red examination of the diethanolamine arylboronates, $ArBO_2(CH_2)_2NH$ (where $Ar = p$-MeC_6H_4—, p-$MeOC_6H_4$—, p-BrC_6H_4—, m-$NO_2C_6H_4$—, and Ph—), led to the formulation shown. There has been the implication that co-ordination of this kind is the route to hydrolytic stability, but there is

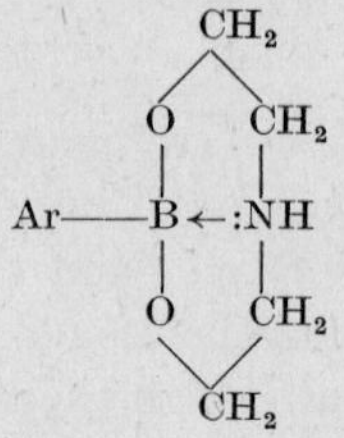

$$
\begin{array}{c}
CH_2 \\
\diagup \diagdown \\
O \quad\quad CH_2 \\
| \quad\quad\quad | \\
Ar\text{——}B\leftarrow\text{:}NH \\
| \quad\quad\quad | \\
O \quad\quad CH_2 \\
\diagdown \diagup \\
CH_2
\end{array}
$$

a danger of over-simplification here, and the principles entailed have been discussed elsewhere (Gerrard, 1956, 1959) (p. 11).

Alkyloxy-exchange can occur; thus the particular alcohol and diethyl phenylboronate gave n-Bu, s-Bu, i-Bu, and octan-2-yl esters.

$$ArB(OR)_2 + 2R'OH \longrightarrow ArB(OR')_2 + 2ROH$$

Of course, one might expect cyclic esters from dihydric alcohols and phenols. Thus Kuivila $et\ al.$ (1954) and Letsinger $et\ al.$ (1954) have prepared a number of such esters by merely mixing the reagents if the resulting ester is less soluble than either reagent. The usual ternary azeotrope (toluene) method afforded the ethylene glycol ester of n-butylboronic acid.

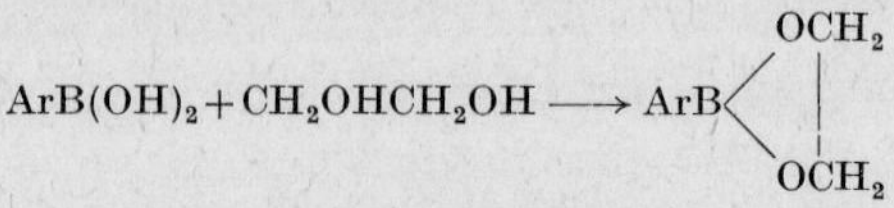

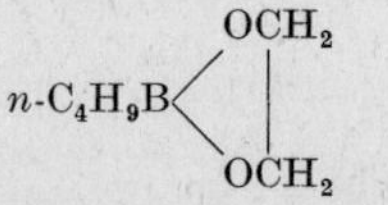

$$
ArB(OH)_2 + CH_2OHCH_2OH \longrightarrow ArB\begin{array}{c} OCH_2 \\ \diagup | \\ \diagdown | \\ OCH_2 \end{array}
$$

$$
n\text{-}C_4H_9B\begin{array}{c} OCH_2 \\ \diagup | \\ \diagdown | \\ OCH_2 \end{array}
$$

The use of the alkyl- or arylboron dichloride and pyridine is also a general procedure for the formation of such cyclic esters.

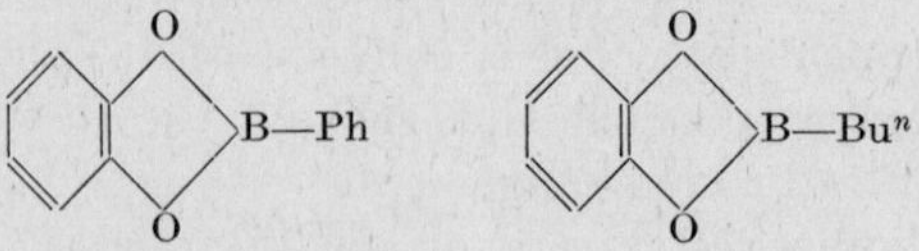

One important property of the esters is their conversion into the corresponding boron dichloride. This may be accomplished by means of boron trichloride or phosphorus pentachloride (Brindley $et\ al.$, 1955; Mikhailov $et\ al.$, 1956; Nielsen $et\ al.$, 1957).

$$PhB(OEt)_2 + 2BCl_3 \longrightarrow PhBCl_2 + 2EtOBCl_2$$

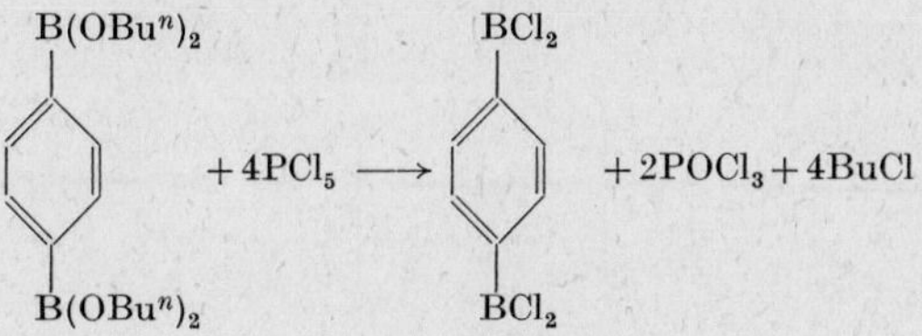

$$+ 4PCl_5 \longrightarrow + 2POCl_3 + 4BuCl$$

m.p. 93–96°C

It is remarkable that phosphorus trichloride, silicon tetrachloride and thionyl chloride were without effect on the B—O—C ester linkage, provided C is of ordinary reactivity (see p. 1).

E. Formation and Properties of Halogeno-Esters

Mutual replacement reactions and the correct stoicheiometry have been used to prepare alkyl- and arylboronic halogeno-esters (Brindley *et al.*, 1956). Reactions occur quickly.

$$n\text{-}BuB(OBu^n)_2 + BCl_3 \longrightarrow n\text{-}BuBCl(OBu^n) + Bu^nOBCl_2$$

n-Butyl *n*-butylchloroborinate
b.p. 48–52°C/2 mm

This member is a colourless fuming liquid, which of course is readily hydrolysed to *n*-butylboronic acid. A similar reaction was observed for the diethyl or di-*n*-butyl esters of phenylboronic acid.

$$PhB(OR)_2 + BCl_3 \longrightarrow PhBCl(OR) + ROBCl_2$$

A better way is to mix the ester with phenylboron dichloride.

$$PhB(OBu^n)_2 + PhBCl_2 \longrightarrow 2PhBCl(OBu^n) \quad n\text{-Butyl phenylchloroborinate,}$$
b.p. 58°C/0.2 mm

Both *n*-butyl phenylchloroborinate and *n*-butyl *n*-butylchloroborinate were stable at 100°C (3 hours), but underwent slow decomposition at 200°C, the arylboron compound being the more stable. A trace of ferric chloride greatly catalysed the decomposition, and at 20°C the reaction was rapid.

$$RBCl(OBu^n) \longrightarrow \tfrac{1}{3}(R.BO)_3 + n\text{-}BuCl$$

As (—)-2-chloro-octane with little loss in rotatory power was obtained by an inversion mechanism from the corresponding ester, it was suggested that the reaction sequence is as follows, and the formation of alkyl chloride is by the SN2 mechanism.

$$\text{R—O—B—Cl} + \text{Catalyst} \longrightarrow \left[\text{R—OB}\right]^{+} + [\text{ACl}]^{-}$$

$$[\text{ACl}]^{-} \rightleftharpoons \text{A} + \text{Cl}^{-} \longrightarrow \left[\text{—C—O—B}\right]^{+} \longrightarrow \text{Cl—C} + \tfrac{1}{3}(\text{OBR}')_3$$

Inverted

However, from the more detailed study of a number of phenylchloro-borinates (Dandegaonker *et al.*, 1957), it is clear that the second step can in certain circumstances involve the intermediate carbonium cation. The *iso*butyl ester was stable at 160–165°C (65 hours); but in the presence of ferric chloride, *iso*butyl, and *tert.*-butyl chloride were formed rapidly at 20°C; the benzyl ester decomposed at 20°C (24 hours) in the absence of catalyst. Relative rates of decomposition were found to be $\text{R} = \text{Bu}^t > \text{Pr}^i > \text{Et} > \text{Me}$. The phenyl ester did not, of course, give phenyl chloride, and so the easily hydrolysed chlorine remained in the system. Thus the reversible reaction

$$2\text{PhBCl(OPh)} \rightleftharpoons \text{PhBCl}_2 + \text{PhB(OPh)}_2$$

could be made to go to the right by removing phenylboron dichloride in a vacuum.

An alternative method of preparing the esters is by the interaction of phenylboron dichloride with the correct amount of alcohol.

$$\text{PhBCl}_2 + \text{ROH} \longrightarrow \text{PhBCl(OR)} + \text{HCl}$$

The full ester can be made from the chloro-ester by adding another molecule of alcohol or phenol. The isolation of mixed esters, $\text{ArB(OR)OR}'$, is rendered difficult by redistribution on distillation.

Compounds with both alkoxyl and bromine attached to boron are relatively unstable and difficult to isolate as a rule. Therefore the formation of *n*-butyl phenylbromoborinate by the cleavage of *n*-butyl *tert.*-butyl ether is of interest (Abel *et al.*, 1957).

$$\text{PhBBr}_2 + \text{ROR}' \longrightarrow \text{PhBBrOR}' + \text{RBr}$$

$$\text{Also } \text{PhBBr}_2 + n\text{-BuOH} \longrightarrow \text{PhBBrOBu}^n + \text{HBr}$$

n-Butyl phenylbromoborinate is a colourless liquid, b.p. 67°C/0.15 mm, which undergoes slow decomposition to phenylboronic anhydride and *n*-butyl bromide at 100°C. The reactivity to alcoholysis and hydro-lysis is similar to that of the chloro-esters.

Related compounds in the boron trifluoride system have been studied (Brindley *et al.*, 1956).

VI. Vinylboronic Acid

Dibutyl vinylboronate, otherwise referred to as dibutyl ethylene-boronate, $CH_2=CH—B(OBu^n)_2$, has received some recent attention by Matteson (1959) in connection with radical catalysed additions to the double bond. Preparation of the ester in 50% yield is claimed by Normant et al. (1959), but description of technique is rather vague. Following details given by Ramsden et al. (1957) for the preparation of vinylmagnesium chloride, Matteson added this compound in T.H.F. to trimethyl borate in ether at $-60°C$ in an apparatus flushed with nitrogen. The resulting system was treated with a mixture of aqueous phosphoric and hydrochloric acids, and a small amount of phenothiazine to prevent polymerization on contact with air. The acid was extracted by n-butanol, water being removed by freezing at $-70°C$, and then esterification was effected by the azeotropic removal of water with the excess of n-butanol. The ester, b.p. 35–40°C/0.08 mm, was obtained in 72% yield.

Interaction with compounds with the form XY, where $X=CCl_3$, $Y=Cl$; $X=CCl_3$, $Y=Br$; or $X=n$-$C_6H_{13}S$, $Y=H$, appears to involve an intermediate free radical, stabilized by C—B π-bonding.

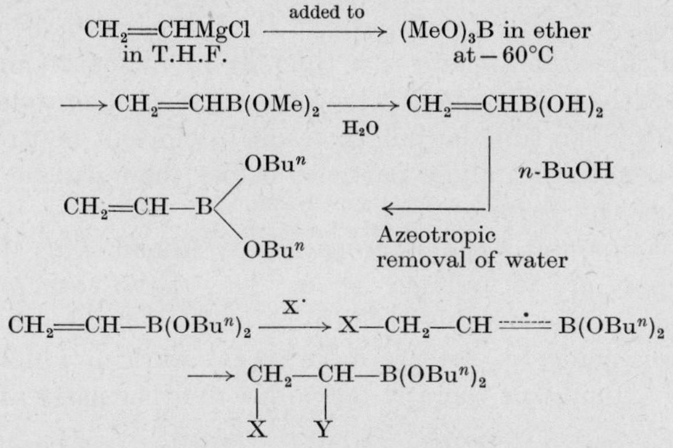

$$CH_2=CHMgCl \text{ in T.H.F.} \xrightarrow{\text{added to}} (MeO)_3B \text{ in ether at} -60°C$$

$$\longrightarrow CH_2=CHB(OMe)_2 \xrightarrow{H_2O} CH_2=CHB(OH)_2$$

$$CH_2=CH—B\begin{matrix} OBu^n \\ \\ OBu^n \end{matrix} \xleftarrow[\substack{\text{Azeotropic} \\ \text{removal of water}}]{n\text{-BuOH}}$$

$$CH_2=CH—B(OBu^n)_2 \xrightarrow{X\cdot} X—CH_2—CH \cdots B(OBu^n)_2$$

$$\longrightarrow \underset{\underset{X}{|}\quad \underset{Y}{|}}{CH_2—CH—B(OBu^n)_2}$$

Thus esters are obtained of those boronic acids, containing a functional group in the α-position, previously inaccessible.

VII. Applications of Boronic Acids

A number of alkyl boronic acids such as nonyl and dodecyl are commercially available in the U.S.A., for research and development purposes. It is suggested that they are: bacteriastatic, fungistatic, surface active and useful as gasoline additives, polymer components and stabilizers for aromatic amines.

Such properties might be of value in the control of slime organisms in the manufacture of paper, and to minimize bacterial damage in paint, leathers, and polymers. They are dehydrated to the anhydrides, $(RBO)_3$, which are oil soluble and are potential oil additives for stabilization against oxidation and polymerization. There are patent specifications, e.g. by Darling (1955) and by Arimoto (1955), for the use of alkylboronic acids and esters in motor fuel (see also Scott *et al.*, 1958). Scintillators containing organo-boron compounds are referred to by Nikitina *et al.* (1958).

The potential uses of organo-boron compounds as biologically active materials is attracting considerable attention in the U.S.A., and there is much sponsored work in progress. Owing to the ease of hydrolysis of many organic compounds of boron, some of the compounds most available at present for routine biological screening would seem to be the alkyl- and arylboronic acids. There are signs of intense activity in this connection.

Of special interest is the use of stable organo-compounds of boron in neutron capture therapy of brain tumours, using slow thermal neutrons. Boron has a high neutron cross section, and therefore neutron capture is very efficient. When introduced into the blood-stream "boron" diffuses rapidly to all parts of the body except healthy parts of the brain, for which there is a time-lag of about 20 minutes. It quickly enters the brain tumours, however. Irradiation by slow thermal neutrons during the time-lag might result in efficient neutron capture and cell destruction by alpha particles before there can be significant damage to healthy parts.

Types of compound recently prepared by Gilman *et al.* (1957, 1958) for this purpose are formulated below. They have been prepared by the interaction of the suitable aryllithium and tributyl borate. A reference has been made (p. 208) to more recent work in which attempts are made to extend this therapy technique to other parts of the body.

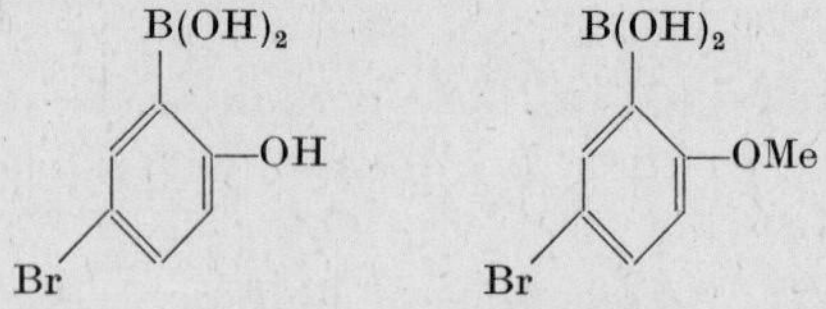

VIII. Summary

Attachment of one alkyl or one aryl group to boron leaves a bifunctional boron atom apart from co-ordination. The functional linkages appear to have the same general properties as they have when in trifunctional boron compounds such as boron trichloride,

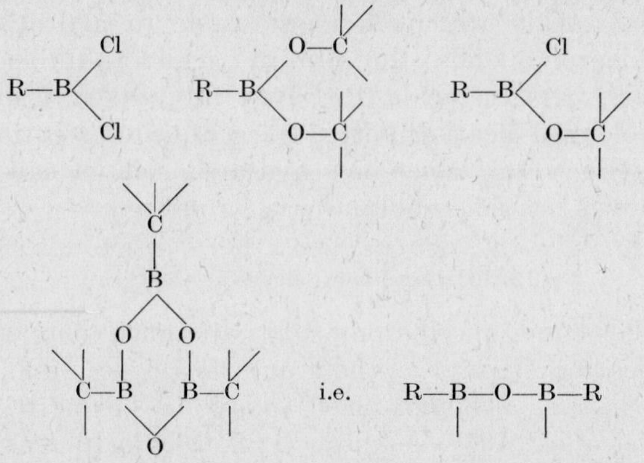

$$R—B\begin{matrix}Cl\\ \\Cl\end{matrix} \qquad R—B\begin{matrix}O—C\\ \\O—C\end{matrix} \qquad R—B\begin{matrix}Cl\\ \\O—C\end{matrix}$$

$$C$$
$$|$$
$$B$$

$$C—B \quad B—C \qquad \text{i.e.} \qquad R—B—O—B—R$$
$$O$$

alkyl borates, aryl borates and esters of metaboric acid. There is insufficient detail to enable an accurate assessment of the influence of the R group on the degree of reactivity of the remaining two bonds. So far as the carbon of the B—O—C linkage is concerned, the factors determining the degree of reactivity of the attached (central) carbon atom, and the reactions which occur according to this reactivity, are the same as for the trifunctional systems. In the chemistry just discussed, the B—C bond has not been reactive. However, this is a simplification which must now be removed.

A. C—B Bond Cleavage

The severance of the C—B bond in arylboronic acids may be referred to as deboronation of the aryl system, or dearylation of the boron part, and the procedure was looked at early in the development of the organic chemistry of boron (Ainley et al., 1930; Challenger et al., 1934). Hydrolytic cleavage is effected by heating with water, or acids or alkalis.

$$PhB(OH)_2 + H_2O \longrightarrow C_6H_6 + B(OH)_3$$
(50% NaOH or conc. HCl under pressure)

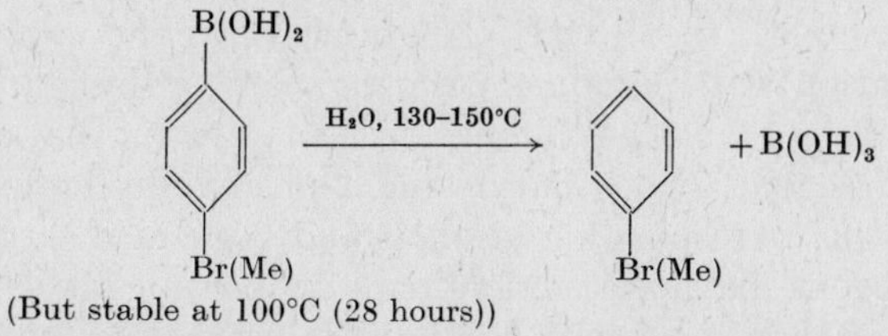

$$+ B(OH)_3$$

(But stable at 100°C (28 hours))

4-Dimethylamino- and 4-methoxy-α-naphthylboronic acids were readily deboronated under milder conditions (Snyder et al., 1948). Benzylboronic acid resembles the aryl acids.

Alkylboronic acids are much more stable to hydrolytic cleavage than are arylboronic acids; the n-propyl acid was stable to water at 140°C (7 hours), and the acids (Pr^n, Bu^n, Bu^i, n-pentyl, n-hexyl) were stable to prolonged heating with strong aqueous solutions of alkali and acids (HBr, HI). When the hydrated sodium salt of n-butylboronic acid was heated, n-butane was formed.

$$n\text{-BuB(OH)}_2\text{(Sodium salt)} \xrightarrow{\text{heat}} n\text{-C}_4\text{H}_{10}$$

On the other hand, alkylboronic acids are much more susceptible to oxidative cleavage than are their aryl analogues, and are readily autoxidized in air, although they appear to be more stable when moist (Snyder et al., 1938; Johnson et al., 1938). Increase in electron release from the alkyl group favours autoxidation; t-butyl-, and benzylboronic acids are oxidized even in moist air, and reactivity is in the order i-Bu > n-Bu.

$$RB(OH)_2 + \tfrac{1}{2}O_2 \longrightarrow ROB(OH)_2 \xrightarrow{H_2O} ROH + B(OH)_3$$

Arylboronic acids, 1-furyl- and 1-thienylboronic acids are stable in the air.

Hydrogen peroxide is the reagent normally used to oxidize both alkyl- and arylboronic acids to boric acid for the quantitative estimation of boron in such compounds.

The following scheme has been suggested (Kuivila, 1954).

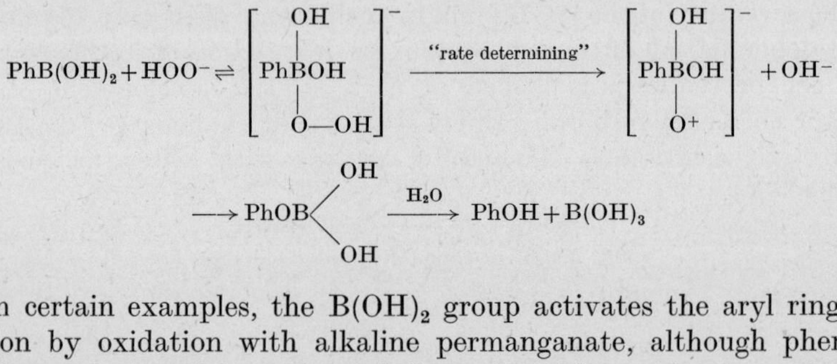

$$PhB(OH)_2 + HOO^- \rightleftharpoons \left[\begin{array}{c} OH \\ | \\ Ph\overset{|}{B}OH \\ | \\ O-OH \end{array}\right]^- \xrightarrow{\text{"rate determining"}} \left[\begin{array}{c} OH \\ | \\ Ph\overset{|}{B}OH \\ | \\ O^+ \end{array}\right]^- + OH^-$$

$$\longrightarrow PhOB\begin{array}{c} OH \\ \diagup \\ \diagdown \\ OH \end{array} \xrightarrow{H_2O} PhOH + B(OH)_3$$

In certain examples, the $B(OH)_2$ group activates the aryl ring for fission by oxidation with alkaline permanganate, although phenyl-, m-phenetyl-, m-tolyl-, and p-tolylboronic acids are in the main unaffected by the reagent, and n-butyl- and 2-phenylethylboronic acids are oxidized to the corresponding alcohols and boric acid.

Whereas halogens have no action on alkylboronic acids, in the presence of water they react with arylboronic acids thus:

$$ArB(OH)_2 + X_2 + H_2O \longrightarrow ArX + HX + B(OH)_3$$

1-Furyl- and 1-thienylboronic acids react similarly.

A volumetric method for estimating arylboronic acids depends on the reaction with bromine (Melnikov *et al.*, 1938). The reaction showed a positive salt effect (Kuivila *et al.*, 1951, 1952, 1954), was catalysed by bases and retarded by acids, and was evidently an electrophilic displacement.

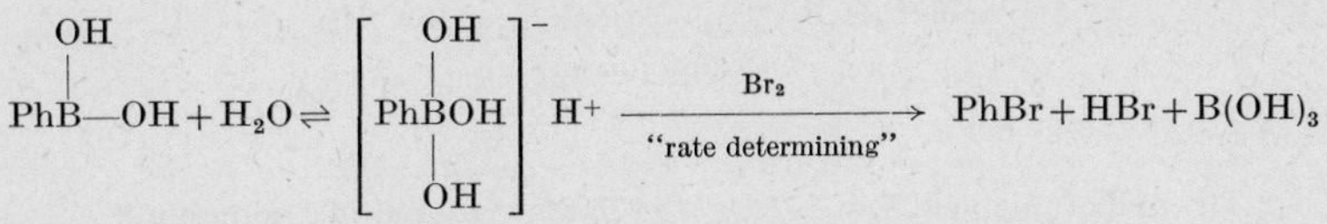

$$\text{PhB---OH} + \text{H}_2\text{O} \rightleftharpoons \left[\begin{array}{c}\text{OH} \\ | \\ \text{PhBOH} \\ | \\ \text{OH}\end{array}\right]^{-} \text{H}^{+} \xrightarrow[\text{``rate determining''}]{\text{Br}_2} \text{PhBr} + \text{HBr} + \text{B(OH)}_3$$

An intermediate of the type shown was indicated.

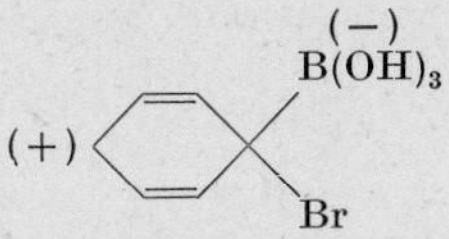

Observations on the rate of iodinolysis of *p*-methoxyphenylboronic acid led to similar interpretations.

A special example, due to Nielsen *et al.* (1957), is the following:

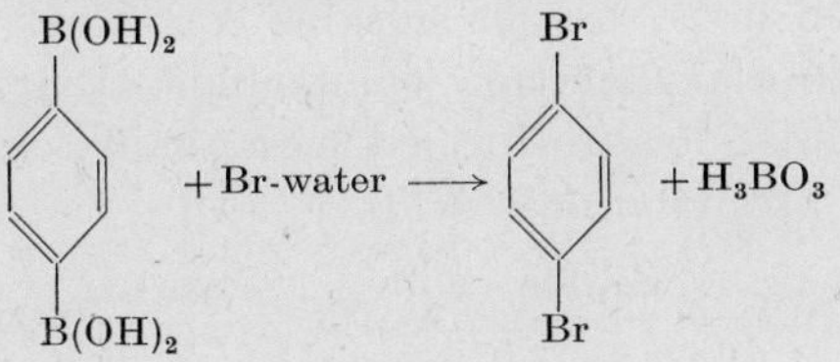

B. Hydrolytic C—B Bond Cleavage in the Presence of Metal Salts

Aqueous solutions of salts of metals of certain B groups of the periodic classification react with arylboronic acids (Ainley *et al.*, 1930).

$$\text{PhB(OH)}_2 + 2\text{CuX}_2 + \text{H}_2\text{O} \longrightarrow \text{PhX} + \text{Cu}_2\text{X}_2 + \text{HX} + \text{B(OH)}_3$$

(For copper, silver, zinc, cadmium, mercury, and thallium, but not for beryllium, magnesium, calcium chlorides or bromides.) Copper sulphate gave benzene, biphenyl, and a trace of phenol, whereas cupric acetate gave cuprous oxide and the binuclear hydrocarbon with phenylboronic, *o*-nitrophenylboronic, and *m*-nitrophenylboronic acids (Holzbecher, 1952). 2-Furyl- and 2-thienylboronic acids reacted with cupric chloride to give the corresponding 2-chloroheterocycle.

Alkylboronic acids did not react with cupric salts.

The following reactions have also been observed by Michaelis *et al.* (1882, 1894), Ainley *et al.* (1930), and Johnson *et al.* (1938).

$$PhB(OH)_2 + AgNO_3 \longrightarrow PhB\!\!-\!\!OH \longrightarrow C_6H_6 + Ag_2O + B(OH)_3$$
$$\underset{OAg}{|}$$

$$o\text{-}MeC_6H_4B(OH)_2 + AgNO_3 \longrightarrow MeC_6H_5$$
$$\text{(Ammoniacal)}$$

2-Furylboronic acid

2-Thienylboronic acid $\quad + AgNO_3 \longrightarrow$ Parent heterocyclic compound
$\qquad\qquad\qquad\qquad$ (Ammoniacal)

Alkylboronic acid $\qquad + AgNO_3 \longrightarrow$ hydrocarbon
$\qquad\qquad\qquad\qquad$ (Ammoniacal)

$$PhB(OH)_2 + MX_2 + H_2O \longrightarrow PhMX + HX + B(OH)_3$$
$$\text{(Cd, Zn)} \qquad\qquad \Big|\; H_2O$$
$$\qquad\qquad\qquad\qquad\longrightarrow C_6H_6$$

$$ArB(OH)_2 + HgCl_2 \longrightarrow ArHgCl + HCl + B(OH)_3$$

Where Ar = Ph, $o\text{-}MeC_6H_4$, 2-furyl, 2-thienyl.

In general, alkylboronic acids do not react with mercuric salts; but benzylboronic acid does. Under suitable conditions arylboronic acids are converted into diarylmercury compounds. Diarylthallium halides or arylthallium dihalides are formed with thallic chloride which does not react with *n*-propylboronic acid.

$$2ArB(OH)_2 \xrightarrow{\;TlX_3,\ H_2O\;} Ar_2TlX \xrightarrow{\;TlX_3,\ H_2O\;} ArTlX_2$$

CHAPTER VI

THE ATTACHMENT OF TWO HYDROCARBON GROUPS TO BORON

I. Attachment of Alkyl Groups

DIRECT attachment of two alkyl groups to boron has received little attention, although there are references (see later) to the approach to compounds having two attached groups from those having three groups so attached; meaning that in these systems, three groups are attached by a synthetic process such as the Grignard reaction, and then one group is removed.

Dimethylboron fluoride is referred to (Wiberg* et al., 1948) as being prepared from metal alkyls, and bis-2-chlorovinylboron chloride (Borisov, 1951) has been obtained from the corresponding zinc alkyl and boron trihalide.

$$BX_3 + ZnR_2 \longrightarrow R_2BX + ZnX_2$$

Referring to the di-n-butyl system, for which standard techniques have been worked out by Gerrard et al. (1957), n-butylmagnesium bromide in ether is added to an ethereal solution of boron trifluoride at $-70°C$ in an atmosphere of nitrogen. After aqueous treatment and hydrolysis with 3N hydrochloric acid, di-n-butylborinic anhydride, b.p. $137°C/9$ mm, is obtained.

$$2Bu^nMgBr + BF_3 \longrightarrow Bu^n_2BF + MgBr_2 + MgF_2 \xrightarrow{H_2O}$$

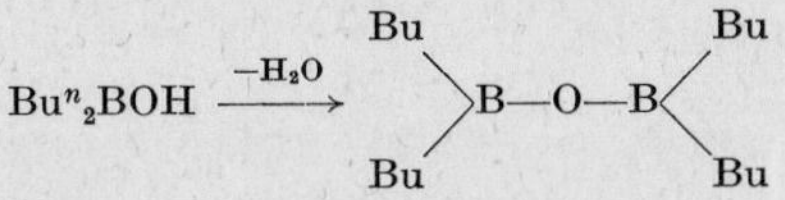

$$Bu^n_2BOH \xrightarrow{-H_2O} \underset{Bu}{\overset{Bu}{>}}B-O-B\underset{Bu}{\overset{Bu}{<}}$$

This seems to be an improvement on previous techniques (Letsinger et al., 1954; Rothstein et al., 1952), and the anhydride is an excellent starting material for the preparation of di-n-butylboron derivatives. It is not unlikely that such reactions are of general applicability, bearing in mind the reservations mentioned (p. 102).

* Cited by Goubeau, F.I.A.T. Review of German Science, Vol. I, p. 218.

Esterification is readily carried out by using alcohol in excess, and distilling off the water as an azeotrope; but, as usual, t-butanol does not react.

$$(Bu^n{}_2B)_2O + 2ROH \longrightarrow 2Bu^n{}_2BOR + H_2O$$

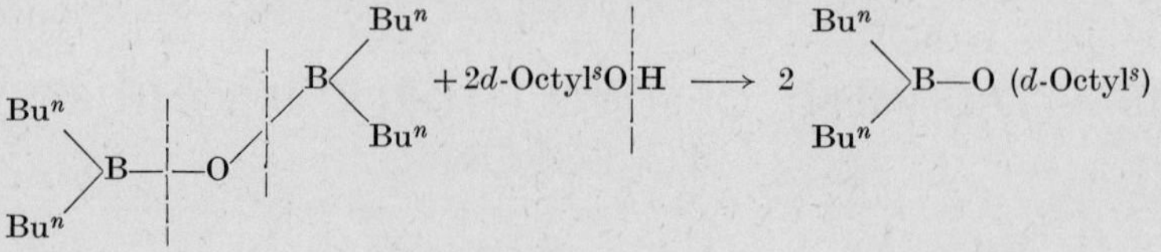

Interaction with optically active 2-octanol showed that the esterification involved boron–oxygen fission, for there was no loss of optical activity or change in configuration.

The anhydride is readily hydrolysed, and removal of volatile matter at $25°C/18$ mm gives gelatinous viscous di-n-butylboronic acid, which becomes the anhydride after being held for several hours at $20°C/0·2$ mm.

$$(Bu^n{}_2B)_2O \; \underset{-H_2O}{\overset{H_2O}{\rightleftharpoons}} \; 2Bu^n{}_2BOH$$

The esters are easily hydrolysed, and the monohalide is readily prepared by the following reactions (X=Cl or Br). Preparation of the chloride by phosphorus pentachloride is rather more convenient; but

$$(Bu^n{}_2B)_2O + BX_3 \longrightarrow 2Bu^n{}_2BX + BOX$$

$$3BOX \longrightarrow B_2O_3 + BX_3$$

neither thionyl chloride nor hydrogen chloride reacted with the anhydride.

$$(Bu^n{}_2B)_2O + PCl_5 \longrightarrow 2Bu^n{}_2BCl + POCl_3$$

Di-n-butylboron chloride is a highly reactive material, b.p. $72–73°C/18$ mm, being quickly oxidized in air, and the corresponding bromide has b.p. $53–55°C/0.5$ mm. These halides are, of course, quickly hydrolysed to the dialkylborinic acid, and readily give esters, especially in the presence of pyridine.

$$Bu^n{}_2BCl + ROH \longrightarrow Bu^n{}_2BOR + HCl$$

$$Bu^n{}_2BCl + t\text{-}BuOH + C_5H_5N \longrightarrow Bu^n{}_2BOBu^t + C_5H_5NHCl$$

Another method of preparing the chloride is by interaction between the ester, $Bu^n{}_2BOBu^n$, and phosphorus pentachloride, and there is mention of a number of other examples (Mikhailov et al., 1956).

An entirely different approach to the dialkylboron halide, and thence to the acid and derivatives, is by the process of attaching three hydrocarbon groups to the boron (see p. 92) followed by removal of one.

This is a process of boron–carbon fission, or dealkylation of boron, and the following reactions have been described.

$$R_3B + HX \longrightarrow R_2BX + RH$$

$$R_3B + X_2 \longrightarrow R_2BX + RX$$

$$2R_3B + BX_3 \longrightarrow 3R_2BX$$

(X=Cl or Br, R=alkyl).

$$Bu^n_3B \xrightarrow[\text{at } 110°C \text{ for 22 hours.}]{\text{HCl passed in}} Bu^n_2Cl + Bu^nH \text{ (Booth } et \text{ } al., 1952)$$

Nearly quantitative yield

$$Bu^n_3B \xrightarrow{\text{HBr}} Bu^n_2BBr + Bu^nH \text{ (Johnson } et \text{ } al., 1938)$$

$$Bu^n_3B \xrightarrow{\text{HI}} Bu^n_2BI + Bu^nH \text{ (Skinner } et \text{ } al., 1953)$$

Dimethylboron chloride and bromide, and diethylboron chloride and bromide have been similarly prepared by Wiberg* et al. (1948). Di-n-butylboron bromide was prepared by the action of bromine on heated tri-n-butylboron (Johnson et al., 1938), and the corresponding chloride was obtained in yields of 90% or better by passing boron trichloride into tri-n-butylboron at 160°C (McCusker et al., 1957; Buls et al., 1957).

One hydrocarbon group can be removed from a trialkylboron by carefully regulated oxidation, and such oxidative fission will be discussed later; for the immediate purpose the following examples are given.

$$2Bu^n_3B + O_2 \xrightarrow{\text{moisture}} 2Bu^n_2BOBu^n$$

$$Et_3B + RCHO \longrightarrow Et_2BOCH_2R + C_2H_4$$

(RCHO=such as CCl_3CHO, PhCHO, p-ClC_6H_4CHO) (Meerwein et al., 1936)

$$Et_3B + RCH_2OH \longrightarrow Et_2BOR + C_2H_6$$

Bromal and triethylboron gave 2,2-dibromovinyl diethylborinate and 2,2,2-tribromoethyl diethylborinate.

Carboxylic acids (acetic and p-chlorobenzoic acids) have been used to remove an alkyl group (Meerwein et al., 1936).

$$Et_3B + RCOOH \longrightarrow \begin{array}{c} Et \\ \diagdown \\ B-O-C-R + C_2H_6 \\ \diagup \quad \parallel \\ Et \quad\quad O \end{array}$$

Acyl diethylborinate.

Tri-n-butylboron and t-butyl hypochlorite at −80°C gave products including n-butyl dibutenylborinate, as shown by Johnson et al. (1938).

* Cited by Goubeau, F.I.A.T. Review.

Esters have been prepared by the Grignard or lithium reagents. For example, addition of allylmagnesium bromide (2 moles) to trimethylborate (1 mole) gave methyl dialkylborinate (Rothstein *et al.*, 1952), and *n*-butyl di-*n*-octylborinate was similarly prepared, as it also was from *n*-octyllithium by Brindley *et al.* (1955).

$$2n\text{-}C_8H_{17}MgBr + (Bu^nO)_3B \longrightarrow (n\text{-}C_8H_{17})_2BOBu^n$$

$$2n\text{-}C_8H_{17}Li + (Bu^nO)_3B \longrightarrow (n\text{-}C_8H_{17})_2BOBu^n$$

Of course, these esters are readily hydrolysed to the acids which can then be converted into derivatives, directly or *via* the anhydride.

Letsinger *et al.* (1954) have shown that a peculiar diborinate can be prepared by the interaction of a chelate orthoborate with a Grignard reagent.

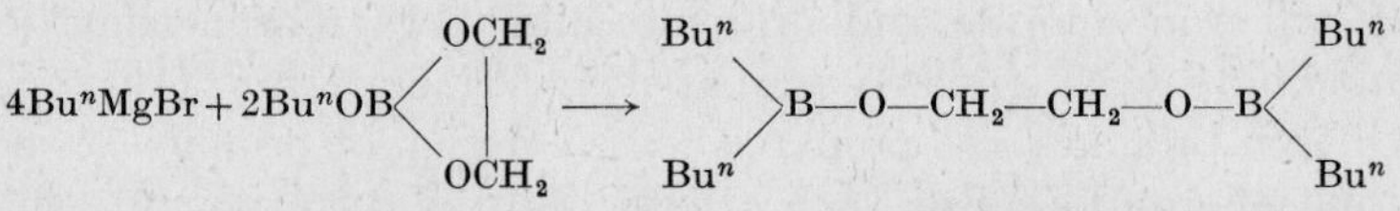

b.p. 133–144°C/1 mm

Simple dialkylboronic acids have been obtained by the hydrolysis-oxidation of alkylated boranes in which two alkyl groups are attached to one boron atom, and it is thus shown how the methyl groups are distributed in the molecule (Schlesinger *et al.*, 1935, 1936).

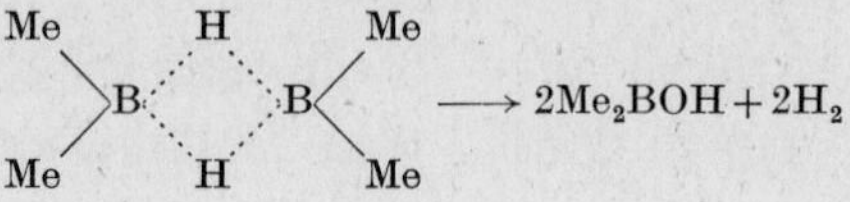

A. Properties of the Dialkylboron Halides

The following schemes illustrate the reactions carried out on specific halides. Ammonia and amines give aminodialkylborons.

$$Bu^n_2BCl + NH_3 \longrightarrow Bu^n_2BNH_2$$

$$Bu^n_2BCl + EtNH_2 \longrightarrow Bu^n_2BNHEt$$

Hydrogen chloride is removed as ammonium or amine hydrochloride by adding the base in excess (Booth *et al.*, 1952).

In phosphinoborine investigations to be described later, phosphino-dimethylboron was prepared by the interaction of triethylamine complex of dimethylboron bromide and phosphine (Burg *et al.*, 1953).

$$Me_2BBr + Et_3N \longrightarrow 1{:}1 \text{ complex} \xrightarrow{PH_3} Me_2BPH_2 + Et_3NHBr$$

Aminodimethylboron (Becher *et al.*, 1952) and dimethylaminodimethylboron (Burg *et al.*, 1954) have been similarly prepared.

$$Me_2BCl + 2NH_3 \longrightarrow Me_2BNH_2 + NH_4Cl$$

$$Me_2BCl + 2Me_2NH \longrightarrow Me_2BNMe_2 + Me_2NH_2Cl$$

Di-*n*-butylboron chloride has afforded the corresponding compounds.

By redistribution (or mutual replacement) reactions, Wiberg *et al.* (1937) showed that dimethylboron chloride gave trimethylboron and boron trichloride; but the trichloride could not be obtained in the butyl system (McCusker *et al.*, 1957) because the equilibrium lay far to the left at temperatures up to 180°C (see discussion on redistribution of groups, p. 102).

$$2R_2BCl \rightleftharpoons RBCl_2 + R_3B \quad (R = n\text{-Bu}, s\text{-Bu}, \text{ or } i\text{-Bu})$$

The question of redistribution or mutual replacement (metathetical) reactions in boron compounds is a complex one, and generalizations may not be confidently made. So much depends on the nature of the R group, the way it is attached to boron, whether directly or through oxygen or nitrogen, and upon experimental conditions. The matter will be referred to again later.

Dealkylation has been effected by interaction with hydrogen chloride in the presence of aluminium chloride (Booth *et al.*, 1952), thus affording a route to alkylboron dihalides other than by primary attachment of only one alkyl group.

$$Bu^n_2BCl + HCl \xrightarrow{\text{AlCl}_3} Bu^nBCl_2 + C_4H_{10}$$

Other reactions are as follows.

$$3Pr^n_2BI + 3SbF_3 \longrightarrow 3Pr^n_2BF \xrightarrow{\text{Redistribution}} 2Pr^n_3B + BF_3 \text{ (Long } et\,al.,\ 1953)$$

$$Bu^n_2BCl \xrightarrow{\text{Na in liquid NH}_3} Bu^n_3B + \text{(solid product) (Auten } et\,al.,\ 1952)$$

$$Me_2BBr + (Me_3Si)_2O \longrightarrow Me_2BOSiMe_3 + Me_3SiBr$$

The reaction between hexamethylsiloxane and dimethylboron bromide (Wiberg *et al.*, 1953) is related to a field of research having considerable possibilities, and more recent and systematic work will be discussed later (Gerrard *et al.*, 1958, 1960).

Hydrocarbonation of boron in dialkylboron halides has been done by means of metal alkenyls, of which vinylsodium and propenyllithium may be cited (Parsons *et al.*, 1954).

$$RCH{=}CHM + Me_2BBr \longrightarrow Me_2BCH{=}CHR$$

An outstanding function of certain boron compounds is their coordination with bases, and in Table VIII representative information is recorded for the dialkylboron halides.

TABLE VIII

Compound	Melting point °C	Reference
Me_2BFNH_3	51	Wiberg et al. (1951)
Me_2BFNH_2Me	68	Wiberg et al. (1951)
$Me_2BFNHMe_2$	21	Wiberg et al. (1951)
Me_2BFNMe_3	33	Wiberg et al. (1951)
Me_2BClNH_3	—	Wiberg et al. (1948)
Me_2BClNH_2Me	—	Wiberg et al. (1947)
$Bu^n_2BClNEt_3$	—	Stone et al. (1955)

II. Attachment of Aryl Groups

Diphenylboron chloride was first prepared by Michaelis et al. (1894, 1901) in low yield from diphenylmercury and phenylboron dichloride; and the corresponding bromide was made from boron tribromide and diphenylmercury. By hydrolysis the acid was readily obtained; tendency to ready dehydration was soon discovered, and indeed it

$$HgPh_2 + PhBCl_2 \longrightarrow Ph_2BCl$$
$$2HgPh_2 + BBr_3 \longrightarrow Ph_2BBr$$
$$\left. \right\} \xrightarrow{H_2O} Ph_2BOH \text{ Diphenylborinic acid}$$

was p-tolylborinic anhydride, and not the acid, that was obtained by a similar procedure.

The Grignard technique has normally been used to attach two aryl groups to boron, and König et al. (1930) and Melnikov et al. (1936, 1938) prepared a number of diarylborinic acids by the addition of triisobutyl borate in ether to the Grignard reagent at room temperature. A standard procedure is to add the Grignard reagent to an ethereal solution of tri-n-butyl borate at about −60°C, followed by

$$2ArMgBr + B(OBu^i)_3 \longrightarrow Ar_2BOBu^i \xrightarrow{H_2O} Ar_2BOH$$
$$\text{(Not isolated)}$$

isolation of the diarylborinic acid after the usual hydrolysis. Diarylboronic acids (Ph, p-MeC_6H_4), isolated as anhydrides, were prepared by adding tri-n-butyl borate (1 mole) (Neu, 1955) to the Grignard reagent (2 moles), instead of the other way round. Povlock et al. (1958) have used trimethoxyboroxine with aryl Grignard reagents.

The essential points in the chemistry of diarylborinic acids and derivatives are illustrated in the following schemes (see Melnikov *et al.*, 1936, 1938; Letsinger *et al.*, 1955).

$$Ar_2BOH + X_2 + H_2O \longrightarrow ArB(OH)_2 + ArX + HX$$

$$ArB(OH)_2 + X_2 + H_2O \longrightarrow B(OH)_3 + ArX + HX$$

(X = Cl, Br, or OH)

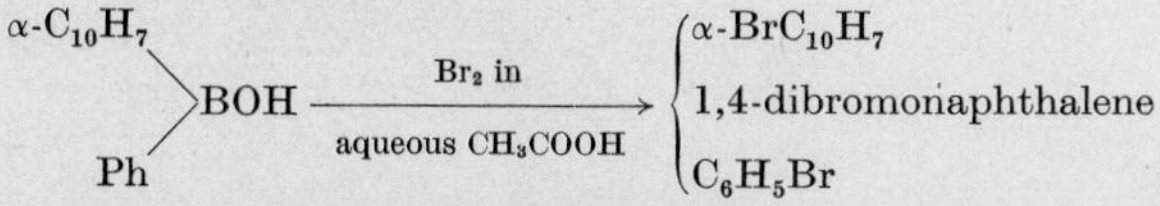

$$\alpha\text{-}C_{10}H_7 \diagdown \atop Ph \diagup BOH \xrightarrow[\text{aqueous } CH_3COOH]{Br_2 \text{ in}} \begin{cases} \alpha\text{-}BrC_{10}H_7 \\ 1,4\text{-dibromonaphthalene} \\ C_6H_5Br \end{cases}$$

Thus dearylation of boron, or deboronation of the aryl ring, according to one's viewpoint, is effected by halogen.

Dearylation is also effected by heating the acids to a sufficiently high temperature.

$$(\alpha\text{-}C_{10}H_7)_2BOH \xrightarrow[\text{6 hours at 120-130°C}]{Heat \text{ for}} \begin{array}{c} C_{10}H_8 \\ (\alpha\text{-}C_{10}H_7BO)_3 \end{array}$$

$$\alpha\text{-}C_{10}H_7 \diagdown \atop Ph \diagup BOH \xrightarrow[\text{with aqueous ethanolic solution} \atop \text{of 2-(dimethylamino) ethanol}]{Heat} \begin{cases} C_{10}H_8 \\ PhB(OH)_2 \end{cases}$$

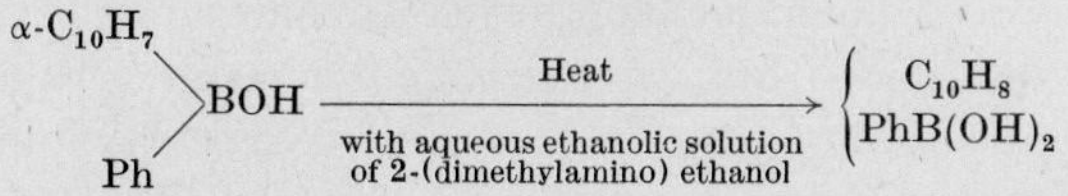

A detailed study has been made on the diphenylborinic acid system by Abel *et al.* (1956, 1957, 1958). The acid itself was prepared by the Grignard procedure, and converted into the anhydride $(Ph_2B)_2O$, m.p. 116°C, by agitating it at 20°C/0.05 mm for 2 hours, a better procedure than that entailing heating to 200°C.

The chloride, Ph_2BCl, b.p. 98°C/0.1 mm, and the bromide, Ph_2BBr, b.p. 132°C/0.4 mm, have been obtained by the ready interaction of the anhydride and the correct boron trihalide, but it must be pointed out that boron trifluoride is without similar action.

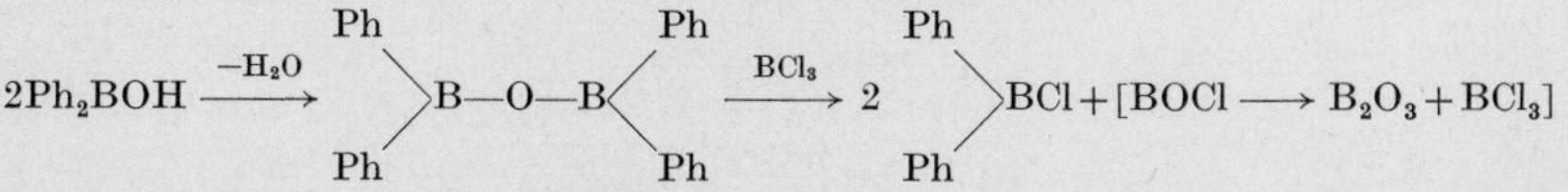

$$2Ph_2BOH \xrightarrow{-H_2O} \underset{Ph}{\overset{Ph}{\diagdown}}B\text{---}O\text{---}B\underset{Ph}{\overset{Ph}{\diagup}} \xrightarrow{BCl_3} 2 \underset{Ph}{\overset{Ph}{\diagdown}}BCl + [BOCl \longrightarrow B_2O_3 + BCl_3]$$

The anhydride had to be heated with phosphorus pentachloride or the pentabromide to afford the corresponding diphenylboron halide. As in the examples of esters of phenylboronic acid, interaction of an ordinary ester of diphenylborinic acid with phosphorus pentahalide (Cl, Br) or boron trihalide (Cl, Br) also affords the diphenylboron halide.

$$Ph_2BOR + PCl_5 \longrightarrow Ph_2BCl + RCl + POCl_3$$

Esters, Ph_2BOR, are prepared by azeotropic removal of water from a mixture of an alcohol and the acid (or anhydride), assisted where

4

necessary by an inert solvent, or may be prepared from the chloride, in the presence of pyridine, of course, when the alcoholic carbon atom is very reactive, such as in t-butanol. The 2-octyl ester and the 1-phenylethyl ester have been obtained without change in configuration, and without loss in optical activity, thus again showing B—O and not C—O fission. Similarly, hydrolysis in alkaline solution leads also to retention of configuration and rotatory power.

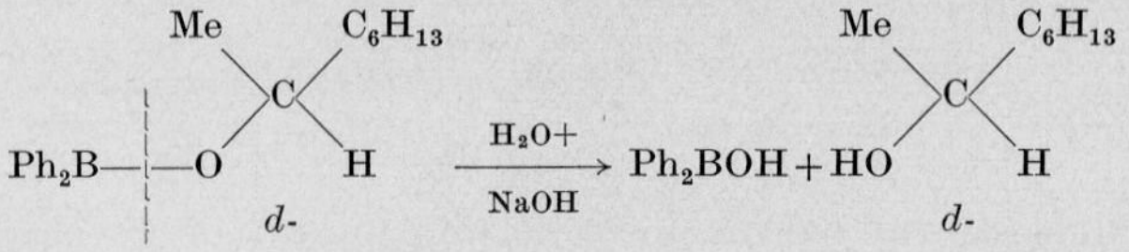

The halides are very readily hydrolysed, and in general so are the esters (see Mikhailov *et al.*, 1956). Neither the 1-butyl nor the 2-butyl ester reacted with hydrogen chloride (10 hours) at 20°C, but both the t-butyl and the 1-phenylethyl esters did and afforded the alkyl chloride, in accordance with the principles discussed (p. 1). Preponderantly inverted 1-chloro-1-phenylethane, with considerable loss in rotatory power, was obtained from that particular ester. When hydrogen bromide was passed into the 2-octyl ester at 20°C (4 hours) there was no loss of ester; but when the latter was mixed with liquid hydrogen bromide at −80°C, and the mixture allowed to remain in a sealed tube at 20°C (25 days), 2-bromooctane having preponderantly inverted configuration (but with some loss in optical activity) was obtained. The following schemes show the important types of reaction.

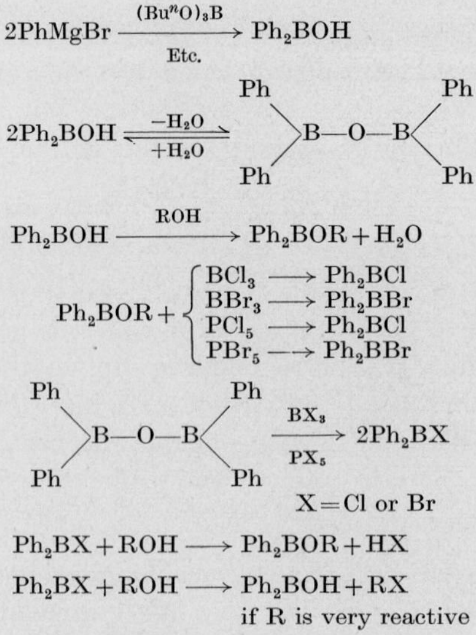

Although the borinic esters are usually easily hydrolysed so far as the B—O—C linkage is concerned, the C—B bond is not by any means so susceptible to hydrolysis; C—B fission, however, is accomplished by heating diphenylborinic acid at 175°C. Of course, in the presence of added water, complete dephenylation occurs.

$$Ph_2BOH \xrightarrow{175°C} PhH + (PhBO)_3$$
$$\text{Benzene} \quad \text{Phenylboronic anhydride}$$

$$Ph_2BOH \xrightarrow[175°C]{H_2O} 2PhH + (HO)_3B$$

At 200°C esters of diphenylborinic acid show some stability, and slowly undergo redistribution of OR groups rather than dephenylation by elimination of olefin; but pyrolysis of n-octyl diphenylborinate at 340°C gave phenylboronic anhydride, benzene, and a mixture of oct-2- and-3-ene (Abel et al., 1958).

$$Ph_2BOOct^n \xrightarrow{340°C} PhH + C_8H_{16} + (PhBO)_3$$

Mutual replacement occurred when diphenylboron halides were heated at 200°C with a boron trihalide, and when heated with hydrogen

$$Ph_2BX + BX_3 \longrightarrow 2PhBX_2$$

bromide at 100°C in a sealed tube, diphenylboron bromide gave benzene and boron tribromide. Cleavage also occurred when the bromide was heated with bromine at 200°C in a sealed tube.

$$Ph_2BBr + 2HBr \longrightarrow 2PhH + BBr_3$$

$$Ph_2BBr + Br_2 \longrightarrow PhBr + PhBBr_2$$

It has been suggested that the initial step is the formation of an ion pair $H^+[Ph_2BBr_2]^-$, $Br^+[Ph_2BBr_2]^-$, $[BX_2]^+[Ph_2BX_2]^-$, which undergoes reaction by either or both of two mechanisms illustrated in one example.

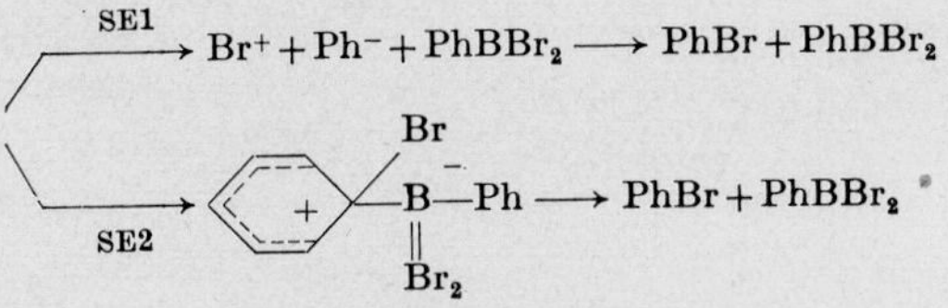

Ethers were not split by the diphenylboronhalides, thus showing low electrophilic reactivity of the boron atom, nor was there any reaction with sodium in an attempt to form tetraphenyldiboron, Ph_2B—BPh_2.

III. Ethanolamine Esters

2-Aminoethyl esters of diarylborinic acids are comparatively resistant to hydrolysis and oxidation; and as they can usually be recrystallized from water, they constitute a means to the isolation and purification of the acids after preparation by the Grignard or lithium procedure. Thus the Grignard reagent (2 moles) (Ph, α-naphthyl) was added to tri-n-butyl borate (1 mole) at $-60°C$, and after hydrolysis, an ethanolic solution of 2-ethanolamine was added to precipitate the 2-aminoethyl ester (Letsinger et al., 1955). o,o'-Dibenzyldilithium and tri-n-butyl borate gave the interesting structure shown in the diagram, pointing to the formula given.

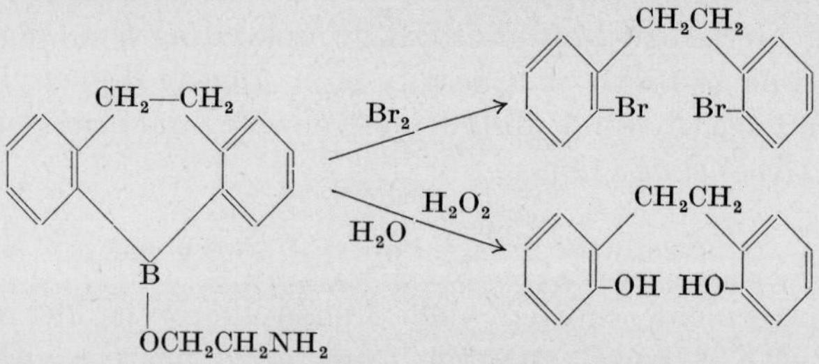

2-Aminoethyl di(p-bromophenyl)borinate is also described, and a suggested formula for the 2-ethylamino-ester is as follows.

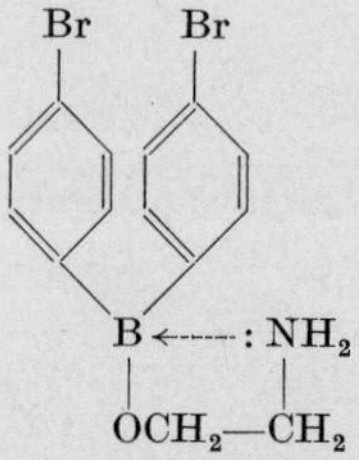

Mixed diarylborinic acids (and esters) have been formed by interaction of a dialkyl arylboronate (e.h. Ph, α-naphthyl) and the corresponding arylmagnesium bromide (Letsinger et al., 1954; Torssell, 1955).

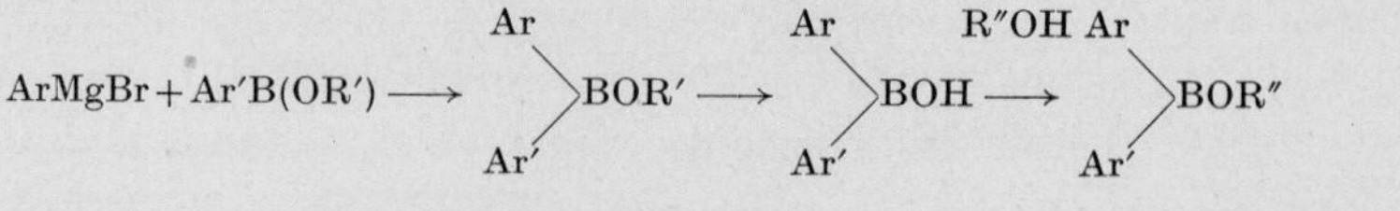

An alkylmagnesium bromide in place of an aryl one gives the mixed alkyl-arylborinic acid and esters.

Lawesson (1956) has reported on diethanolamine esters of alkylborinic acids.

In a series of papers by Neu (1955–1958) the applications of diphenylborinic anhydride and the acid as reagents for the detection of such compounds as hydroxyflavones and long-chain cation active compounds of ampholytes and amine salts are discussed. Diphenylborinic acid is considered as an analytical reagent.

IV. Stabilization by Chelation

Stabilization of boron by internal co-ordination to nitrogen is demonstrated by means of the 2-ethylamino-esters just mentioned. Gerrard *et al.* (1958) have reported that outstanding examples of co-ordination to oxygen are afforded by 2-ethoxycarbonyl-1-methyl-vinyl di-*n*-butyl- and diphenylborinate. These are formed by the interaction of di-*n*-butylboron chloride and diphenylboron chloride with ethyl acetoacetate.

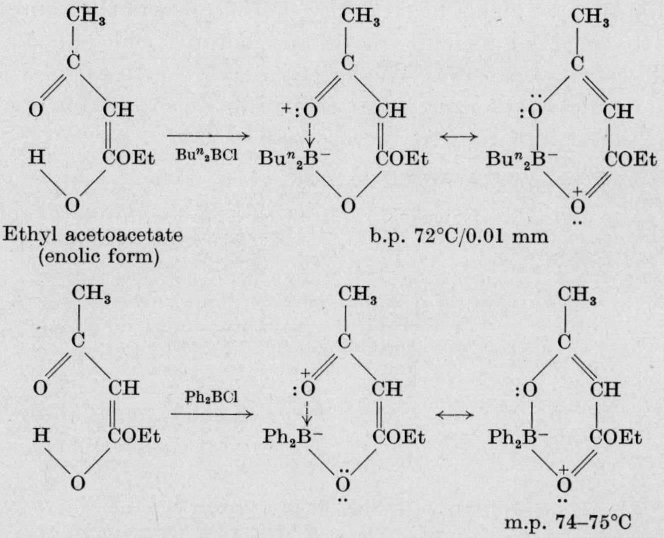

Ethyl acetoacetate
(enolic form)   b.p. 72°C/0.01 mm

m.p. 74–75°C

Whereas the dibutylboron compound was much more stable to oxidation and hydrolysis than the usual alkyl di-*n*-butylborniate, and the diphenylboron compound was stable to water and 3N-HCl during 5 hours, it seems probable that chelation itself is insufficient to account for the stability of these compounds, as amine complexes of borinates are still easily hydrolysed, as are chelated acyloxyboron compounds (see p. 231). The main cause of stability seems to be in the aromatic character of the ring; again showing that mere co-ordination does not necessarily reduce hydrolytic propensity to any marked extent.

CHAPTER VII

THE ATTACHMENT OF THREE HYDROCARBON GROUPS TO BORON

I. Formation of Trialkyl- and Triarylborons

THREE hydrocarbon groups (Me and Et) were first attached to boron by means of zinc alkyls, but diphenylmercury was apparently incapable

$$3ZnR_2 + 2B(OMe)_3 \longrightarrow 2BR_3 + 3Zn(OMe)_2$$

of arylating boron beyond the diphenylboron stage, Frankland *et al.* (1860, 1862) and Michaelis (1901) being associated with this early work. Quantitative yields of trimethyl boron and triethyl boron were reported to be given by zinc alkyls and boron trichloride, or boron trifluoride (Stock *et al.*, 1921; Wiberg *et al.*, 1937). An early procedure for a Grignard reagent and boron trifluoride was used for a number of trialkyl- and arylborons by Krause *et al.* (1921).

$$BF_3 + 3RMgX \longrightarrow BR_3 + 3MgXF$$

The following compounds were prepared by this technique (Table IX).

TABLE IX

Trisubstituted Borons and their Properties

$B(R)_3$ R=	Boiling point °C/mm	Melting point °C
Me	−21.8	—
Et	95	—
Pr^n	60/20	—
Pr^i	33–35/12	—
Bu^n	90–91/9	—
Bu^i	86/20	—
Bu^t	71/12	—
$Pentyl^i$	119/14	—
Hex^n	97/0.002	—
*cyclo*Hexyl	194/15	98–100
Ph	203/15	137
$PhCH_2$	229–232/13	47
o-MeC_6H_4	208/12	67–69
m-MeC_6H_4	218–222/12	59–60
p-MeC_6H_4	233–234/12	142–144
p-$MeOC_6H_4$	—	128
o-PhC_6H_4	—	201–203
$2,5$-$Me_2C_6H_3$	221/12	146–147
α—$C_{10}H_7$	—	206–207

There are modifications in technique, of course, and there is no report of a systematic examination of operational permissive and compelling factors. Tri-*iso*propylboron and tri-*t*-butylboron (Brown *et al.*, 1945; Long *et al.*, 1953) are prepared in 60–65% yields, and tri-*n*-propylboron in 50–60% yields, by using the diethyl ether complex of boron trifluoride instead of the gas. Whereas trimethylboron was isolated in only 10% yield with the Grignard reagent in diethyl ether, a much improved yield was obtained by Brown *et al.* (1944, 1945), when the reagent was added to boron trifluoride di-*n*-butyl etherate. Triethylboron was prepared by adding the di-*n*-butyl etherate of boron trifluoride *to* the Grignard reagent in *n*-butyl ether.

Trimethylboron in 87% yield was obtained from methylmagnesium bromide (or iodide) and boron trichloride in diethyl ether (Wiberg *et al.*, 1937; Smith *et al.*, 1951), and tri-*n*-butylboron was obtained in 70% yield. By the addition of boron tribromide in benzene to a benzene solution of the Grignard reagent, largely freed from diethyl ether, trimethylboron was obtained in 87% yield. Tri-*n*-butylboron was obtained in 50% yield from the Grignard reagent and trimethyl borate (Johnson *et al.*, 1938). Partly alkylated boron halides have been tried; tri-*n*-butylboron was obtained from allylmagnesium bromide and di-*n*-butylboron chloride, a result presumably involving redistribution of groups in allyldibutylboron (see later). Secci (1958) has examined the EtMgI + BF₃ system as a means of producing ^{14}C labelled alkylborons.

Organolithium compounds have been effectively used by Torssell (1954; see also report in *Chem. and Eng. News*, 28 April, 1958, p. 56).

$$
\begin{array}{ccc}
\text{H}_2\text{C} \!-\!\!-\! \text{CH}_2 & & \text{H}_2\text{C} \!-\!\!-\! \text{CH}_2 \\
| \qquad\quad | & +\,\text{PhBF}_2 \longrightarrow & | \qquad\quad | \\
\text{H}_3\text{C} \quad\; \text{CH}_2 & & \text{H}_2\text{C} \quad\; \text{CH}_2 \\
| \qquad\quad | & & \diagdown\;\diagup \\
\text{Li} \quad\; \text{Li} & & \text{BPh}
\end{array}
$$

$$
\begin{array}{ccc}
\quad\; \text{CH}_2 & & \quad\; \text{CH}_2 \\
\diagup\;\diagdown & & \diagup\;\diagdown \\
\text{CH}_2 \;\; \text{CH}_2 & +\,\text{PhBF}_2 \longrightarrow & \text{CH}_2 \;\; \text{CH}_2 \\
| \qquad\; | & & | \qquad\; | \\
\text{CH}_2 \;\; \text{CH}_2 & & \text{CH}_2 \;\; \text{CH}_2 \\
| \qquad\; | & & \diagdown\;\diagup \\
\text{Li} \quad\; \text{Li} & & \text{BPh}
\end{array}
$$

Vinylsodium or propenyllithium and dimethylboron bromide gave the corresponding mixed alkyl-alkenylboron which redistributed, as shown by Parsons *et al.* (1954). Trialkenylborons are described by Parsons *et al.* (1957) and by Topchiev *et al.* (1958).

$$\text{Me}_2\text{BCH}\!\!=\!\!\text{CHR} \longrightarrow \text{MeB(CH}\!\!=\!\!\text{CHR)}_2 + \text{Me}_3\text{B}$$

Mixed trialkylborons have been prepared by addition of an alkyllithium in diethyl ether to n-butyl di-n-butylborinate in ether at $-70°C$ (Mikhailov *et al.*, 1956). The mixture was saturated with hydrogen chloride, filtered to remove lithium chloride, and the filtrate washed with aqueous sodium hydroxide. The organic layer was freed from alcohol and water, and azeotropically dried with benzene. With n-butyllithium, tri-n-butylboron, b.p. $109–110°C/23$ mm (38% yield), with n-propyllithium, di-n-butyl-n-propylboron, b.p. $76–80°C/9$ mm (50% yield), and with ethyllithium, ethyl-di-n-butylboron, b.p. $67–69°C/9$ mm (26% yield) were obtained.

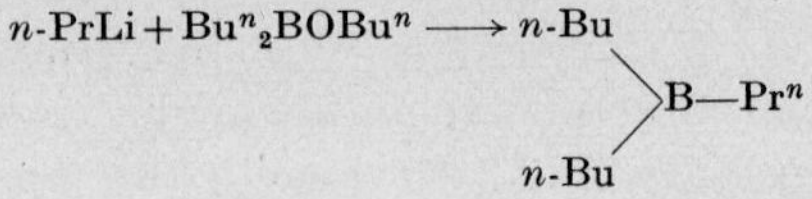

$$n\text{-PrLi} + Bu^n_2BOBu^n \longrightarrow \begin{array}{c} n\text{-Bu} \\ \diagdown \\ \phantom{n\text{-Bu}}B\text{—}Pr^n \\ \diagup \\ n\text{-Bu} \end{array}$$

Mixed trialkylborons have been prepared by exchange with Grignard reagents (see report in *Chem. and Eng. News*, 28 April, 1958, p. 56).

$$EtMgBr + Bu^n_3B \longrightarrow Bu^n_2BEt \text{ or } Bu^nBEt_2$$

In an attempt to get 1,4-bis(di-n-butylboryl)-butane and the analogous pentane derivative, the bifunctional Grignard and di-n-butylboron chloride afforded the boron ring compound.

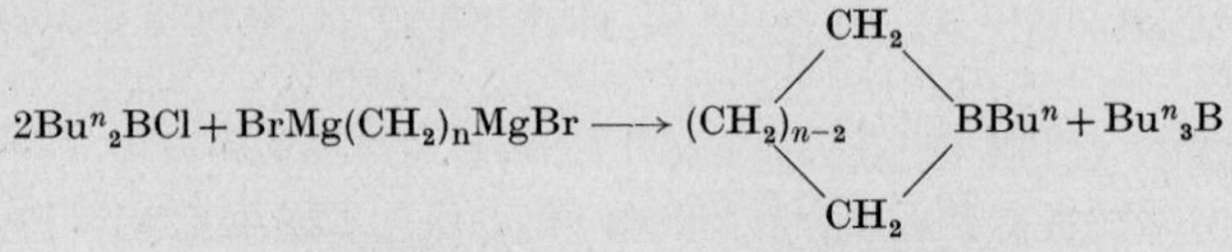

$$2Bu^n_2BCl + BrMg(CH_2)_nMgBr \longrightarrow (CH_2)_{n-2} \begin{array}{c} \diagup CH_2 \diagdown \\ BBu^n + Bu^n_3B \\ \diagdown CH_2 \diagup \end{array}$$

$$(n=4 \text{ and } 5)$$

$$2Bu^n_2BCl + BrMg(CH_2)_4MgBr \longrightarrow \begin{array}{c} CH_2\text{—}CH_2 \\ | \diagdown \\ | BBu^n + Bu^n_3B \\ CH_2\text{—}CH_2 \diagup \end{array}$$

$$2Bu^n_2BCl + BrMg(CH_2)_5MgBr \longrightarrow \begin{array}{c} \diagup CH_2\text{—}CH_2 \\ CH_2 \diagdown \\ BBu^n + Bu^n_3B \\ \diagdown CH_2\text{—}CH_2 \diagup \end{array}$$

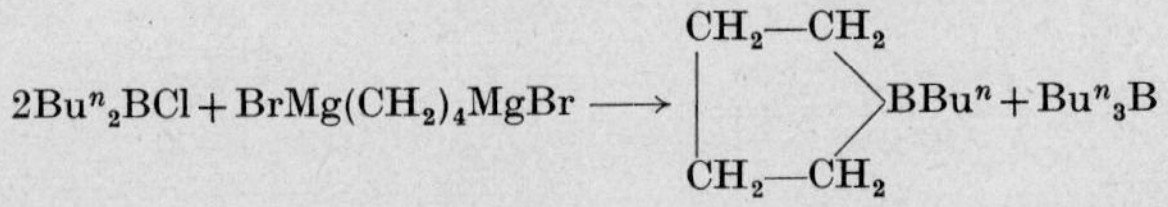

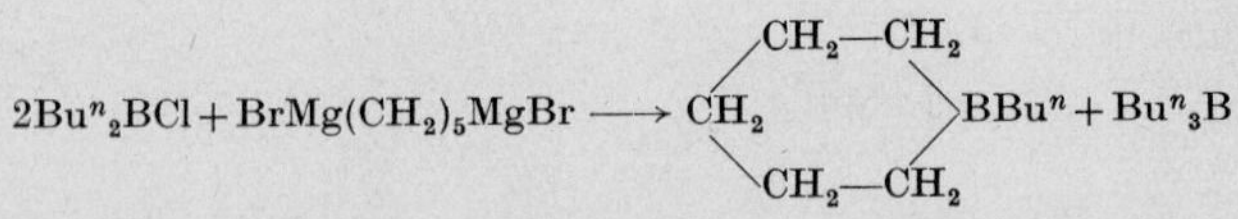

It is necessary to state at this stage that there is controversy about the production of mixed trialkyl or triarylborons, owing to the prevalence of mutual replacement or redistribution. Indeed, some believe that it is difficult, if not impossible, to isolate such compounds in a pure state. This point will be taken up specially later (see p. 102), because it relates to a considerable number of types of organic compound of boron.

A. Aluminium Alkyls

Trialkylaluminium compounds and alkylaluminium sesquihalides are an economic and convenient source of alkyl groups.

One method for the commercial production of triethylboron is by the interaction of the ethylaluminium sesquihalide and trimethylborate (Ashby, 1959).

$$Et_3Al_2Cl_3 + (MeO)_3B \longrightarrow Et_3B$$

Alternative procedures have been investigated, and a suggested procedure for those alkylborons containing alkyl groups for which the aluminium reagent is available (Me, Et, Bu^i) involves the interaction of trimethoxyboroxine (trimethoxyboroxole).

$$3Et_3Al + (MeOBO)_3 \longrightarrow 3Et_3B + Al(OMe)_3 + Al_2O_3$$
(in mineral oil at
room temperature)

$$3Et_3Al_2Cl_3 + (MeOBO)_3 \longrightarrow 3Et_3B + Al(OMe)_3 + Al_2O_3 + 3AlCl_3$$

With reference to trialkylaluminium as a reagent for the preparation of trialkylborons, see Köster (1958) and Zakharkin et al. (1959).

Reference must be made to a more direct procedure for the production of trialkylborons by the interaction of an alkyl chloride (Me, Et) and a boron halide (F, Cl, Br) when passed over aluminium at 350°C, or zinc at 325–350°C (Hurd, 1948). 2-Methylpropene and diborane are stated to give tri-*iso*butyl and tri-*t*-butylborons at 100°C (24 hours); but doubt has been cast on the existence of tri-*t*-butylboron. Benzene and diborane gave triphenylboron at 100°C (12 hours); but such procedures as described are not of practical significance. Related to these reactions is an attempt to obtain phenylborane ($PhBH_2$) by reduction of phenylboron dichloride with lithium aluminium hydride; triphenylboron was isolated, presumably as a result of mutual replacement in the phenylboranes (Nielsen et al., 1957). It is claimed that this is the first instance of positive evidence of mutual replacement in an arylborane.

$$3PhBCl_2 \xrightarrow{\text{LiAlH}_4} 3PhBH_2 \longrightarrow Ph_3B + B_2H_6$$

The recent production of a series of trialkylborons, R_3B, where R = such as $CH_2\!\!=\!\!CHCH_2CH_2$, or $MeOCH_2CH_2CH_2$, or $PhOCH_2CH_2CH_2CH_2$, is of special interest because of the influence of the functional part of R, as shown by Lyle et al. (1956).

B. Cyclic Boron–Carbon Compounds

Interest in the development of boron–carbon heterocycles is quickening; in addition to the methods already mentioned, for example those by Letsinger *et al.* (1955) and Torssell (1954), the pyrolysis of trialkyl-borons and the hydroboration of dienes and trienes have been shown by Köster *et al.* (1957, 1959) and Winternitz (1959) to yield heterocycles. Hydroboration of *cyclo*dodeca-1,5,9-triene has been carried out by Greenwood *et al.* (1960), using triethylamine–borane in light petroleum. Perhydro-9b-boraphenalene was formed as a colourless liquid, readily oxidized, and capable of forming a fairly stable complex with pyridine, although it shows no co-ordination with T.H.F.

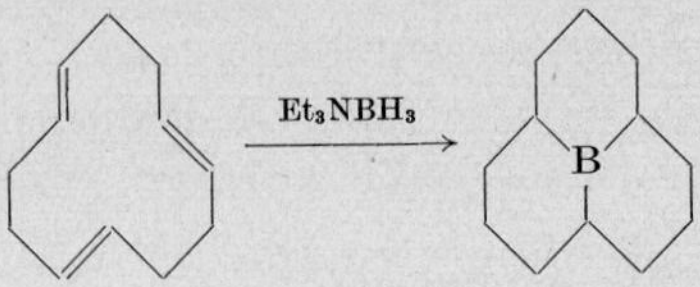

II. Properties of Trisubstituted Borons

Most of the fundamental work on trialkylborons has been done with trimethylboron, and that on triarylborons, with triphenylboron. So far as they have been investigated, trialkyl- and triarylborons are monomeric. The absence of dimerization in trimethylboron is attributed to stabilization by trigonal hyperconjugation. High value of the force constants for trimethylboron point to absence of the resonance found in the boron halides, and the molecule is considered to be planar. The following physical measurements have been made (see Table X).

TABLE X

Trisubstituted Borons—Physical Measurements

Measurement	Reference
Trouton's constant (Me, Pr^n, Pr^i)	Bamford, 1946
Raman spectrum (Me)	Goubeau *et al.*, 1952; Siebert, 1952
Raman spectrum (Et)	Blau *et al.*, 1957
Ultra-violet spectrum (Me)	Goubeau *et al.*, 1952
Force constants	Goubeau *et al.*, 1952; Siebert, 1952
Nuclear quadruple spectra for ^{10}B and ^{11}B (Me, Et)	Dehmelt, 1952, 1953; Das, 1957
Parachor (Et)	Laubengayer *et al.*, 1941
Electron diffraction	Levy *et al.*, 1937
Heat of combustion	Long *et al.*, 1949
Potential barrier values	French *et al.*, 1946; Shamin-Ahmad, 1952
Dipole moment	Curran *et al.*, 1957
Electric mobility (Me)	Ferguson *et al.*, 1957
Infra-red spectrum	Stewart, 1956

Tri-α-naphthylboron appears to exist in two rotational isomeric forms, and it crystallizes from benzene with solvent of crystallization (Krause et al., 1930; Brown et al., 1948). A sodium derivative of it is mentioned by Moeller et al. (1959).

Thermal stability has been of interest; heated for 2 days under reflux at 90–100°C/5 mm tri-n-butylboron gave butylene, whereas at 125–130°C/10 mm (10 days), trans-isobutylene and di-n-butyldiborane were formed (Rosenblum, 1955). It was reported that triethylboron decomposes at 100°C (Stock et al., 1921), that it is insoluble in cold water (Frankland et al., 1860), but is decomposed by hot water, although the products were not identified (Meerwein et al., 1936).

Oxidation reactions have been studied. Whereas trimethylboron and triethylboron are spontaneously combustible in air, and in oxygen form explosive mixtures, higher members and the triarylborons are more stable (Frankland, 1862; Johnson et al., 1938). Although the triarylborons, Ph, p-MeC$_6$H$_4$—, and MeOC$_6$H$_4$—, fume in air (Krause et al., 1931), tri-α-naphthylboron is quite stable (Brown et al., 1948).

A kinetic study (Bamford et al., 1946) of the homogeneous oxidation showed that the rate for tri-n-propylboron was too quick for measurement, and the slower rate for trimethylboron was attributed to the strengthening of the C—B bond by hyperconjugation. Effects of controlled oxidation were noticed by Frankland (1862); but as oxidative fission of the carbon–boron bond is of special interest, this is discussed in another section.

Interactions of trialkylborons with aldehydes and alcohols have been examined by Meerwein et al. (1936), and the interactions with carboxylic acids have been used in connection with hydroboronation procedure (p. 154).

$$Et_3B + RCHO \longrightarrow Et_2BOCH_2R + C_2H_4$$
$$(R = CCl_3CHO, PhCHO \text{ or } p\text{-}ClC_6H_4CHO)$$
$$Et_3B + RCH_2OH \longrightarrow Et_2BOCH_2R + C_2H_6$$

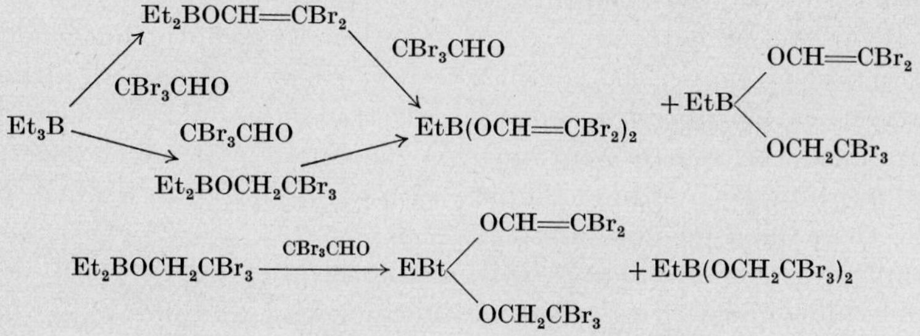

$$Et_3B + R.COOH \longrightarrow Et_2BOOCR + C_2H_6$$
$$(R = \text{acetic or } p\text{-}ClC_6H_4COOH)$$

Whereas triphenylboron heated with even large excess of methanol afforded only the phenyl diphenylborinate, tri-α-naphthylboron gave naphthalene and trimethyl borate (Rondestvedt et al., 1955).

Interactions with boron trichloride and hydrogen chloride have had significance in earlier procedures for making the alkylboron chlorides.

$$Bu^n_3B + 2BCl_3 \xrightarrow[\text{AlCl}_3]{110°C} 3Bu^nBCl_2$$
$$10\% \text{ yield, pure}$$

There was an improvement by a two-step procedure.

$$Bu^n_3B + HCl \xrightarrow{110°C} Bu^n_2BCl + C_4H_{10}$$
$$\text{B.p. } 173°C$$

$$2Bu^n_2BCl + HCl \xrightarrow{110°C} Bu^nBCl_2 + BCl_3 + C_4H_{10}$$
$$\text{B.p. } 108°C \text{ (and also butylene)}$$

At 180–210°C hydrogen chloride gave the dichloride, Bu^nBCl_2, but there were fractionation difficulties.

The matter of toxicity has been considered by Hill et al. (1958).

A. Co-ordination with Bases

The co-ordination compounds between trialkyl- and triarylborons on the one hand, and bases on the other, has been of considerable interest in connection with the fundamental work on the intervention of steric factors into the order of strength of a series of bases, particularly nitrogen bases, as discussed by Brown, H. C., in many papers. Electronic factors influence the electron density on the nitrogen atom of the base, and also the acid strength (electrophilic strength) of the boron atom. Steric factors influence the availability of the electron density on the nitrogen atom, and also have an effect on the ease with which the boron system can become tetrahedral in the co-ordination compound.

Thus trimethylboron forms definite complexes Me_3BNR_3, where R = H, or Me, or both, and where R = H or Et or both; but when the base is triethylamine, the complex is very easily dissociated, although with respect to proton acceptance, triethylamine is a stronger base than any of the others mentioned. α-Picoline is a stronger base than pyridine; but the pyridine complex with trimethylboron is much more stable than the α-picoline complex.

Similarly, trimethylboron will not form a complex with di-*iso*-propylamine, nor with 2-*t*-butylpyridine, nor with 2,2-dimethylpyridine. On the other hand, neither tri-*iso*propylboron and trimethylamine, nor tri-*t*-butylboron and trimethylamine complex. Trimethylboron formed a complex at each of the nitrogen atoms in ethylenediamine and 1,3-diaminopropane (Goubeau *et al.*, 1955). When the ethylene-diamine complex was heated at 250°C in a bomb, methane and the corresponding "diborazene" were formed: $CH_4 + Me_2BNHCH_2CH_2NHBMe_2$.

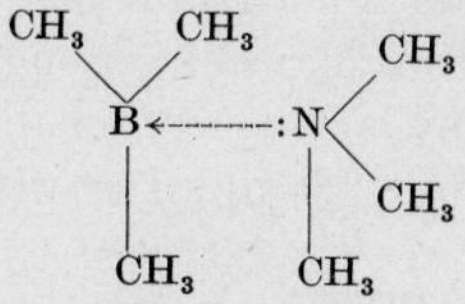

Form complex

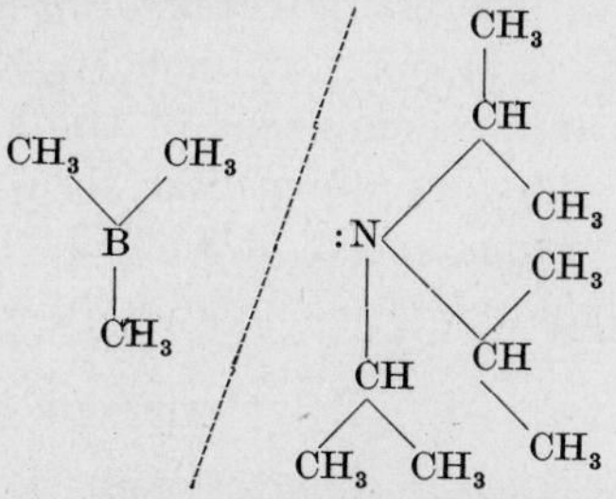

No complex formed, due to steric hindrance.

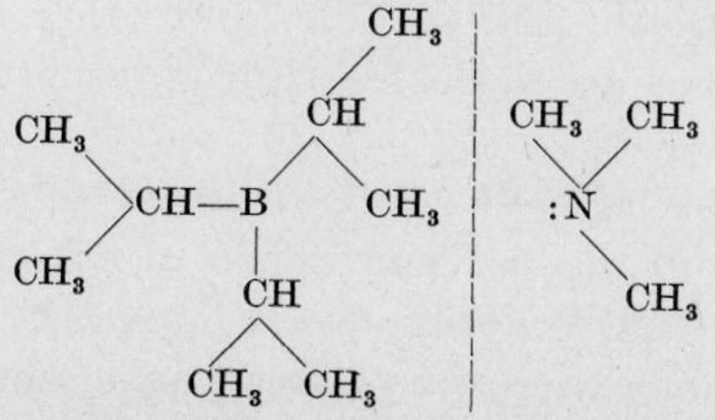

No complex formed, due to steric hindrance 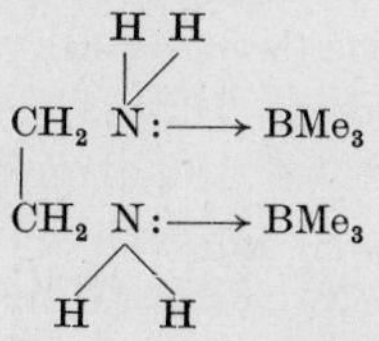   Dicomplex

Steric factors (Sujishi *et al.*, 1954) accounted for the non-formation of a complex between trimethylboron and $(SiH_3)_2NCH_3$ or $(SiH_3)_3N$; and the complex between $(SiH_3)Me_2N$ and trimethylboron was less stable than that between the latter and Me_3N.

Heat of dissociation of the complexes of trimethylboron and NH_3, $MeNH_2$, $EtNH_2$, Pr^nNH_2, Bu^nNH_2 are all about -7.26 ± 0.21 kcal/mole. Taft (1953) and Graham *et al.* (1956) have commented on the relative stabilities of certain co-ordination compounds relating to borane and trimethylboron.

There do not appear to be any examples of co-ordination of tri-alkylborons to oxygen in organic compounds.

III. Suggested Applications of Trisubstituted Borons

As with all definite types of organic boron compounds, energetic attempts have been made to find applications of trialkyl and triaryl-borons. The alkylborons have been examined as polymerization catalysts. The alkylborons are mentioned in connection with jet fuels, because of their high flame speed, wide range of inflammability, and high blow-out velocity. Their hydrolytic stability is good in neutral, acid, and alkaline solution; but they are in the main easily oxidized.

Triethylboron and higher members are in general soluble in hydrocarbons, but insoluble in water, and are miscible with most organic solvents.

The liquid adduct formed by triethylboron and sodium hydride should have interesting catalytic properties, and the oxidation of trialkylborons to primary alcohols (p. 154) has remarkable potentialities.

Certain amine complexes of trialkylborons have been suggested as diesel oil additives (Carpenter, 1957); they act as cetane improvers, increasing flame speed of fuel–air mixture. Complexes with higher amines might serve as sludge inhibitors.

IV. Polymerization by Trialkylborons

Kolesnikov et al. (1957, 1958) showed that tri-n-butylboron functioned as a catalyst in the polymerization of such vinyl monomers as acrylonitrile, methyl methacrylate, and styrene, and other examples of polymerization by alkylborons are discussed (Furukawa et al., 1957, 1958; Ashikari, 1958).

Pressures of 100 atmospheres at 50°C were required to effect the production of polyethylene from the polymerization of ethylene catalysed by tributylboron in toluene (Kolesnikov et al., 1957).

Traces of oxygen appear to be necessary for polymerization, and Bawn et al. (1959) have carried out kinetic studies on the polymerization of methyl methacrylate in benzene at 25.0°C and 48.2°C, by using different concentrations of tributylboron in the presence of oxygen in controlled amounts. It appears that the oxygen converts a part of the trialkylboron into a peroxide (see Abraham et al., 1957, 1959, and p. 115), which reacts with unoxidized trialkylboron to form initiating free radicals.

Kolesnikov et al. (1957) found that tri-n-butylboron was a very active catalyst for the polymerization of methyl methacrylate, the reaction being carried out in toluene at 60°C, the concentration of catalyst being 2 moles per cent of monomer. About 60% of the monomer was converted into polymethyl methacrylate in 3 hours. Polystyrene was obtained in 20% yield in 2 hours by the catalytic conversion of the monomer at 60°C in the presence of 2 moles per cent of n-butylboron. For acrylonitrile, the catalyst was but slightly active, and required an activator, which could be boron trifluoride etherate.

Furukawa et al. (1957) used triethylboron to initiate the polymerization of such vinyl monomers as vinyl acetate, vinyl chloride, vinylidene chloride, methacrylic ester, acrylic ester, and acrylonitrile. With vinyl acetate, for example, in n-hexane, the solution became turbid immediately, and reaction temperature had to be controlled at 20°C by cooling. The conversion was 42% in 2 hours, and average molecular weight of the polymer was about 2×10^4. It is believed that triethylboron acts as an anionic initiator. Vinyl chloride is polymerized at moderate rates even at $-30°C$ by triethylboron. Furthermore, because the trialkylboron is stable to water, it appears possible to carry out polymerization in water suspension.

Ashikari (1958) has studied the polymerization of acrylonitrile with triethylboron, and has found the reaction to occur readily at comparatively low temperature, and rapidly at room temperature, and is best carried out in a solvent such as hexane. Vinyl chloride was also polymerized rapidly by tri-n-butylboron, and with emulsion polymerization, the yield was about 90%, and the molecular weight large. Copolymerization was also investigated.

Alkyl borates catalyse the interaction of diazomethane with alcohols to give the corresponding methyl ether (Meerwein $et\ al.$, 1931), and such reactions, catalysed by alkylborons and alkyl borates, have been studied in some detail by Bawn $et\ al.$ (1958). Triethylboron was used to catalyse diazomethane in the presence of n-butanol, to give polymethylene in 60% yield, and a small amount of ethyl n-butyl ether was also formed. In the presence of ethanol, with tri-n-butyl borate as catalyst, small quantities of methyl n-butyl ether, ethyl n-butyl ether, and n-propyl n-butyl ether were obtained in addition to polymethylene. The following is the mechanism suggested, and conditions

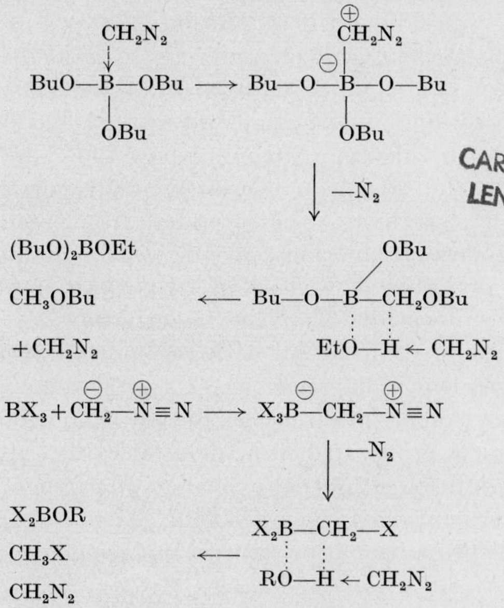

$$
\begin{array}{ccc}
 & \overset{CH_2N_2}{\underset{\downarrow}{|}} & \overset{\oplus}{\overset{CH_2N_2}{\underset{}{|}}} \\
BuO{-}B{-}OBu & \longrightarrow & Bu{-}O{-}\overset{\ominus}{B}{-}O{-}Bu \\
 & | & | \\
 & OBu & OBu
\end{array}
$$

$$\downarrow -N_2$$

$$(BuO)_2BOEt$$

$$CH_3OBu \qquad \qquad \overset{OBu}{\overset{\diagup}{\longleftarrow Bu{-}O{-}B{-}CH_2OBu}}$$

$$+CH_2N_2 \qquad \qquad EtO{-}H \leftarrow CH_2N_2$$

$$BX_3 + \overset{\ominus}{CH_2}{-}\overset{\oplus}{N} \equiv N \longrightarrow X_3\overset{\ominus}{B}{-}CH_2{-}\overset{\oplus}{N} \equiv N$$

$$\downarrow -N_2$$

$$X_2BOR$$

$$CH_3X \qquad \longleftarrow \qquad X_2B{-}CH_2{-}X$$

$$CH_2N_2 \qquad \qquad RO{-}H \leftarrow CH_2N_2$$

for the preparation of aliphatic ethers of chain lengths varying from methyl to polymethylene are thereby indicated.

Diazomethane will then react with X_2BOR, and the process will continue.

V. Mutual Replacement of Groups in Trialkylborons

Remarks made from time to time infer that the isolation of mixed trialkylborons, $RR'R''B$, or RR'_2B, is rarely accomplished (Krause *et al.*, 1931; Parsons *et al.*, 1954), because of a redistribution of the groups to give homogeneous trialkyl borons. This means that C—B bonds are being broken, even in the examples of the homogeneous

$$3RR'R''B \longrightarrow R_3B + R'_3B + R''_3B$$

trialkylborons, for although the diagnostic factors available in following the changes in the mixed trialkylborons are absent, there are no grounds for denying that if it happens in the one case, it happens in the other. The point is obviously an important one, and some attention has been given to an elucidation of factors involved in some recent work by McCusker *et al.* (1955, 1957, 1958) and Hennion *et al.* (1957, 1958, 1959).

Attempts to prepare tri-*t*-butylboron from *t*-butylmagnesium chloride and boron trichloride, or trifluoride, gave only tri-*iso*butylboron in good yield. Evidence of character was obtained from refractive index, density, infra-red spectra, and oxidation (see Brown *et al.*, 1956) to *iso*butanol. It was doubted if three *t*-butyl groups can be attached to boron, and the authenticity of materials so named is a moot point. Tri-*s*-butylboron can be prepared from *s*-butylmagnesium bromide and boron halides, so long as operations do not involve elevated temperatures. It can be distilled in a vacuum below 150°C, but at atmospheric pressure, distillation at 214°C causes complete conversion into tri-*n*-butylboron. Furthermore, the product from *iso*propylmagnesium bromide and boron trichloride appears with a boiling range of 148–155°C and is probably a mixture of tri-*iso*propylboron, tri-*n*-propylboron and the intermediate mixed trialkylborons.

Not only have we to contend with the mutual exchange of groups, but also consequent isomerization. A troublesome point will be to decide whether isomerization occurs during attachment of the carbon atom of the alkyl group to boron, or afterwards, as a result of conditions which determine mutual exchange of partners.

The workers mentioned suggest a four-centre system to account for the so-called disproportionation (mutual replacement) (mutual exchange of partners).

The exchange is supposed to occur through the overlap of the sigma orbital of the R—B bond of one molecule and the vacant p-orbital of the boron atom of the other molecule. Undue emphasis appears to be placed on the availability of the unoccupied p-orbital on the boron

$$\begin{array}{c} R \\ \diagup \\ B \\ \diagup \\ R \end{array} \begin{array}{c} R' \\ \diagdown \diagup \\ B \\ \diagdown \\ R \end{array} \begin{array}{c} R \\ \diagdown \\ \diagup \\ R' \end{array} \longrightarrow R_3B + R'_2BR$$

atom, and insufficient attention is given to the precise degree of reactivity of the "central" carbon atom of the alkyl group, i.e. the reactivity of the C—B bond with special reference to the electronic condition of C. This is especially noticeable when a comparison of mutual replacement tendencies in different types of compound, R_3B, R_2BX, $(RBO)_3$, etc., are compared.

The isomerization of t-butyl to isobutyl in these systems has been examined further. Whereas one t-butyl group can clearly be attached to boron and give t-butylboronic anhydride or t-butylboron dichloride, attempts to attach three such groups by using three moles of Grignard reagent for one mole of boron trifluoride or trimethyl borate gave only di-isobutyl-t-butylboron, and when boron trichloride was used, or when one mole of t-butylmagnesium chloride was added to one mole of di-isobutylboron chloride, only tri-isobutylboron was found.

$$3t\text{-BuMgCl} + \left\{ \begin{array}{c} BF_3 \\ \text{or} \\ (MeO)_3B \end{array} \right\} \longrightarrow i\text{-Bu}_2BBu^t$$

$$2t\text{-BuMgBr} + t\text{-BuBCl}_2 \longrightarrow i\text{-Bu}_2BBu^t$$

$$2i\text{-BuMgBr} + t\text{-BuBCl}_2 \longrightarrow i\text{-Bu}_2BBu^t$$

$$t\text{-BuMgCl} + i\text{-Bu}_2BF \longrightarrow i\text{-Bu}_2BBu^t$$

$$3t\text{-BuMgCl} + BCl_3 \longrightarrow i\text{-Bu}_3B$$

$$t\text{-BuMgCl} + i\text{-Bu}_2BCl \longrightarrow i\text{-Bu}_3B$$

Di-isobutyl-t-butylboron could be distilled under 60°C without redistribution or isomerization; but at higher temperatures tri-isobutylboron was formed. Two t-butyl groups could not be attached, and whereas t-butylmagnesium chloride and boron trifluoride gave t-butylboron difluoride, when boron trichloride was used instead of the trifluoride, not even one t-butyl group could be attached. The suggestion that the major difference between fluorine and chlorine is a matter of size, and therefore the difference in behaviour is due to steric factors, is probably an oversimplification; for polarizability might also be significant here.

An explanation of the rearrangement in the alkyl group during attachment process is not easily found. One suggestion is illustrated by the following scheme.

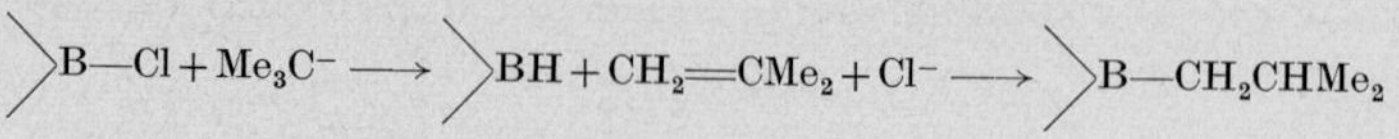

$$\text{>B—Cl} + \text{Me}_3\text{C}^- \longrightarrow \text{>BH} + \text{CH}_2\text{=CMe}_2 + \text{Cl}^- \longrightarrow \text{>B—CH}_2\text{CHMe}_2$$

A distillable trialkylboron having three different alkyl groups has been prepared by interaction of t-butylmagnesium chloride and n-amylboron difluoride to give n-amylisobutyl-t-butylboron, the same compound being obtained by interaction of t-butyl-di-n-amylboron and isobutylmagnesium bromide (Clark $et\ al.$, 1958). Of interest, too, is the fact that the t-butyl-di-n-amylboron, b.p. 47.5°C/0.14 mm was prepared by the interaction of t-butylboron dichloride and n-amylmagnesium bromide.

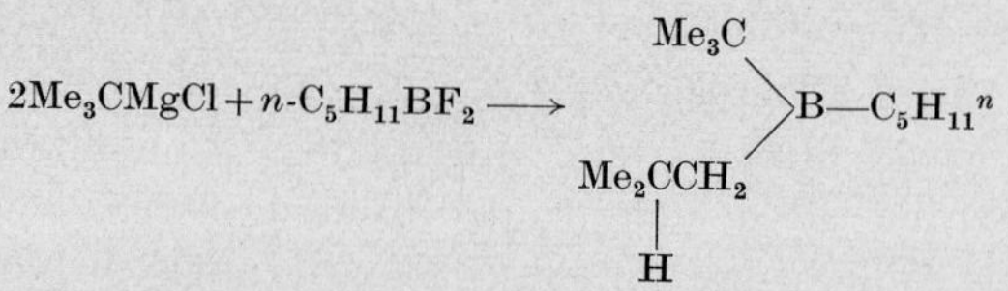

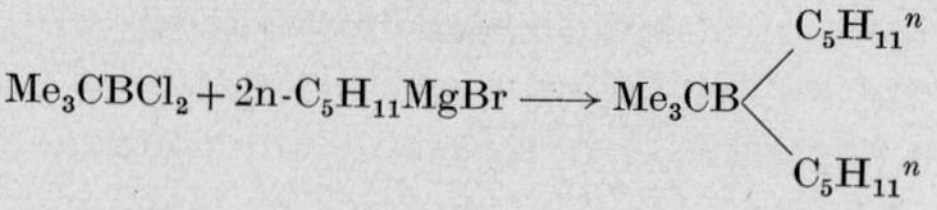

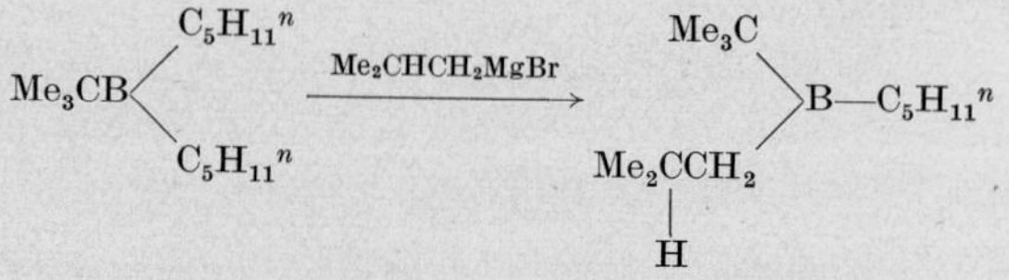

$$2\text{Me}_3\text{CMgCl} + n\text{-C}_5\text{H}_{11}\text{BF}_2 \longrightarrow \quad \begin{array}{c} \text{Me}_3\text{C} \\ \text{>B—C}_5\text{H}_{11}{}^n \\ \text{Me}_2\text{CCH}_2 \\ | \\ \text{H} \end{array}$$

$$\text{Me}_3\text{CBCl}_2 + 2n\text{-C}_5\text{H}_{11}\text{MgBr} \longrightarrow \text{Me}_3\text{CB} \Big\langle \begin{array}{c} \text{C}_5\text{H}_{11}{}^n \\ \text{C}_5\text{H}_{11}{}^n \end{array}$$

$$\text{Me}_3\text{CB} \Big\langle \begin{array}{c} \text{C}_5\text{H}_{11}{}^n \\ \text{C}_5\text{H}_{11}{}^n \end{array} \xrightarrow{\text{Me}_2\text{CHCH}_2\text{MgBr}} \begin{array}{c} \text{Me}_3\text{C} \\ \text{>B—C}_5\text{H}_{11}{}^n \\ \text{Me}_2\text{CCH}_2 \\ | \\ \text{H} \end{array}$$

This appears to be a convenient place to refer briefly to redistribution tendencies in compounds related to trialkylborons, for such tendencies have been found to be in the following order (see references just cited).

$$\text{RR}'_2\text{B} > [\text{R}_2\text{B}]_2\text{O} > \text{R}_2\text{BF} > \text{R}_2\text{BCl} >$$
$$\text{R}_2\text{BOR}' > [\text{RBO}]_3 > \text{RBX}_2 \quad (\text{X=halogen})$$

It is stated that the alkylboron dihalides prepared from boron halides and boronic anhydrides (trialkylboroxoles, or trialkylborines) are resistant to redistribution up to a temperature of 180°C. This stability is attributed to the influence of the doubly bonded resonance forms

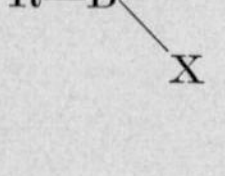

$$\text{R—B} \overset{\overset{\oplus}{\ominus}X}{\underset{X}{\Big\langle}}$$

in the four-centre reactions mentioned. It is contended that the over-lap of the sigma orbital of the R—B bond of one molecule and the vacant p-orbital of the boron atom of the other molecule cannot occur because of the low availability of the vacant orbital consequent on the back–co-ordination from halogen. This explanation is much too simple. There is little or no evidence that any back–co-ordination from halogen to boron noticeably reduces the electrophilic function of boron. On the other hand, the inductive effect of the halogen can materially reduce electron density on the "central" carbon atom of the alkyl group, and so hinder the nucleophilic function of this group required for attachment to a foreign boron atom.

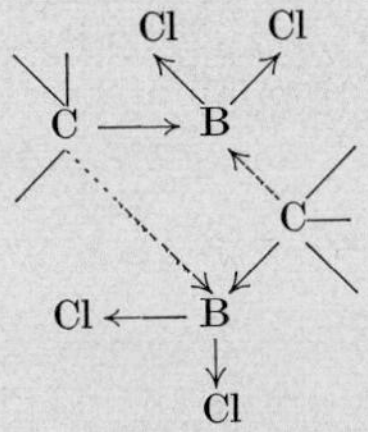

$$
\begin{array}{ccc}
 & Cl & Cl \\
 & \nwarrow & \nearrow \\
C & \longrightarrow & B \\
\diagup & & \\
 & & C- \\
Cl \longleftarrow & B & \\
 & \downarrow & \\
 & Cl &
\end{array}
$$

Steric requirements are also involved, and these are different for the four-centre system and the three-centre "end-on" system as illustrated by the example of the neopentyl group (Gerrard *et al.*, 1947, 1950). The points could be tested by a systematic change in the electron-releasing power of the alkyl group, especially by the introduction of halogen in the group.

Likewise the stability of trialkylboroxoles (boroxines, trimers of boronic anhydrides), as typified by tri-*n*-butylboroxole, up to tempera-ture of 190°C is attributed to the effect of back–co-ordination from oxygen to boron. The situation here is entirely different from the

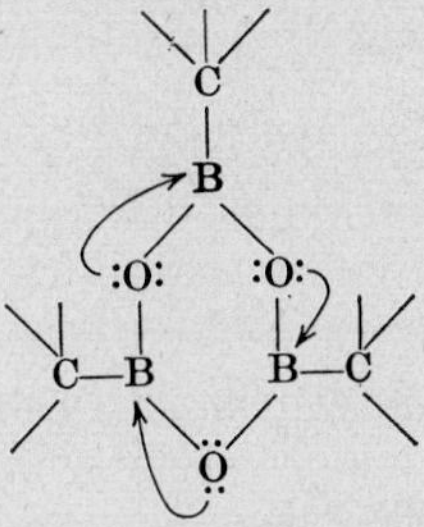

$$
\begin{array}{ccc}
 & C & \\
 & | & \\
 & B & \\
:O: & & :O: \\
C-B & & B-C \\
 & O: &
\end{array}
$$

example of the dihalide, because there is an established strong effect which can materially reduce the electrophilic function of boron, as is clearly shown by the fact that trialkyl borates will not co-ordinate with pyridine (see later). This back–co-ordination will tend to increase the nucleophilic function of the alkyl carbon atom; but the determining condition here is the lowered electrophilic function of boron, *and* the lowered nucleophilic function of oxygen for the exchange shown in the diagram. Above 200°C, tri-*n*-butylboroxole shows slight

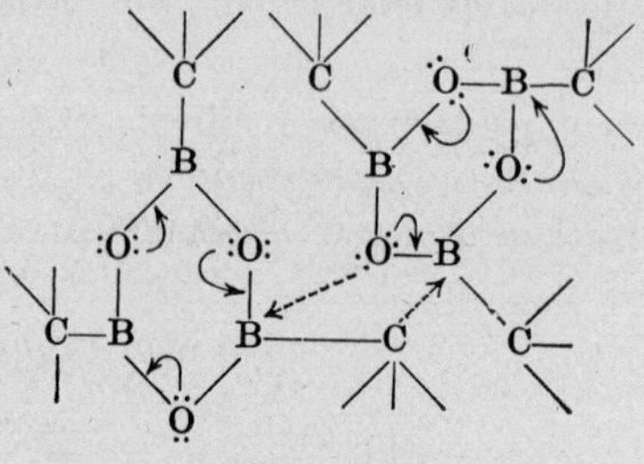

redistribution to boric oxide and tri-*n*-butylboron depicted as an equilibrium system with a point of equilibrium well to the side of the boroxole.

$$[Bu^nBO]_3 \underset{}{\overset{\text{Above 200°C}}{\rightleftharpoons}} Bu^n{}_3B + B_2O_3$$

In the examples of dialkyl esters of alkylboronic acids, which have been made by interaction of trialkyl borates and Grignard reagents, little success attended efforts to obtain pure specimens. Low-pressure fractionation of di-*n*-butyl *n*-butylboronate gave on continued operation tri-*n*-butylboron and tri-*n*-butyl borate.

$$(Bu^nO)_2BBu^n \xrightarrow[\text{Redistribution}]{\overset{\text{Long fractional}}{\text{distillation}}} (Bu^nO)_3B + Bu^n{}_3B$$

Boron trichloride and trialkylborons undergo mutual exchange of alkyl and chlorine at 160°C and afford dialkylboron chlorides in yields of about 90%. These chlorides do not show redistribution effects at temperatures below about 150°C. Distillation at higher temperatures results in the separation of alkylboron dichlorides and trialkylborons.

$$2R_2BCl \longrightarrow R_3B + RBCl_2$$

It is argued that resonance stabilization of alkylboron dichlorides is greater than for the dialkylboron chlorides, because there are two atoms of chlorine in each molecule, and this accounts for the lack of reactivity of the dichlorides with respect to redistribution. One might ask, however, why boron trichloride, in which stabilization must, on the same argument, be all the greater, should react with trialkylborons as stated above.

Di-*iso*butylboron fluoride was stable when distilled at 49°C/52 mm; at 123°C/760 mm *iso*butylboron difluoride and tri-*iso*butylboron were formed, pointing to a readier redistribution than occurred with the corresponding dialkylboron chlorides. The view adopted was that the smaller fluorine atom allows the easier formation of the transition state than the larger chlorine atom.

Borinic anhydrides are apparently suspect, for in an attempt to purify a specimen of dibutylborinic anhydride by fractional distillation, redistribution occurred, resulting in the formation of tributylboron and butylboroxole.

$$3R_2B—O—BR_2 \longrightarrow 3R_3B + (RBO)_3$$

According to workers in this field of research (see references cited), the borinic anhydrides undergo redistribution more readily than the esters, R_2BOR'.

Some help in determining the structure of the butyl group in such systems comes from the study of proton magnetic resonance (Davies et al., 1959), and already some twenty boron compounds containing i-Bu and t-Bu groups have been examined.

Goubeau et al. (1957) have studied the thermal decomposition (400–600°C) of trimethyl boron, and have obtained a number of products such as $(BC_2H_5)_4$, $(BCH)_x$ and, remarkably enough, compounds formulated B_2OMe_4 (for example) containing oxygen due to attack on the glass reaction tube.

Further references to redistribution reactions are made by Buls et al. (1957), Solomon et al. (1958), and Clifford (1957).

Chapter VIII

SPECIAL TETRAVALENT BORON COMPOUNDS DERIVED FROM TRIPHENYLBORON

I. Introduction

THE method most commonly used for the attachment of three aryl groups to boron is by the interaction of boron trifluoride–ether complex and aryl magnesium halide (Krause *et al.*, 1921, 1922; Wittig *et al.*, 1949).

Triphenylboron can be distilled from the reaction mixture and purified by crystallization from ether under nitrogen. It has m.p. 142°C, and b.p. 203°C/15 mm. It is sensitive to oxygen and is dephenylated readily by alcohol. Rondestvedt *et al.* (1955) have investigated alcoholysis, and give the example of methanolysis.

$$Ph_3B + EtOH = Ph_2BOEt + C_6H_6$$

See recent work by Neu (1959). Tetramethylammonium triphenyl-hydroxyboron, $Me_4N^+[Ph_3B.OH]^-$, is formed from triphenylboron and tetramethylammonium hydroxide (Fowler *et al.*, 1940), the salt crystallizes with water or alcohol. By fusion of triphenylboron with sodium hydroxide, the corresponding sodium salt is formed, $Na^+[Ph_3B.OH]^-$, crystallizing from ether with solvent of crystallization. It is soluble in water, is slowly hydrolysed, acetic acid precipitates triphenylboron, and ammonium chloride gives the complex, $Ph_3B.NH_3$.

$$Na[Ph_3B.OH] + NH_4Cl = NaCl + Ph_3B.NH_3 + H_2O$$

A corresponding compound $Na[Ph_3BCN]$ is more stable to acids than is the hydroxy compound, and aqueous solutions are neutral; the Li, Na, K, NH_4 salts are soluble, the Rb salt slightly so, and the Cs salt is insoluble (70 mg Cs in 100 c.c. water at 0°C) (Wittig, 1951). The sodium salt is used for estimating pharmaceutical chemicals containing nitrogen (Schultz *et al.*, 1955).

Acceptor properties of triphenyl boron are further illustrated by the reaction with metallic sodium in ether in dry and inert atmosphere (Krause *et al.*, 1937). Wittig *et al.* (1956) have investigated the addition of triphenylmethylsodium to butadiene in the presence of triphenyl-boron.

$$Ph_3B + Na \longrightarrow Ph_3B^-Na^+$$

The triphenylboron sodium is diamagnetic, and the dimeric $[Ph_3B . BPh_3] = Na_2$ (Li Chu, 1953) is of interest as an example of B—B bonding. Considerable ion-association is indicated by low conductance in ethereal solution. The compound Ph_3PBPh_3 is discussed by Wittig et al. (1955).

Sodium triphenylborohydride is obtained by the interaction of triphenylboron–sodium and methanol, the salt (and the lithium salt,

$$2Ph_3B^- . Na^+ + MeOH \longrightarrow Na^+[Ph_3B \ OMe]^- + Na^+[Ph_3BH]^-$$

too) being also obtained by the addition of triphenylboron to sodium hydride (or lithium hydride) in ether. The lithium salt is also obtained (see papers by H. C. Brown et al.) by heating lithium hydride together with triphenyl boron at 180°C until the melt solidifies. The borohydride is fairly readily hydrolysed with water, but hydrogen is vigorously evolved on treatment with acids.

$$Na^+Ph_3BH^- + H_3O^+ \longrightarrow Na^+ + H_2 + H_2O + Ph_3B$$

II. Tetraarylboron Salts

In very recent years intense interest has been centred on sodium tetraphenylboron, $Na^+[BPh_4]^-$, and a very considerable number of papers have appeared during the last 6 or 7 years. The special incentive was originally the application of this salt as a reagent for potassium, because the potassium salt, $K^+[BPh_4]^-$, is insoluble and constitutes by far the best means of estimating potassium (see Geilmann et al., 1953). Naturally, all the techniques appertaining to volumetric, gravimetric, electrometric, polarographic analysis, macro, semimicro, and micro, have been investigated for this determination. This work is still going on; but interest in tetraarylboron salts has widened and deepened very considerably and has become a new and separate chapter in the organic chemistry of boron. It would be extremely confusing in the small space available to make detailed reference to each paper which has been published, and so a literature list with brief indication of subject matter is given as an appendix.

Arylation of boron beyond the triaryl stage is effected by the use of the requisite excess of arylmagnesium bromide in conjunction with the boron trifluoride–ether complex. The solution contains the compound $MgBr^+BAr_3^-$, and to isolate the sodium salt, where $Ar = Ph$, water is added, the magnesium precipitated as carbonate by the addition of sodium carbonate, and by addition of sodium chloride in excess, sodium tetraphenylboron is obtained (Wittig et al., 1951). In amended instructions, the reference to removal of magnesium is deleted, and the process then simply involves the addition of the residue, after removal of solvent, to water, followed by saturation with sodium chloride. The salt is soluble in ether, chloroform, and water. A later method of purification entails recrystallization from ether–*cyclo*hexane.

Addition of phenyl–lithium to an ethereal solution of triphenylboron gives lithium tetraphenylboron, as shown by Wittig *et al.* (1949), and the salt crystallizes from ether as the octa-etherate, but the ether is relinquished in a vacuum. The salt is soluble in water, and in alcohol, insoluble in benzene, *cyclo*hexane, and carbon tetrachloride. It is stable in boiling water, but is decomposed by acids at about 80°C.

Addition of trimethylamine hydrochloride to an aqueous solution of the sodium salt gives a precipitate of the ammonium salt, $Me_3NH^+[BPh_4]^-$, which decomposes by dephenylation at 200°C in a stream of nitrogen; and after removal of the base and benzene in the carrier gas, triphenylboron can be obtained in 90% yield.

$$Me_3\overset{+}{N}H[BPh_4]^- \xrightarrow{\ 200°C\ } Me_3N + C_6H_6 + Ph_3B$$

Similarly, the insoluble ammonium salt affords triphenylboron by dephenylation; but, on the other hand, the tetramethylammonium

$$NH_4BPh_4 \xrightarrow{\ 240°C\ } NH_3 + C_6H_6 + Ph_3B$$

salt is stable to 340°C. Some electrochemical properties of the salt Bu_4NBPh_4 have been examined by Accascina *et al.* (1959).

There is a rapid, quantitative, and analytical reaction with mercuric chloride, although most recent work shows sources of error in the example of the potassium salt dissolved in acetone.

$$NaBPh_4 + 4HgCl_2 + 3H_2O \longrightarrow 4PhHgCl + NaCl + 3HCl + B(OH)_3$$

Certain metal chlorides cause dephenylation.

$$2LiBPh_4 + 2CuCl_2 \xrightarrow{\ In\ Et_2O\ } 2CuCl + 2NaCl + 2Ph_3B + Ph_2$$

$$2NaBPh_4 + MCl_2 + 6H_2O \xrightarrow{\ Water\ } M(OH)_2 + 2NaCl + 2PhB(OH)_2 + C_6H_6$$
$$\text{where } M = Mn^{++}, Fe^{++}, Co^{++} \text{ or } Ni^{++}.$$

In ether, $M + Ph—Ph + Ph_3B$ are obtained.

In contrast to the boron anion typified by sodium tetraphenylboron, $Na^+BPh_4^-$, is the interesting example of a boron cation provided by the structure referred to as α,α'-dipyridyldiphenylboronium perchlorate, which has been prepared by Davidson *et al.* (1959, see also 1958) by the interaction of diphenylboronium perchlorate and α,α'-dipyridyl in nitromethane.

III. Sodium Tetraphenylboron as an Analytical Reagent

The use of sodium tetraphenylboron for the gravimetric precipitation of potassium, rubidium and caesium as corresponding salts quickly emerged. The sodium salt is completely dissociated in water, giving a faintly acid solution, but is undissociated in acetone except at high dilution.

With reference to the determination of potassium, a number of papers appeared in sequence in which operational details were discussed. Influential factors such as pH, temperature of precipitation, temperature of filtration, preparation of solution of reagent, the stability of this, the nature of the wash solution, temperature of drying, and interference or not of other metal ions were considered.

The simple technique involved the precipitation of potassium tetraphenylboron in acetic acid solution at 60–70°C, followed by filtration through an A.1 porcelain filter, washing with water, and drying at 120°C (1 hour) (Raff et al., 1951). The potassium salt is less soluble than potassium perchlorate. Silver and thallium and other heavy metals interfere, and alkaline earth metals may be removed by prior treatment with sodium carbonate, or E.D.T.A. (ethylenediaminetetraacetate, disodium salt).

A micro-procedure was devised (Flaschka et al., 1952, 1953, 1955), and analysis of biological ash involves determination of potassium by the tetraphenylboron technique (Spier, 1952). Further observations were reported (Rüdorff et al., 1952, 1953, 1954), and a procedure for a titrimetric finish was introduced. Titration of BPh_4^- with Ag^+ in acetone with eosin indicator has been carried out; but the silver nitrate must be standardized by the tetraphenyl ion. The potassium salt is soluble in acetone, but insoluble in water. The silver salt is sparingly soluble in both. Rubidium is similarly determined titrimetrically.

Precipitation factors had to be carefully considered (Kohler, 1953); it is stated that the precipitate is coarser at room temperature, and that 0.1 N mineral acid is better than acetic acid. It is desirable to use pure sodium tetraphenylboron, and although the reagent is stable in acid at room temperatures, it is decomposed in 20 minutes at 80°C. The sodium salt, 1.5 g in 250 c.c of water, may be purified just before use by stirring (5 minutes) with pure alkali free aluminum hydroxide (0.5–1.0 g). The solution is filtered, the first 20 c.c of filtrate being refiltered or rejected. This gives a 0.6% solution.

It is stated that students got better results with this method for potassium than experienced analysts did with the older methods.

The early observations on the precipitation of ammonium and amine salts mentioned in the foregoing naturally led to extended studies of compounds, such as alkaloids, containing basic nitrogen. A procedure for the determination of alkaloids was described, entailing a gravimetric (Schultz, 1952) and a volumetric (Schultz, 1953) technique. A titrimetric method for atropine, pyramidone, antipyrine, and pyridine gave good and rapid results (Rüdorff et al., 1952–1954), and sodium tetraphenylboron has been used for the isolation of choline and acetyl choline (Marquardt et al., 1952). Triethanolamine tetraphenylboron has been described (Neu, 1954), and microscale m.ps of salts from 30 analgesics and alkaloids have been determined by Fischer et al. (1953).

The interaction of sodium tetraphenylboron with nitrogen-bearing medicinals listed in *Pharmacopoeia Helvetica* V and supplements I and II, and also of some others, has been examined (Aklin et al., 1956). Precipitations occurred according to the basic strength of the nitrogen, but results showed a need for re-examination of reaction conditions. For one thing, desiccation temperature should be no higher than 80°C. Good results were obtained in the gravimetric determination of papaverine chloride and strychnine nitrate after details of procedure had been worked out. The "Kalignost" ($NaBPh_4$) method is believed to be suitable as a pharmacopoeia method, but it is desirable to examine alkaloidal salts individually in order to establish incontestable assay procedures.

Precipitation of quaternary ammonium salts has also been examined (Wittig et al., 1956). Jansons et al. (1958) have reported on the volumetric determination of aromatic and heterocyclic nitrogen-containing compounds, and intense interest is being developed in the determination of thallium as the tetraphenyl boron salt (Wendlandt, 1957; Sirotina et al., 1957; Veiss et al., 1959; Mukherji et al., 1959).

IV. Tetraphenylboron Derivatives of Amides

It has been suggested that an amine must have a basic dissociation constant of at least 10^{-11} to undergo qualitative precipitation as a tetraphenylboron salt, and it is now found that some organic acid amides, usually considered neutral, give insoluble salts. Zief et al. (1959) have examined this system from the point of view of biologically active substances.

$$\left[R-\overset{\overset{\displaystyle O}{\|}}{C}-\underset{\underset{\displaystyle H}{|}}{N}\diagup^{R_2}_{R_3} \right]^{+} Ph_4B^{-}$$

Electron-attracting groups reduce electron density on nitrogen, and hinder salt formation. Thus N-vinyl-, N-allyl pyrrolidone, and acrylamide do not form salts, but in N-allyl-γ-hydroxybutyramide the electron attraction effect of the allyl group is balanced by the OH group, and a salt is formed. Polyvinylpyrrolidone (molecular weight, 50,000) does form a derivative. Increase in electron density caused by proximity of a methyl group is felt in N-methylpyrrolidone-2, which forms a derivative, whereas 5-methylpyrrolidone-2 and unsubstituted pyrrolidone-2 do not.

In antibacterial and antifungal activity, N-methylpyrrolidinium tetrapenhylboron compares favourably with C.P.C. (cetylpyridinium chloride) in activity against Gram-positive bacteria and fungi (active at 1–10 μg/ml).

The following melting-points of the tetraphenylboron compounds are recorded: acetamide, 180–182°C, dimethylacetamide, 119–120°C, formamide, 187–190°C, dimethylformamide, 117–121°C, N-methylpyrrolidone-2, 148–150°C, N-allyl-γ-hydroxybutyramide, 127–130°C, and polyvinylpyrrolidone, > 250°C.

CHAPTER IX

OXIDATIVE FISSION OF CARBON–BORON BONDS

THE oxidative fission of the carbon–boron bond has become a subject of detailed investigation. Controlled autoxidation of diethylborinic acid and its esters was one of the earliest processes to be investigated in the organic chemistry of boron (Frankland, 1862, 1876) and was formulated thus:

$$Et_2BOH \xrightarrow{\frac{1}{2}O_2} EtB(OH)OEt \qquad Et_2BOEt \xrightarrow{\frac{1}{2}O_2} EtB(OEt)_2$$

Trialkylborons are similarly converted into the boronic ester, and thence to the acid, and further autoxidation leads to the trialkyl borate, and thence to boric acid (Meerwein et al., 1936).

$$R_3B + O_2 \longrightarrow RB(OR)_2 \xrightarrow{H_2O} RB(OH)_2$$

$$R = Et,\ Pr,\ Pr^i,\ Bu^n,\ Bu^i,\ Bu^t,\ CH_2Bu^i$$

Alkenyl- and aryl-borons are usually more oxidatively stable than the corresponding alkyl compounds. In 1-bromonaphthalene, triphenylboron reacts fairly quickly with oxygen at room temperature to afford phenyl diphenylborinate, whereas tri-1-naphthylboron is more stable, and trimesitylboron is remarkably so. Likewise, trivinylboron is stable at room temperature, whereas methyldivinylboron is autoxidized. The following workers have referred to oxidative fission of the carbon–boron bond: Krause et al. (1921, 1931), Snyder et al. (1938), Johnson et al. (1938), Bamford et al. (1946), Brown et al. (1957), Parsons et al. (1957), McCusker et al. (1955).

Although dialkylboron compounds are usually fairly quickly oxidized by air, monoalkylboron compounds are more stable, and with the exception of the oxidatively reactive secondary and tertiary alkyl-boron difluorides, alkylboron dihalides can be more stable to oxygen at ordinary temperatures. Esters of alkylboronic acids have a certain stability; but the acids themselves are more susceptible, secondary and tertiary alkylboronic acids being rapidly autoxidized. However, the primary and secondary alkyl compounds appear to be stable in water.

Monoarylboron compounds are more stable to oxidation than are diarylboron ones (Torssell, 1955).

Trialkylboroxoles, $(RBO)_3$, are quantitatively oxidized to metaborates (ROBO), slowly when R is primary, rapidly when R has a tertiary structure, and the liberation of iodine from potassium iodide, and the polymerization of styrene, led to the conclusion that the peroxide intermediate formulated as the monomer was probably formed (Grummitt, 1942).

$$R-B{=}O \xrightarrow{\ O_2\ } \underset{\overset{\cdot\cdot}{O_2}}{R-B{=}O} \xrightarrow{\hspace{2cm}} ROB{=}O$$

As monomer:	Peroxide	As monomer:
alkylboronic anhydride	intermediate	Alkyl metaborate

As trimer: As trimer:

$\left\{\begin{array}{l}\text{trialkylboroxole}\\\text{trialkylboroxine}\end{array}\right\}$ $\qquad\qquad$ $\left\{\begin{array}{l}\text{trialkoxyboroxole}\\\text{trialkoxyboroxine}\end{array}\right\}$

Note. Evidence points to the trimeric structure, and names are to be taken as equivalent.

Furthermore, it has been supposed (Petry *et al.*, 1956) that the autoxidation of trimethylboron gave the peroxide, Me_2BOOMe.

The peroxidic effect has been investigated in detail by Davies *et al.* (1958, 1959) and Abraham *et al.* (1959), and the following scheme for $R=Bu^n$ has been formulated.

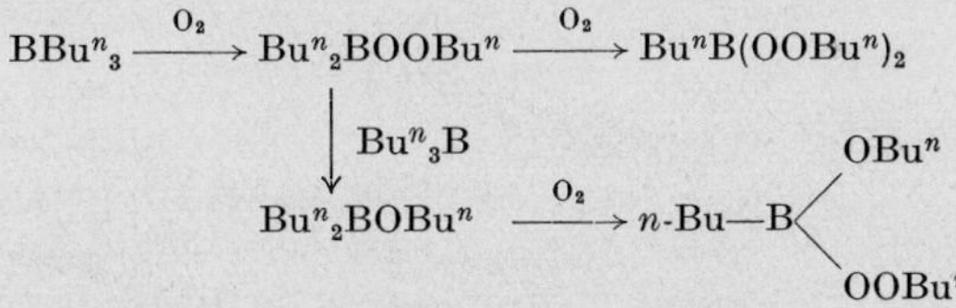

$$BBu^n_3 \xrightarrow{\ O_2\ } Bu^n_2BOOBu^n \xrightarrow{\ O_2\ } Bu^nB(OOBu^n)_2$$

$$\downarrow{\scriptstyle Bu^n_3B}$$

$$Bu^n_2BOBu^n \xrightarrow{\ O_2\ } n\text{-}Bu-B{\big\langle}\begin{array}{l}OBu^n\\[4pt]OOBu^n\end{array}$$

It is a matter of acute interest how the atom of oxygen gets between the carbon and boron atoms. The mechanism is believed to involve a 1 : 3 shift of alkyl or (aryl) from boron to oxygen. Examples of tri(alkylperoxy)boron compounds are afforded by the reaction of boron

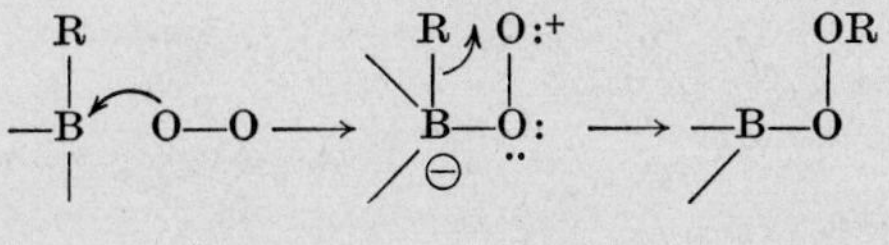

$$\underset{|}{\overset{R}{\underset{|}{-B}}}{\overset{\curvearrowleft}{}}O-O \longrightarrow \overset{R\,{\nearrow}\,\overset{+}{O}:}{\underset{/\ominus}{\underset{|}{B-O}:}} \longrightarrow \overset{OR}{\underset{/}{\underset{|}{-B-O}}}$$

trichloride with *t*-butyl hydroperoxide and with *n*-butyl hydroperoxide, and *o*-nitrophenyl chloroboron compounds reacted similarly.

$$3ROOH + BCl_3 \longrightarrow 3HCl + B(OOR)_3$$

R = t-Bu, b.p. 60–70°C/10⁻³ mm

R = n-Bu, b.p. 50–60°C/10⁻³ mm

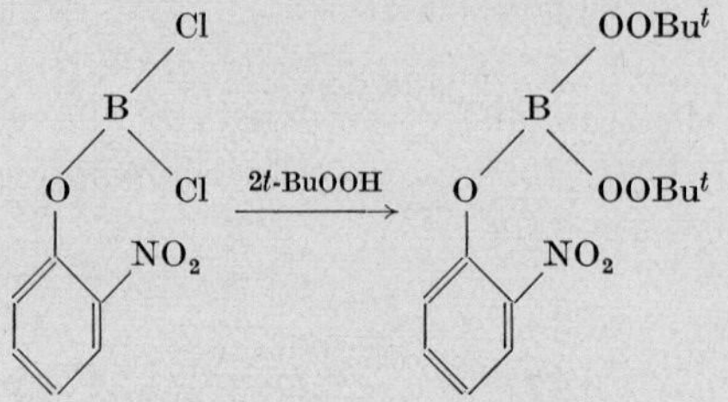

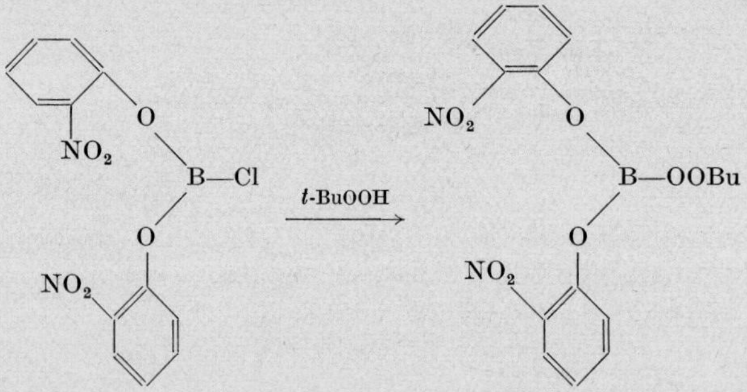

These peroxides react with water or alcohols to give the original alkyl hydroperoxide.

Tetraacetyl diborate and t-butyl hydroperoxide gave di(t-butylperoxy) boron hydroxide.

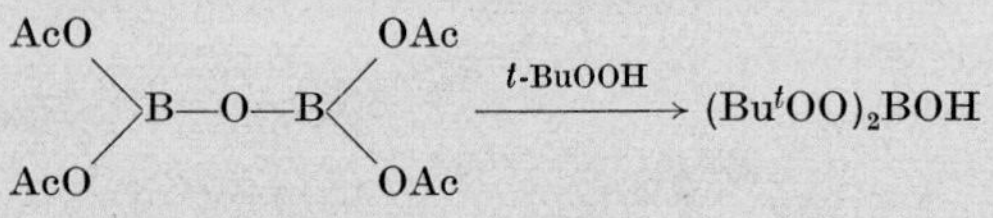

Oxidative dearylation of phenylboronic acid, by hydrogen peroxide (Ainley et al., 1930) has been studied kinetically (Kuivila et al., 1957), and the scheme has been formulated as follows, and this scheme has received support (Davies et al., 1958, 1959) by showing that isotopically normal phenol was obtained by the interaction of

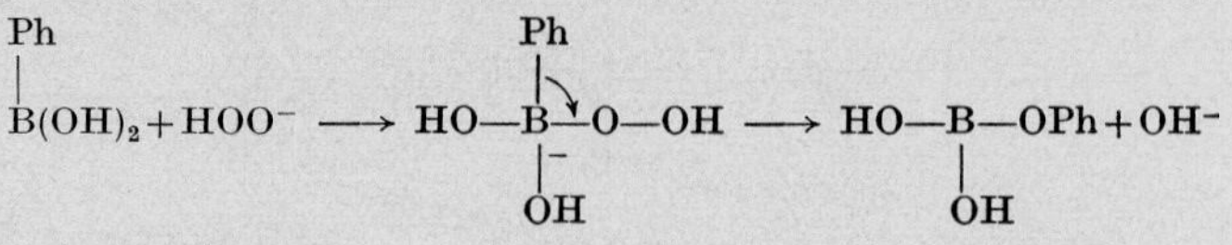

isotopically normal hydrogen peroxide in ¹⁸O water with phenyl-boronic acid containing ¹⁸O. The peroxy-compounds, RB(OOR)₂ or

R_2BOOR could not be obtained by nucleophilic substitution, because of this 1 : 2 shift.

$$C_6H_{13}BF_2 + Bu^tOO^- \longrightarrow F_2BOC_6H_{13} + Bu^tO^-$$

Chemical evidence points to the following autoxidation scheme for tri-*iso*butylboron.

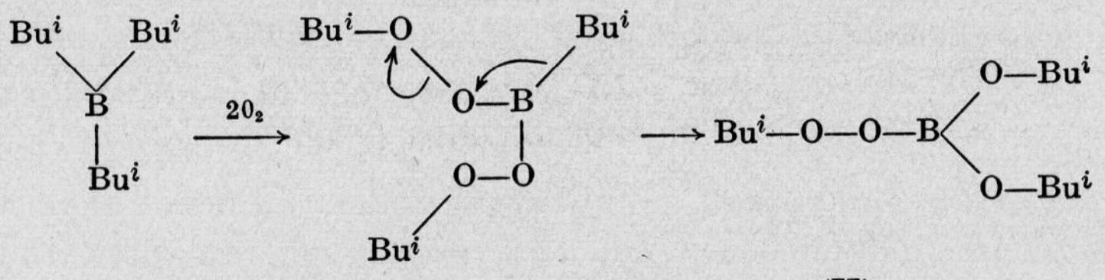

(I)   (II)

Proton magnetic resonance has now been harnessed by Davies *et al.* (1959) to investigate the structure of (I) and (II). Comparison of the spectrum of freshly prepared compound (I) with that of the material observed 2.5 hours later, showed that at room temperature a considerable extent of rearrangement from (I) to (II) had occurred.

CHAPTER X

HYDRIDO-COMPOUNDS OF BORON

I. Introduction

BORON hydrides, and alkali metal borohydrides, are not in themselves organic compounds of boron; but their production and application are so interwoven with the recent developments in such chemistry that a brief discussion specifically on the boron hydrides is imperative.

II. Boron Hydrides

A. Historical Development

In 1881 the gas evolved by the action of hydrochloric acid on magnesium boride was found to contain the boron hydride in too small amount for accurate study, and on doubtful grounds the formula BH_3 was assumed (Jones et al., 1881), and later accepted without analysis by another worker (Sabatier) in 1891, who concluded that the compound was decomposed by caustic potash. The Proceedings of the Chemical Society of London, 1901, contains a preliminary note by Ramsay et al. on this system.

Foundations of borohydride chemistry are due mainly to Stock and his co-workers (Stock, 1933). After spending a year in Moissan's laboratory in Paris, Stock returned to the University of Berlin in 1901 with the intention of studying the chemistry of the hydrides of boron, because it was expected that being a neighbour of carbon in the periodic table, boron might form a variety of interesting compounds other than merely boric acid and borates. He continued work in the Technische Hochschule of Breslau, and publication began to appear about 1911. From then on Stock and his co-workers published a series of papers, mainly in Berichte, terminating about 1936, by which time other schools were in the process of development.

Considerable care was required in the preparation of magnesium boride from magnesium and boric oxide. Furthermore, as the magnesium contained silicon, manipulation of the already small yield of "boron hydrides" was still further hampered by the presence of silicon hydrides, which had to be separated from the boron compounds. An elaborate vacuum technique had to be devised, to ensure that the

118

system was completely enclosed. Complicated fractionations involving condensations by liquid air had to be practised. About 1916, Stock turned rather more to the silicon hydride system, and later, taking advantage of the considerably increased experience with the vacuum technique, returned to the boron hydrides.

The crude mixture from the primary interaction of the magnesium boride and hydrochloric acid contained the tetraborane, B_4H_{10}, a very small amount of the pentaborane, B_5H_9, and rather more of the hexaborane, B_6H_{10}, with traces of the decaborane $B_{10}H_{14}$. The lowest hydride in the crude gas was B_4H_{10}, which could be converted into the diborane, B_2H_6, by warming.

Certain physical and chemical properties of the individual hydrides were investigated, and it should be pointed out that the task was made much more difficult by the fact that tap grease is readily attacked and had to be rigorously excluded. Furthermore, the procedure of preparation was so involved that only very small quantities of specimens were available for study.

Between 1917 and 1930 some 45 papers appeared on the structure of the boron hydrides, discussed by almost as many different authors.

About this time two new schools of research into the chemistry of boron hydrides were in the process of being created. Wiberg published several papers with Stock and on his own, and subsequently continued investigations in Germany, establishing for himself a considerable school of research in Munich. The second school was created by Schlesinger in America, who with Burg devised an entirely different and greatly improved technique for the production of diborane and the other hydrides obtained from it.

Schlesinger and Burg (1931) described the preparation of diborane by the following procedure. Pure hydrogen was passed through boron trichloride at $-40°C$, the resulting mixture then being led through a 12–15-kilovolt discharge at 10 mm. Hydrogen chloride, chlorodiboranes, unused boron trichloride were condensed from hydrogen and fractionated to remove hydrogen chloride. The fractionation column was then operated at 2 atmospheres, and at $0°C$ reflux to convert chloroboranes into diborane which is quickly removed. Diborane was purified by fractionation from the other boranes.

$$6B_2H_5Cl \longrightarrow 5B_2H_6 + 2BCl_3$$

These workers could make ten times as much diborane in a week as Stock could produce in six weeks by the magnesium boride method. It was thus that avenues were quickly opened for the rapid development of hydrido-chemistry of boron, and the name of H. C. Brown must be strongly associated with this development which has led boron hydrides and metal borohydrides into startling actual and potential industrial applications. Stock and Sütterlin (1934) found that boron tribromide was a better reagent than boron trichloride, because of the easier decomposition of the bromoborane, and subsequent easier separation of the products. This modification of the discharge technique became more attractive when an improved method of preparing the tribromide was made available by Gamble et al. (1940).

During the 1936–1942 period, Wiberg in Germany and Schlesinger and Burg in U.S.A. pursued experimental investigations into the methylated boranes, borazole and its derivatives. One to four of the hydrogen atoms in diborane can be indirectly replaced by methyl

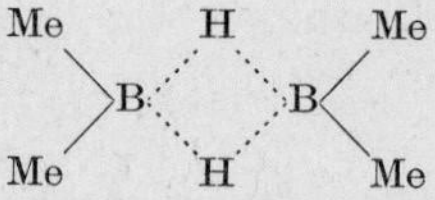

Tetramethyldiborane

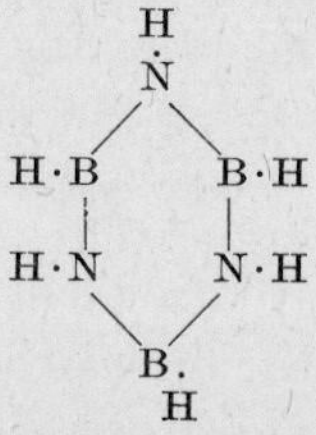

borazole

groups, provided not more than two methyl groups are on the same boron atom. When three methyl groups are on one boron atom the B—B bond cannot exist, and the compound becomes trimethylboron. The work of this period has been reviewed by Schlesinger et al. (1942), Bauer (1942), Bell et al. (1948), and Wiberg (1948), and was again referred to in part in a later review by Stone (1955).

During the early 1940s, under a National Defense Research Program, Schlesinger, Brown and Burg gave the hydrido- and organo-chemistry of boron an entirely different course which rendered the earlier methods, including the sparking method, of producing boron hydrides completely obsolete. Owing to the usual restriction, much of this work was not published until 1953, when it was described in a series of papers in the *Journal of the American Chemical Society*.

By means of diborane prepared as described, a number of metal borohydrides were prepared (Schlesinger et al., 1939, 1940, 1943, 1953),

$$Al_2Me_6 + 4B_2H_6 \longrightarrow 2BMe_3 + 2Al(BH_4)_3$$

$$3BeMe_2 + 4B_2H_6 \longrightarrow 2BMe_3 + 3Be(BH_4)_2$$

$$LiEt + 2B_2H_6 \longrightarrow BEt_3 + 3LiBH_4$$

and by the same reaction, trialkylborons. Later, in the search for volatile compounds of uranium, the borohydride, $U(BH_4)_4$ was attractive; but available methods for the preparation of diborane and hence borohydrides were inadequate for producing satisfactory quantities of uranium(IV) borohydride. During the course of this work the interaction of lithium hydride with boron trifluoride etherate to afford diborane was discovered, and so too was lithium aluminum hydride, $LiAlH_4$, the use of which as a reducing agent has given rise to over 1700 published accounts, a record likely to be far exceeded by the more recently publicized sodium borohydride, $NaBH_4$.

$$LiH + BF_3,OEt_2 \longrightarrow B_2H_6$$
$$3LiAlH_4 + 4BCl_3 = 3LiCl + 3AlCl_3 + 2B_2H_6$$

B. Preparative Reactions

The following reactions are amongst those available for the preparation of diborane.

$$2BCl_3 + 6H_2 \xrightarrow{\text{electric discharge}} B_2H_6 + 6HCl$$
$$BCl_3 + 3H_2 + 2Al \longrightarrow B_2H_6 + 2AlCl_3$$
$$3LiAlH_4 + 4BF_3 \longrightarrow 2B_2H_6 + 3LiF + 3AlF_3$$
$$3NaBH_4 + 4BF_3 \xrightarrow{\text{ether}} 2B_2H_6 + 3NaBF_4$$
$$6LiH + 8BF_3 \xrightarrow{\text{ether}} B_2H_6 + 6LiBF_4$$
$$6LiH + 2BCl_3 \longrightarrow B_2H_6 + 6LiCl$$
$$6NaH + 2BCl_3 \longrightarrow B_2H_6 + 6NaCl$$
$$3NaBH_4 + BCl_3 \longrightarrow 2B_2H_6 + 3NaCl$$

The ($NaBH_4$–BCl_3) system appears to be prominent amongst those favoured for technical production, but the development of the (NaH–BCl_3) system, and especially the (BCl_3–H_2) one, would offer considerable overall advantages.

Schlesinger et al. (1953) obtained diborane in 87% yield by adding boron trifluoride–ether complex to an ethereal solution of lithium hydride, and sodium hydride may be similarly used, although it is essential to have the metal hydrides in a finely divided state. Mikheeva et al. (1956) declare that diborane of exceptional purity is prepared from lithium hydride and the trifluoride etherate, and other methods are mentioned by Köster et al. (1957). Diborane is made as a regular supply in quite a number of laboratories these days; but there is a certain variation in the reaction selected for preparation, and the apparatus used is essentially the Stock one, with whatever modification improvement in glass and vacuum techniques have allowed.

During the last two years, boron hydrides have attracted even more acute attention than formerly; on the one hand because of trends towards the plant-scale industrial production of diborane and higher boranes for exploitation in the development of high-energy rocket fuels and "boron polymers", and on the other because of further work leading to the more precise formulation of their electronic structures which afford outstanding examples of a two-electron three-centre bond.

C. Structure and Properties

Recent work on the structures of boron hydrides has been reviewed by Longuet-Higgins (1957), who refers to the contributions of Price on rotational infra-red lines, Shoolery (1955) on nuclear magnetic resonance, and Lipscomb (1954) on the structures of the higher boranes.

Diborane is diamagnetic and colourless, the electro-negativities of boron and hydrogen are almost equal, and the two boron atoms are 1.78 Å apart. It will be recalled that Pitzer (1945) compared the bridge between the two boron atoms with the double bond in ethylene, and designated it the *protonated double bond*, meaning that four electrons are engaged in binding the two atoms of boron, and the two protons are located at the centres of maximum electron density.

Protonated double bond

Three-centre bond

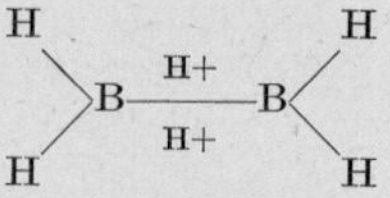

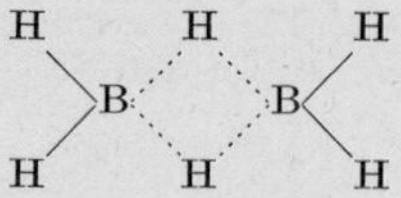

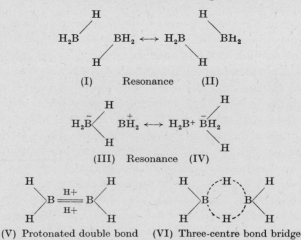

The three-centre bond has been more recently favoured. In the diagram each full line represents a normal two-electron, two-centre bond, thereby accounting for eight electrons, and leaving four electrons to form two three-centre bonds, each involving the two boron nuclei and

(I) Resonance (II)

(III) Resonance (IV)

(V) Protonated double bond

(VI) Three-centre bond bridge

one hydrogen nucleus, and indicated by the broken lines. However, Dickens *et al.* (1957) (cf. Hamilton, 1956) believe that in this picture insufficient allowance has been made for electron correlation. It is felt that the different structures (some involving resonance) which can be formulated are approximations to the truth, taking into account in a different degree the mutual effect of the electrons.

In tetraborane, B_4H_{10}, each atom of boron is surrounded tetrahedrally by four atoms.

Each full line represents an ordinary two-centre bond, involving altogether seven bonds and fourteen electrons. The remaining eight electrons are assigned to four three-centre bonds, each being represented by *two* dotted lines B----H----B.

The pentaborane, B_5H_{11}, as distinct from the pentaborane, B_5H_9, has the structure involving 8B—H two-centre bonds (full lines) using sixteen electrons, and of the remaining ten, six are occupied in forming three three-centre hydrogen bridges (dotted lines) and four are engaged in three-centre bonding (dotted lines) of the boron frame.

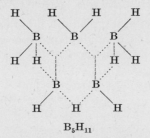

B_5H_{11}

Isotopic analysis of gaseous boron hydrides by neutron absorption (Hamlen *et al.*, 1956), infra-red study of the exchange of deuterium between decaborane and diborane (Kaufman *et al.*, 1956), the solubility of hydrogen in diborane (Hu *et al.*, 1956), the mass spectrometric appearance potential study of diborane (Margrave, 1957), effect of hyperconjugation on the strength of effect of bases (McDaniel, 1957), mass spectrum and vapour pressure of pentaborane-d_9 (Shapiro *et al.*, 1957), and nuclear magnetic resonance study of B_2D_6–B_5H_9 exchange reaction (Koski, 1957), are amongst the subjects of topical interest.

A comprehensive tabulation of simple thermodynamic functions and related data for all known hydrides, including B_2H_6, B_4H_{10}, B_5H_9,

B_5H_{11}, B_6H_{10}, B_6H_{12} and $B_{10}H_{14}$, has been published (Elson et al., 1956) and includes 213 references.

A series of papers, I, II, III, IV, V, by Bauer (1956–1958), on the energetics of boranes, has appeared. In part V (1958) Bauer discusses the interconversion of the hydrides and refers to the "valence cone" of Platt (1954), the "smoothed potential" theory of Arnold (1954), and to the topological approach by Dickerson et al. (1957).

The kinetics of pyrolysis of diborane has naturally received attention. In one investigation (Clarke et al., 1951) the rate was observed to be depressed by addition of hydrogen; but unaffected by an equivalent amount of nitrogen, or by an increase in surface. The reaction was believed to be a radical-type polymerization process, complicated by quasi-reversible dehydrogenation. In another investigation (Bragg et al., 1951) the rate of increase of total pressure was measured as a function of temperature and initial pressure. The rate of formation of hydrogen was measured, and the results confirm the value of 1.5 for the order, and indicate an activation energy of 25.5 ± 0.5 kcal/mole. The reaction was followed by the mass spectrometer, and in this way the concentrations of several species were measured. According to Beachell et al. (1956) the sorption of diborane, diborane (deuterium) and trimethylboron by palladium and charcoal is purely physical.

D. Alkylated and Arylated Boranes

By interaction of trialkylborons and diborane, mutual replacement of partners can occur and give alkylated diboranes (Schlesinger et al., 1935).

$$B_2H_6 + BMe_3 \longrightarrow$$

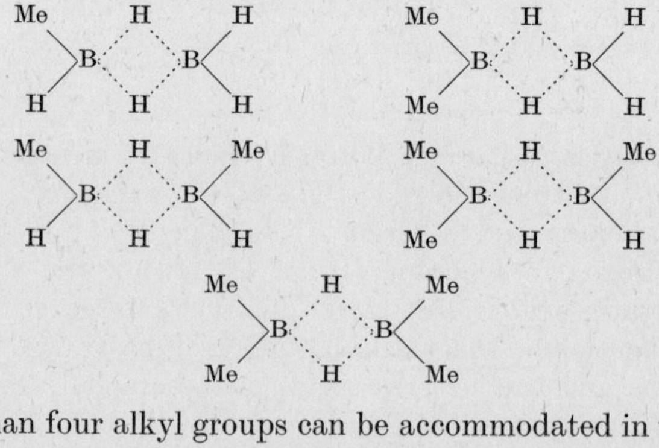

Not more than four alkyl groups can be accommodated in the diborane structure, thus preserving the essential three-centre hydrogen bridges. After equilibrium has been reached at room temperature, the mixture is cooled in order to effect separation in a vacuum system at low temperatures. These alkylated diboranes are, or tend to be, spontaneously oxidized in air. Higher homologues have, of course, been investigated, especially in connection with rocket fuel research projects. Methyldiboranes are gases or liquids according to number of methyl groups present.

The constitution of alkylated diboranes has been determined by hydrolysis to the corresponding acids. Thus with two alkyl groups on the same boron atom, the dialkylborinic acid is formed, whereas when only one alkyl group is so attached, the alkylboronic acid is formed.

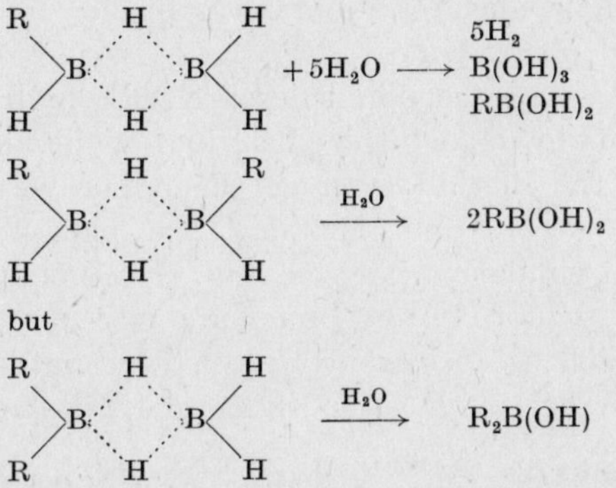

A different approach to the formation of alkylboranes has recently been made by Long *et al.* (1959), in which the trialkylboron is reduced by interaction with lithium borohydride, sodium borohydride or lithium aluminium hydride in the presence of either a hydrogen halide or a boron halide. Another approach is from the dialkylboron halide, R_2BX, or the alkylboron dihalide, RBX_2.

Thus hydrogen chloride, a trialkyl boron, and sodium borohydride at temperatures in the range 150–175°C react according to the following schemes.

$$NaBH_4 + HCl + 3BR_3 \longrightarrow 2B_2H_2R_4 + NaCl + RH$$

$$3NaBH_4 + 3HCl + 5BR_3 \longrightarrow 4B_2H_3R_3 + 3NaCl + 3RH$$

$$NaBH_4 + HCl + BR_3 \longrightarrow B_2H_4R_2 + NaCl + RH$$

$$5NaBH_4 + 5HCl + 3BR_3 \longrightarrow 4B_2H_5R + 5NaCl + 5RH$$

Low-pressure conditions are advisable for R = Me or Et, but atmospheric pressure is more suitable when R = Pr^n or higher. It is not unexpected to learn that aluminium chloride ($AlCl_3 : NaBH_4 = 1 : 12$) catalyses the reaction. As is to be expected, lithium borohydride reacted at lower temperatures, but gave hydrogen instead of hydrocarbon, for example, as follows.

$$2LiBH_4 + 2HCl + 4BR_3 \longrightarrow 3B_2H_2R_4 + 2LiCl + 2H_2$$

A typical reaction when boron trihalide is used in place of hydrogen chloride is illustrated by the following equation, although with sodium borohydride, hydrocarbon is also produced.

$$3LiBH_4 + BCl_3 + 2BMe_3 \longrightarrow 3B_2H_4Me_2 + 3LiCl$$

Lithium aluminium hydride has a certain advantage over sodium borohydride for the preparation of the higher alkyl compounds.

$$LiAlH_4 + 4BPr^n{}_2Cl \longrightarrow 2B_2H_2Pr^n{}_4 + LiCl + AlCl_3$$
(Vapour pressure, 21 mm at 76°C)

Information about arylated diborane is sparse. Pace (1929) claimed to have prepared phenylborane ($PhBH_2$) by reduction of the phenyl-boron diiodide–hydrogen iodide complex with hydrogen in ethanol. However, one would expect the phenylboron diiodide to react with alcohol to give the diethyl ester of phenylboronic acid before any reduction could occur, and certainly the phenylborane could be expected to react similarly. Nielsen *et al.* (1957) could not repeat this work, although by a different procedure (see p. 95) produced tri-phenylboron, which was apparently formed by mutual replacement in phenylborane. Wiberg *et al.* (1958) refer to 1,2-diphenyldiborane.

$$3PhBCl_2 \xrightarrow[\text{in dioxan}]{\text{LiAlH}_4} 3PhBH_2 \longrightarrow Ph_3B + B_2H_6$$

Dimethylborane, Me_2BH, appears not to be known, electronic require-ments leading to the dimeric form as the substituted diborane. Di-methoxyborane, $(MeO)_2BH$, does exist, being formed by the interaction of diborane and methanol.

$$B_2H_6 + 4MeOH \longrightarrow 2(MeO)_2BH + 4H_2$$

It is conceivable that here is a partial satisfaction of the electrophilic nature of boron by back–co-ordination from the electron-rich atoms of oxygen. Of course, as is to be expected, dimethoxyborane under-goes rapid mutual replacement at room temperature, giving diborane and trimethyl borate, and thus affords a route from diborane to tri-alkyl borates.

$$6(MeO)_2BH \longrightarrow B_2H_6 + 4(MeO)_3B$$

Diborane and dimethylamine interact at 200°C, affording dimethyl-aminoborane, Me_2NBH_2, which at room temperature is dimeric $[Me_2NBH_2]_2$, as a white crystalline solid, m.p. 73.5°C, insoluble in water, and not readily attacked by water or hydrogen bromide. Bis-dimethylaminoborane, $(Me_2N)_2BH$, does not dimerise, and has m.p. −45°C, and b.p. 109°C. It readily hydrolyses to bis-dimethylamino-boric acid, $(Me_2N)_2BOH$, and hydrogen.

Recently, Hawthorne (1957) has approached the phenylborane problem differently and has obtained the pyridine complex as a solid, m.p. 80–83°C. Diethyl phenylboronate or triphenylboroxine was mixed with pyridine at $-70°C$ in diethyl ether, and treated with lithium aluminium hydride. Hydrolysis with water then afforded the product. The general nature of this reaction made it even more interesting, for

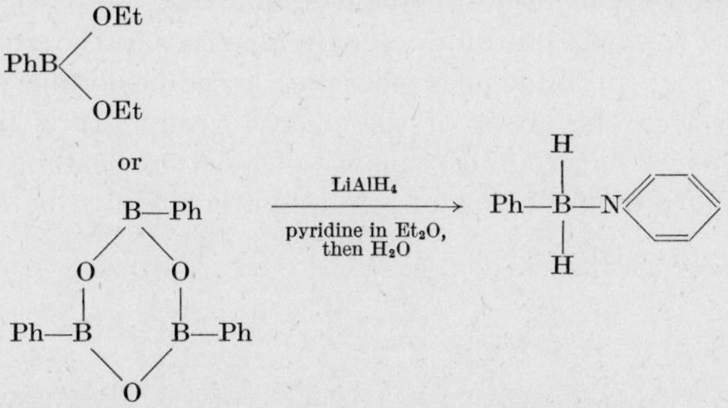

$$PhB\overset{OEt}{\underset{OEt}{<}}$$

or

$$
\begin{array}{c}
\underset{\underset{\displaystyle Ph-B}{}{\diagdown}\underset{\displaystyle O}{|}\ \underset{\displaystyle B-Ph}{|}}{\overset{\displaystyle B-Ph}{\diagup\diagdown}}\\
O \qquad O
\end{array}
\xrightarrow[\substack{\text{pyridine in Et}_2\text{O,}\\ \text{then H}_2\text{O}}]{\text{LiAlH}_4}
Ph-\underset{\underset{\displaystyle H}{|}}{\overset{\overset{\displaystyle H}{|}}{B}}-N\hexagon
$$

it was found that trimethylamine and triethylamine could be used instead of pyridine, and trialkylboroxines could be used in place of phenylboroxine. Furthermore, substituted phenyl compounds were prepared.

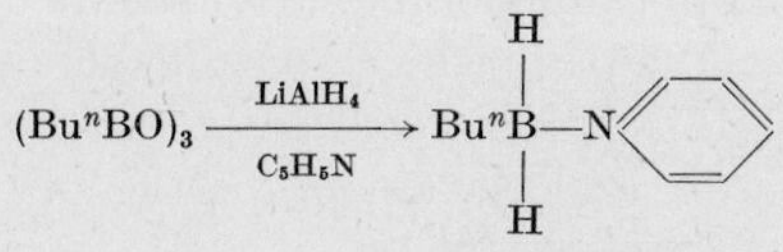

$$(Bu^nBO)_3 \xrightarrow[C_5H_5N]{LiAlH_4} Bu^nB\underset{\underset{\displaystyle H}{|}}{\overset{\overset{\displaystyle H}{|}}{}}-N\hexagon$$

An oil

$$p\text{-ClC}_6H_4B(OEt)_2 \longrightarrow p\text{-ClC}_6H_4-B\underset{\underset{\displaystyle H}{|}}{\overset{\overset{\displaystyle H}{|}}{}}-N\hexagon$$

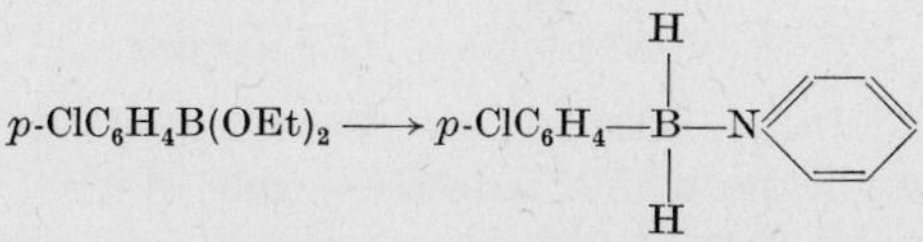

m.p. 61–62°C

Apart from the p-anisyl (yield, 61%) and the phenyl compound (yield, 53%) the yields were normally less than 40%.

The pyridine complexes of the diarylboranes were also obtained.

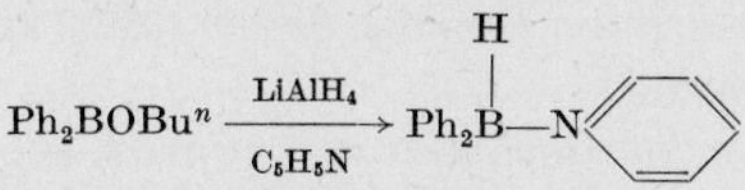

$$Ph_2BOBu^n \xrightarrow[C_5H_5N]{LiAlH_4} Ph_2B\overset{\overset{\displaystyle H}{|}}{}-N\hexagon$$

Also for Ar = $p\text{-ClC}_6H_4$, $p\text{-MeC}_6H_4$ or $p\text{-MeOC}_6H_4$.

It is remarkable that when the reagents were mixed in the absence of pyridine, phenylborane seemed to persist at 0°C for about 20 minutes;

for when pyridine was then added, the pyridine phenylborane was obtained in undiminished yield.

Pyridine borane, $C_5H_5NBH_3$, is commercially available as a reducing agent (see p. 164). The pyridine arylboranes and diarylboranes are much less reactive, and are apparently less reactive to carbonyl groups (but see Barnes *et al.*, 1958), although the diphenylborane compound reduced *iso*butyryl chloride to *iso*butyraldehyde.

Reactivity towards protonic solvents increases in the order pyridine diphenylborane > pyridine phenylborane > pyridine borane, and attracts acute interest in the effect of the phenyl group. It is necessary to know the precise nature and sequence of reactions leading to the final products before the influence of the aromatic ring can be assessed.

$$\underset{\underset{\text{H}}{|}}{\overset{\overset{\text{H}}{|}}{\text{Ph——B——N}}}\hspace{-0.2em}\text{<}\hspace{-0.2em}\bigcirc\xrightarrow[\text{HCl}]{\text{H}_2\text{O}}\text{PhB(OH)}_2+\text{H}_2+\text{C}_5\text{H}_5\text{N}$$

$$\underset{\text{Ph}}{\overset{\text{Ph}\quad\text{H}}{\underset{|}{\text{B——N}}}}\hspace{-0.2em}\text{<}\hspace{-0.2em}\bigcirc\xrightarrow[\text{HCl}]{\text{H}_2\text{O}}\text{Ph}_2\text{BOH}+\text{H}_2+\text{C}_5\text{H}_5\text{N}$$

Referring again to pyridineborane, attention is drawn to some work by Mikheeva *et al.* (1956) in which pyridineborane, $C_5H_5NBH_3$, and quinolineborane, $C_9H_7NBH_3$, were prepared by passing diborane through the dry bases; and certain properties of the complexes were described (see also Mikheeva, 1957).

E. Complexes of Decaborane

Of late, considerable interest has been taken in decaborane, the white crystalline borane which is fairly stable in air.

Hawthorne *et al.* (1960) have discovered an interesting alkoxylation reaction whereby alkoxydecaborane compounds, $B_{10}H_{13}OR$ (R=Me, Et, Pr^n, Bu^n or Ph), are obtained as readily hydrolysed high boiling liquids by the interaction of iodine, and sodium decaborane in the appropriate ether. This constitutes an ether cleavage; but more work has to be done to determine the fate of the other alkyl group. Spectroscopic data were recorded.

The acid function of decaborane towards acetonitrile (Schaeffer, 1957; Reddy *et al.*, 1959), morpholine, piperidine, *iso*quinoline, 4-methylpyridine, 3-methylpyridine, 2-methylpyridine, and pyridine (Mosher *et al.*, 1958) is shown in the formation of complexes with 2 of base to 1 of decaborane. A complex with dimethylamine is reported by Fitch *et al.* (1958). A 2 : 1 complex of 2-bromopyridine and decaborane has been obtained by Burkardt *et al.* (1959) by the vacuum distillation of 2-bromopyridine onto decaborane at −78°C. The mixture was then allowed to warm to room temperature, and after removal of excess of 2-bromopyridine in a vacuum, the product was crystallized from acetone. These workers obtained a 3 : 1 pyridine–decaborane complex, and not a 2 : 1 complex, as brilliant orange crystals having quite remarkable thermal stability (heating to 105°C gave no pyridine).

$$2 \bigcirc_{\mathrm{N}}^{} \mathrm{Br} + \mathrm{B}_{10}\mathrm{H}_{14} \longrightarrow \left[\bigcirc_{\mathrm{N}}^{\mathrm{Br}} \right]_2 \mathrm{B}_{10}\mathrm{H}_{14}$$

$$3 \bigcirc_{\mathrm{N}}^{} + \mathrm{B}_{10}\mathrm{H}_{14} \longrightarrow \left[\bigcirc_{}^{} \mathrm{N} \right]_3 \mathrm{B}_{10}\mathrm{H}_{14}$$

The interaction of decaborane and alcohols has been studied by Beachell *et al.* (1956, 1958), and alkyldecaboranes have been prepared by the Grignard procedure (Siegel *et al.*, 1958; Callaghan *et al.*, 1959).

Zhigach *et al.* (1956) have studied the interaction of alcohols and ketones with pentaborane, and refer to esters such as $(\mathrm{EtO})_2\mathrm{BH}$.

F. High-energy Fuels

It would be difficult to point to a compound which has for so long touched the hearts of pure scientists, and yet caused a greater flutter in Wall Street than diborane.

In the intensive research and development of jets and rockets in connection with guided missiles, the speed and range required called for fuels beyond those usually available. The high heats of combustion of boron hydrides attracted attention just at a time when the production of diborane had emerged from its micro-world.

The United States Government allocated 100 million dollars to the development of large-scale production of boron fuels, and the most amazingly colourful references appeared in the Wall Street and technical press in America. "Unquestionably one of the biggest chemical feats of the decade has been the development of the high-energy boron-based fuels. Almost overnight they have created an industry that will soon have more than 90 million dollars' worth of plant in place." So wrote one of the more reserved writers.

Exotic fuels, Hycal fuels, Zip fuels—these were the fanciful names given to the new materials. There were those who believed that the sputnik rockets were powered by such fuels. "The missile bandwagon really took off last week", wrote another reporter. "On Thursday, capping 10 hectic days, a rocket chemical company activity announced a joint company to develop and produce boron compounds headed for use in high-energy fuels." "Bounding Borax" was referred to in connection with the share market relating to boron fuels.

The precise nature of these fuels was kept strictly secret; but vague references to the alkylated boranes, pentaborane, and decaborane appeared. Sodium borohydride and boron trichloride were mentioned as key chemicals in these projects. As one statement had it, "while security regulations prevent revealing more detailed facts, most prominent in the discussion of solid propellants was decaborane".

Plant was erected, boron trichloride was transported in tank quantities, diborane was transported in cylinders cooled with dry ice, and pentaborane was similarly conveyed.

Then came the shock. The U.S. Government departments cancelled the high-energy boron fuel programme, and the American technical press during the months of September and October, 1959, were full of the matter. What use is to be put to the 5 ton-per-day plants is a question the manufacturers were asking.

III. Metal Borohydrides

A. Introduction

It is relevant to refer to certain borohydrides because they are deeply involved in the organic chemistry of boron. By 1943 metal borohydrides were no longer regarded as chemical oddities and borohydrides of a large number of metals have since been prepared, including mixed salts of the type $MXBH_4$, e.g. $AlF_2(BH_4)$, as well as compounds in which one or more of the hydrogen atoms have been substituted by other groups such as methyl or methoxy.

B. Preparation

The above method of preparation was unsatisfactory for larger amounts, and so attempts were made to discover better methods. When diborane was passed into an ethereal suspension of lithium hydride an exothermic reaction was initiated which finally resulted in an almost quantitative yield of lithium borohydride; but unfortunately this method was not general, and under no prevailing circumstances would sodium hydride react with diborane.

With the increasing awareness of the possible chemical potentialities of these compounds, many attempts were made to find an economic and efficient method for large-scale manufacture (Schlesinger *et al.*,

1949, 1950) of alkali metal borohydrides. Several methods of preparation involving the displacement of the weaker acid trimethyl borate by the stronger acid, borane, were investigated, and although these methods produce sodium borohydride smoothly and rapidly in good yield, the need to use diborane was a severe disadvantage.

$$2NaBH(OCH_3)_3 + B_2H_6 \longrightarrow 2B(OCH_3)_3 + 2NaBH_4$$

$$3NaB(OCH_3)_4 + 2B_2H_6 \longrightarrow 4B(OCH_3)_3 + 3NaBH_4$$

$$3NaOCH_3 + 2B_2H_6 \longrightarrow B(OCH_3)_3 + 3NaBH_4$$

Similarly, diborane will react with potassium methoxyborohydride to yield potassium borohydride, but will not react with potassium methoxide.

The preparation of sodium borohydride without the use of diborane was effected by displacing methoxide ions from sodium methoxyborohydrides by hydride ion acting as a stronger base, and a novel method involved thermal redistribution.

$$3NaH + NaBH(OCH_3)_3 \longrightarrow 3NaOCH_3 + NaBH_4$$

$$4NaH + NaB(OCH_3)_4 \longrightarrow 4NaOCH_3 + NaBH_4$$

$$4NaBH(OCH_3)_3 \longrightarrow NaBH_4 + 3NaB(OCH_3)_4$$

Under the thermal conditions necessary (about 230°C), the sodium tetramethoxyboron underwent redistribution to sodium methoxide and methyl borate, and unless the latter was removed the reaction was inhibited, and the yield of sodium borohydride poor.

$$3NaB(OCH_3)_4 \longrightarrow 3NaOCH_3 + 3B(OCH_3)_3$$

These complications are avoided by slowly adding methyl borate to an excess of a well-stirred mass of powdered sodium hydride at 250°C. Under these conditions a 90–93% yield was obtained, the reaction proceeding *via* the formation of intermediate sodium trimethoxyborohydride.

$$NaH + B(OCH_3)_3 \longrightarrow NaBH(OCH_3)_3$$

The overall reaction is represented by the following equation.

$$4NaH + B(OCH_3)_3 \longrightarrow 3NaOCH_3 + NaBH_4$$

The sodium borohydride is extracted by liquid ammonia or *iso*propylamine (acetone free) (Banus *et al.*, 1951) in which borohydride ion is very soluble and methoxide almost insoluble.

$$4Na + 2H_2 + B(OCH_3)_3 \longrightarrow 3CH_3ONa + NaBH_4$$

This system formed the basis of the described industrial method of making sodium borohydride. Hydrogen was passed under pressure into finely dispersed sodium in high boiling mineral oil to form finely dispersed sodium hydride which was then treated with trimethyl borate. It was for this process that pure trimethyl borate was required.

Ethyl and n-butyl borates may be employed with some advantage, although sodium ethoxide is less stable than the corresponding methoxide and so close control is required to prevent breakdown.

Sodium hydride and boron trioxide heated together under the appropriate conditions will give a yield of sodium borohydride.

$$4NaH + B_2O_3 \longrightarrow 3NaBO_2 + NaBH_4$$

Lithium borohydride may be prepared by interaction of lithium chloride and sodium borohydride in ethylamine.

$$LiCl + NaBH_4 \longrightarrow LiBH_4$$
$$\text{Extracted by ether}$$

Wiberg *et al.* (1958) have described the preparation and properties of lithium monophenylborohydride.

C. Properties of Sodium Borohydride

Sodium borohydride is a white crystalline solid with face-centred cubic lattice determined from X-ray diffraction data (Soldate, 1947). It is probable that the structure consists of tetrahedral $[BH_4]^-$ ions and Na^+ ions. It is stable in dry air at 300°C, but slowly decomposes at 400°C in a vacuum tube. When ignited in a Bunsen flame it burns quietly.

The hydrolysis of sodium borohydride has been extensively studied. A solution is very unstable in water although the resulting hydrolysis tends to increase the stability in increasing the alkalinity. For this reason it dissolves in cold water without extensive evolution of hydrogen and may be partially recovered as the dihydrate ($NaBH_4, 2H_2O$).

Rapid hydrolysis is effected by the addition of acid, by raising the temperature of the solution or by the addition of catalysts, especially salts of iron, cobalt or nickel. Rate of hydrolysis increases with increasing acidity, is dependent upon pH, and is independent of the specific acid used. The addition of alkali to sodium borohydride solutions increases the stability considerably. A solution of the borohydride (0.1 M) in sodium hydroxide (0.1 N) lost about 12.8% of available hydrogen in 6 days at 24°C, whereas when 1.0 N sodium hydroxide is present the loss is only 5.5%. Thus sodium borohydride has its highest stability in strongly basic solution at low temperature.

Complete hydrolysis of sodium borohydride liberates 2.37 litres of hydrogen (STP) per gram.

$$NaBH_4 + 2H_2O \longrightarrow NaBO_2 + 4H_2$$
$$NaBH_4 + 3H_2O + HCl \longrightarrow NaCl + H_3BO_3 + 4H_2$$

Strong acid solutions react to form hydrogen and diborane, and an effective procedure in which concentrated sulphuric acid or methylsulphonic acid is used has been described by Weiss *et al.* (1959).

$$2NaBH_4 + 2HCl \longrightarrow 2NaCl + 2H_2 + B_2H_6$$

Relative solubility in different solvents is an important function of borohydrides (see Table XI). Sodium borohydride is commonly used in ethanol or methanol but undergoes reaction with these solvents.

TABLE XI

The Solubility of Sodium Borohyride

Solvent	Temperature °C	Solubility g/100 g solvent
Methylamine	−20	27.6
Ethylamine	17	20.9
n-Propylamine	28	9.7
*iso*Propylamine	28	6.0
n-Butylamine	28	4.9
Ethylenediamine	75	22
Pyridine	25	3.1
Ethanol	20	4
Methanol	25	25
Water	22	55
*iso*Propanol	25	0.37
	60	0.88
t-Butanol	25	0.11
Diglyme	20	3.3
	40	11.0
	60	2.6

An approx. 1.0 M solution in methanol at 60°C loses 88% of its available hydrogen in 12 min., whereas in ethanol under the same conditions only 54% of the available hydrogen is lost in 6 hours (Brown *et al.*, 1955). *iso*Propanol and *t*-butanol are also solvents, but exhibit no tendency towards the formation of hydrogen under the same conditions.

$$NaBH_4 + 4MeOH \longrightarrow NaB(OMe)_4 + 4H_2$$

The dimethyl ether of diethylene glycol (diglyme) has recently established itself as a useful solvent for sodium borohydride (Brown *et al.*, 1955). Maximum solubility is observed at 40°C, and on cooling a saturated solution crystals of a solvate consisting of an equimolecular proportion of sodium borohydride and diglyme are deposited. On heating a saturated solution to higher temperatures, precipitation of finely divided unsolvated sodium borohydride occurs.

Sodium borohydride has very low solubility in diethyl ether, dioxan, diethyl cellosolve, ethyl acetate and methyl borate.

The process of forming a complex hydride anion generally involves donation of a negative hydride ion [H]⁻ to a specific acceptor molecule.

$$X^+ : H^- + \overset{\displaystyle H}{\underset{\displaystyle H}{\overset{|}{\underset{|}{B}}}} : H \longrightarrow X^+ \left[\overset{\displaystyle H}{\underset{\displaystyle H}{\overset{|}{\underset{|}{H : B}}}} : H \right]^-$$

Hydride ion is not an especially strong electron donor, and thus the acceptor component must necessarily be a strong electron attractor for a complex ion to be formed. Elements with particularly well developed electron-accepting properties are the members of Group III of the periodic table, notably boron aluminium and gallium. Boron is a strong acceptor and so it is not surprising that the borohydride ion is stable. Once the complex is formed all the covalent bonds to the hydrogen atom are equivalent.

The stability of the borohydride ion depends also upon the electronic nature of the associated metal component, with stability decreasing with increasing electron affinity of the component. Thus, the alkaline metal borohydrides are particularly stable, whereas aluminium borohydride, for example, is very unstable and highly reactive.

Similarly, acidification rapidly brings about the decomposition of complex hydride anions, since hydronium ion is strongly electron-attracting.

$$BH_4^- + H_3O^+ \longrightarrow [BH_3] + H_2O + H_2$$

The BH_3 unit is then attacked and decomposed by water. The picture is further complicated because borohydrides with less electropositive metals appear to possess covalent characteristics. Thus aluminium borohydride is a very unstable, low boiling (covalent) liquid. It is possible that the covalent borohydrides have chemical bonding closely related to the boron hydrides and involve "hydrogen bridges" or three-centre bonds (Longuet-Higgins, 1946, 1957).

IV. Borohydrides as Reducing Agents

A. The Effect of Metal Ions and Solvents

The chemistry of the borohydride ion cannot be considered independently of the metal ion with which it is associated; and the solvent is also of significance.

Sodium borohydride is practically insoluble in ether and exhibits poor reducing properties, whilst lithium borohydride is readily soluble in this solvent and forms a capable reducing solution. The difference in reactivity is not entirely due to solubility difference, because both sodium and lithium borohydrides are soluble in *iso*propanol and diglyme, and, whereas solutions of the lithium compound readily reduce esters, the sodium borohydride solutions have low activity towards esters. Furthermore, the order of increasing solubility is *t*-butanol < *iso*propanol < diglyme, but the order of increasing reactivity towards ester reduction is diglyme < *iso*propanol. Reactivity is not a function entirely dependent upon the solubility, but differs according to the solvent and the particular borohydride compound employed. The lithium compound is essentially a more powerful reducing agent than sodium borohydride.

The solubility of sodium borohydride in diglyme is attributed (Brown *et al.*, 1957) to the ability of diglyme to serve as a tridentate chelate in co-ordinating with the sodium ion.

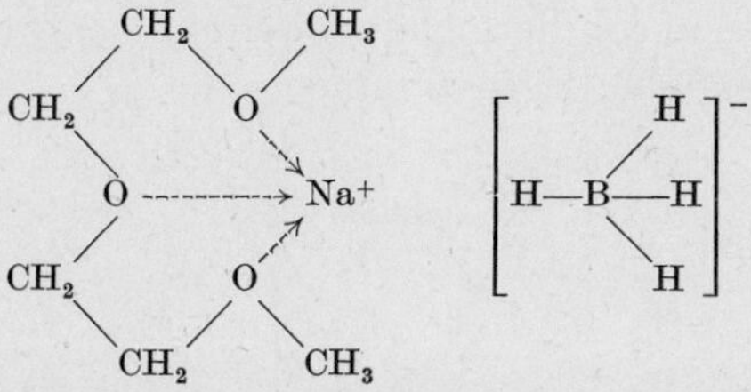

The reduction of ketones by sodium borohydride is faster in *iso*propanol than in diglyme. The rate was studied using *iso*propanol as solvent and the reaction follows second-order kinetics with the rate constant decreasing in the order $Me_2C=O > MeEtC=O > MePr^iC=O > MeBu^tC=O$ (Brown *et al.*, 1955). Reduction of ketones by sodium borohydride in diglyme is much slower, but the reduction is catalysed by water, *iso*propanol and especially by triethylamine. It follows that the solvent influences the rate of reaction and indicates that it must be involved in some manner in the transfer of the hydride ion to the carbonyl group.

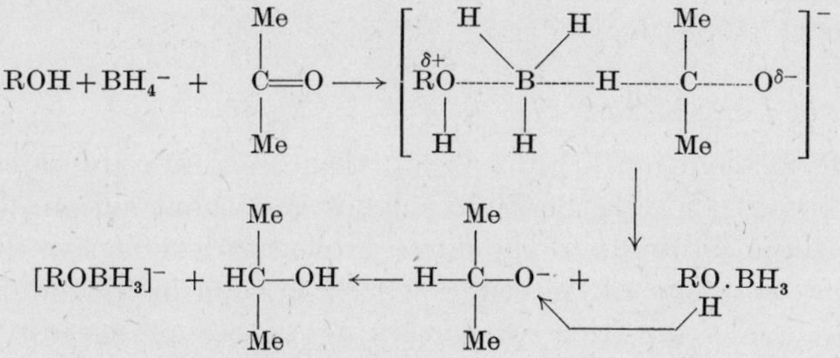

According to this hypothesis the slow reaction in diglyme is attributed to the poor donor properties of the ether oxygen atoms compared with the superior donor properties of the oxygen atom of *iso*propanol. The strong catalytic action of triethylamine is accounted for by the high donor properties of the nitrogen atom of the amine group.

$$Et_3N + BH_4^- \; + \; \underset{\displaystyle Me}{\overset{\displaystyle Me}{C}} = O \longrightarrow \left[Et_3N \cdots \overset{\overset{\displaystyle H \quad H}{\diagdown \diagup}}{\underset{\displaystyle H}{B}} \cdots H \cdots \underset{\displaystyle Me}{\overset{\displaystyle Me}{C}} \longrightarrow O^{\delta-} \right]^-$$

$$Et_3N + H\underset{\displaystyle Me}{\overset{\displaystyle Me}{C}} - OBH_3^- \; \longleftarrow \; Et_3N : BH_3 \; + \; H\underset{\displaystyle Me}{\overset{\displaystyle Me}{C}} - O^-$$

The reduction is practically halted after the transfer of the first hydride ion because of conflicting steric requirements of the *iso*propoxyborohydride ion and the triethylamine.

$$Et_3N + (Me)_2CHOBH_3^- \; + \; \underset{\displaystyle Me}{\overset{\displaystyle Me}{C}} = O \xrightarrow{\text{very slow}} \left[\begin{array}{c} Et \\ Et - N \cdots \overset{\overset{\displaystyle H \diagdown \diagup H}{}}{B} \cdots H \cdots \underset{\displaystyle Me}{\overset{\displaystyle Me}{C}} \longrightarrow O^{\delta-} \\ Et \quad O \\ Me - CH - Me \end{array} \right]^-$$

Just as the nature of the solvent alters the rate of reduction, the presence of various metallic ions substantially alters the reducing power. The reducing character of such solutions is altered considerably by the presence of lithium, magnesium, calcium, gallium, titanium and aluminium halides and by potassium hydroxide, ruthenium methoxide and caesium methoxide. Thus selectivity of reduction can be achieved by employing sodium borohydride in conjunction with a particular solvent and in the presence of a selected metal ion. The reports of Kollonitsch *et al.* (1954, 1955) are significant.

The etherates of lithium borohydride have attracted attention (Burns *et al.*, 1958; Kolski *et al.*, 1958).

1. *Effect of Lithium Ions*

In diglyme solution lithium borohydride readily reduces esters, but sodium borohydride in the same solvent is without action. The addition of lithium bromide to the latter proved to be effective in bringing about the reduction of the ester. Other lithium halides and different solvents have been tried with varying degree of success, and the results obtained again indicate that the nature of the solvent is important as well as the particular lithium halide employed.

2. *Effect of Lithium Chloride in Ethanol Solution*

$$NaBH_4 + LiCl \xrightarrow[-10°C]{Absolute\ EtOH} LiBH_4 + NaCl$$

A metathetical reaction occurs which results in the almost quantitative precipitation of sodium chloride. The essentially alcoholic solution of lithium borohydride reduces ketones, but gives poor yields for the reduction of esters. A prohibitory factor is the low temperature at which the reagent must be used, and at higher temperatures, decomposition of the lithium borohydride occurs.

3. *Effect of Lithium Iodide in Tetrahydrofuran*

The solution is prepared by stirring together equimolecular quantities of sodium borohydride and lithium iodide in tetrahydrofuran.

$$NaBH_4 + LiI \xrightarrow[T.H.F.]{} LiBH_4 + NaI$$

An equilibrium is believed to occur and the metathetical reaction becomes complete during the reduction reaction. This solution is far more stable and provides powerful reduction with good yields. If the lithium iodide is replaced by either lithium chloride or bromide, no reduction will occur in the resulting solution. Similarly, potassium borohydride will interact with lithium chloride in T.H.F. solution to yield a solution containing lithium borohydride (Davis, 1956).

4. *Effect of Lithium Chloride in isoPropanol*

$$NaBH_4 + LiCl \longrightarrow LiBH_4 + NaCl$$

This solution reduces esters in good yield (Brown *et al.*, 1955). Unlike the sodium compound, lithium borohydride is slowly decomposed by *iso*propanol (especially at elevated temperatures), and thus to ensure complete reduction excess of reagent must be employed. Sodium chloride is precipitated quantitatively.

5. *Effect of Lithium Bromide in Diglyme*

The addition of lithium bromide to a solution of sodium borohydride in diglyme resulted in the formation of a fairly powerful and stable reducing solution. Esters are speedily reduced and the corresponding alcohols are obtained in good yield. Lithium chloride is less soluble in diglyme than the bromide and the solution thus obtained exhibits milder reducing properties.

6. *Effect of Alkaline Earth Ions—Hydroxylic Solvents*

Successful reductions have been reported (Kollonitsch *et al.*, 1954) involving the borohydrides of calcium, magnesium and strontium. Calcium and magnesium borohydrides are conveniently prepared in ethanol solution by metathesis at $-50°$ to $-20°C$.

$$2NaBH_4 + CaCl_2 \longrightarrow Ca(BH_4)_2 + 2NaCl$$

$$2NaBH_4 + MgCl_2 \longrightarrow Mg(BH_4)_2 + 2NaCl$$

Calcium and magnesium borohydrides are very sensitive to water, but they provide neutral solutions which are advantageous in the reduction of alkali-sensitive compounds such as sugar derivatives. Difficulty arises when employing ethanol as solvent because low reaction temperatures are essential to prevent decomposition of these borohydrides. The borohydrides of barium and strontium have very similar reducing properties. These alkaline earth borohydrides are soluble in solvents with hydroxyl groups, and reduction can be carried out in the presence of as much as 50% water providing the temperature is kept very low.

7. *Effect of Alkaline Earth Ions—Non-hydroxylic Solvents*

Better reductions are obtained by stirring calcium iodide or magnesium iodide in a suspension of sodium borohydride in tetrahydrofuran (Kollonitsch *et al.*, 1955) at a temperature between 0° and 25°C.

$$2NaBH_4 + MgI_2 \longrightarrow Mg(BH_4)_2 + 2NaI$$

Anhydrous magnesium chloride and bromide are only slightly soluble in diglyme (0.013 M for $MgCl_2$ and 0.078 M for $MgBr_2$ at 100°C) but the addition of equivalent quantities of these solids to diglyme solution of sodium borohydride brings about the reduction of esters. Calcium borohydride appears to possess more powerful reducing properties than the corresponding lithium compound.

8. *Effect of Potassium Ions*

Potassium borohydride solution can be prepared from the sodium compound by the simple expedient of stirring together aqueous solutions of the borohydride and potassium hydroxide (Banus *et al.*, 1954). Potassium borohydride is less reactive than the sodium compound and provides an extremely gentle reducing agent. It is insoluble in benzene hexane, *iso*propylamine, ether, dioxan and tetrahydrofuran. It is slightly soluble in ethanol (0.25 g in 100 g solvent at 25°C), but is quite soluble in water and aqueous–methanol solutions, although being practically insoluble in methanol.

9. Effect of Aluminium Ions

A very powerful reducing agent is prepared by adding anhydrous aluminium chloride to a solution of sodium borohydride in diglyme (Brown *et al.*, 1955, 1956).

This solution is best prepared by mixing a freshly prepared solution of aluminium chloride (2 M) and sodium borohydride (1.0 M) in diglyme in a molar ratio $AlCl_3 : 3NaBH_4$. The precise order of mixing appears to be unimportant. The resulting clear solution possessed powerful reducing properties, whilst being easy to handle and free from any particular hazard. This solution is stable in open air and no violent reaction occurs when poured into water, thus permitting reduction reactions in the absence of the special techniques necessary for the moisture-sensitive reducing agents such as lithium aluminium hydride. An important feature of this solution is its ability to reduce certain functional groups with the exclusion of others, thus a sodium salt of a carboxylic acid is not reduced, whilst under the same conditions, carboxylic acid is reduced at the carbonyl position.

Nitro groups are not attacked and the reduction of a nitro-ester gives good yields of the corresponding nitro-alcohol. Similarly, chloro-esters may be reduced to the chloro-alcohol.

Investigation of the suitability of other solvents showed that T.H.F. was quite satisfactory, while diethyl ether was less satisfactory and dioxan was unsatisfactory. The presence of as little as 20% diglyme resulted in increased rate of reductions in either T.H.F. or diethyl ether.

10. Effect of Other Metal Ions

There has been very little experimentation into the effect of other metals on reducing properties of sodium borohydride (Brown *et al.*, 1956). It appears, however, that these properties are substantially improved by the addition of other polyvalent metal halides such as gallium trichloride and titanium tetrachloride. Titanium tetrachloride is a less active agent than aluminium chloride, and zinc chloride was unsuitable.

B. The Effect of Alkoxy Substituents on the Reducing Properties

Substitution of alkoxy groups for the hydrogen atoms in the borohydride ion has a profound effect upon reducing properties (Brown *et al.*, 1956). The reducing properties depend upon the nature of the alkoxy group, the number of hydrogen atoms substituted, and the solvent. Sodium trimethoxyborohydride is prepared by the reaction between sodium hydride and trimethyl borate in T.H.F. An analogous but slower reaction leads to the formation of sodium triethoxyboro-hydride, whereas the tri-*iso*propoxyborohydride is only obtained in poor yield with considerable difficulty, and sodium tri-*t*-butoxyboro-hydride cannot be prepared by this method. If diglyme is used as the solvent the reaction rate is increased; sodium tri-*iso*propoxyboro-hydride is obtained more speedily, and a slow reaction even occurs between sodium hydride and tri-*t*-butyl borate yielding tri-*t*-butoxy-borohydride.

$$NaH + (MeO)_3B \longrightarrow NaBH(OMe)_3$$

In contrast with sodium borohydride, sodium tri-*iso*propoxyboro-hydride reacts with *iso*propanol to liberate hydrogen, and will reduce ethyl benzoate at a moderate rate.

$$NaBH(OPr^i)_3 + Pr^iOH \longrightarrow NaB(OPr^i)_4 + H_2$$

The greater reactivity of the trialkoxyborohydride is attributed to the greater ease of removing a hydride ion from the weak Lewis acid, alkyl borate (see p. 12); while the transfer of a hydride ion from BH_4^- involves its removal from the strong Lewis acid, "borane". In accordance with this hypothesis the reducing tendencies of a series of tri-substituted borohydride ions $(XO)_3BH^-$ should be related to the strength of the parent Lewis acid $(XO)_3B$.

In T.H.F. both sodium trimethoxy and triethoxyborohydrides underwent redistribution, whereas the tri-*iso*propoxy and tri-*t*-butoxy compounds were stable in this solvent.

$$2NaBH(OMe)_3 \rightleftharpoons NaBH_2(OMe)_2 + NaB(OMe)_4$$

Perhaps one should refer back to what has been stated about re-distribution in trialkylborons and related compounds.

1. *Reductions involving Trialkoxy-substituted Sodium Borohydride*

Whereas aldehydes, ketones, acid chlorides and acid anhydrides are readily reduced at 35°C by sodium trimethoxyborohydride (Brown *et al.*, 1953), esters and nitrile groups are slowly reduced at 100–140°C. Conjugated C=C remains unaffected; the reagent is without action on carboxylic acids. Nitro groups do not react at low, but do so at elevated, temperatures. Considerable selectivity of reduction is achieved by this reagent in T.H.F. at low temperature ($-80°C$) whereby acid chlorides are reduced to the corresponding aldehyde.

$$RCOCl + NaBH(OMe)_3 \longrightarrow RCHO + NaCl + (MeO)_3B$$

The yield obtained is fairly low (25–40%), and the primary alcohol and methyl ester (i.e. RCH_2OH and RCO_2Me) are concurrently formed.

Sodium trimethoxyborohydride reductions are normally carried out in T.H.F. solution (see Table XII for examples), but the reduction characteristics may be altered by employing different solvents. The reagent is insoluble in diethyl and n-butyl ether, but is moderately soluble in dioxan.

TABLE XII

Reductions by Sodium Trimethoxyborohydride in Tetrahydrofuran

Compound	Product	Yield %
Benzaldehyde	Benzyl alcohol	78
Acetophenone	1-Phenylethanol	82
Cinnamaldehyde	Cinnamyl alcohol	79
Benzoyl chloride	Benzyl alcohol	66
Ethyl benzoate	Benzyl alcohol	33

2. General Review

Trialkoxy borohydrides and tetraalkoxy boron compounds have been prepared and are found to be stable solids. Mono and dialkoxyborohydrides have not been isolated, although they are believed to be formed as intermediates during the reduction by a metal borohydride, the mechanism of the reduction being postulated as proceeding *via* the formation of mono-, di-, tri-alkoxyborohydrides, and tetraalkoxylboron compounds.

Methanol and ethanol (but neither *iso*propanol nor *t*-butanol) react with sodium borohydride to liberate hydrogen and are converted into the corresponding tetraalkoxyboron compound (Brown *et al.*, 1956).

$$NaBH_4 + 4MeOH \longrightarrow Na(MeO)_4B + 4H_2$$

Preliminary results have shown that trialkoxyborohydrides are more powerful reducing agents than sodium borohydride.

Tetramethoxy and tetraethoxylboron compounds of lithium, sodium, potassium, calcium, zinc and thorium have been prepared by the direct combination of the borate ester and the metal alkoxide and by metathesis between the tetraalkoxyboron compounds and metal halides.

$$B(OR)_3 + NaOR \longrightarrow NaB(OR)_4$$

C. Application of Borohydrides as Reducing Agents

1. Mechanism

It has been postulated that the mechanism arising from the reduction by borohydride of a carbonyl group terminates with the formation of $1:4$ adduct (Chaikin *et al.*, 1949; Brown *et al.*, 1956). Hydrolysis of this adduct then yields the reduction products.

$$4R_2C{=}O + NaBH_4 \longrightarrow NaB(OCHR_2)_4.$$

Then, $$NaB(OCHR_2)_4 + H_2O + 2NaOH \longrightarrow Na_3BO_3 + 4R_2CHOH$$

The postulate infers that the adduct is formed in four successive steps, and that the first adduct formation is rate-determining, all subsequent adducts being formed much faster. This would be in accord with the weakening Lewis acid function of boron.

$$R_2C{=}O + BH_4^- \longrightarrow R_2CHOBH_3^-$$

$$R_2CHOBH_3^- + R_2C{=}O \longrightarrow (R_2CHO)_2BH_2^-$$

$$(R_2CHO)_2BH_2^- + R_2C{=}O \longrightarrow (R_2CHO)_3BH^-$$

$$(R_2CHO)_3BH^- + R_2C{=}O \longrightarrow (R_2CHO)_4B^-$$

TABLE XIII

Reduction of Aldehydes and Ketones with Sodium Borohydride

Aldehyde/Ketone	Product	Yield %
Reductions in Aqueous Solution		
Acetonyl acetone	Hexane-2,5-diol	86
n-Butyraldehyde	*n*-Butanol	85
Chloral hydrate	2,2,2,-Trichloroethanol	61
Crotonaldehyde	Crotyl alcohol	85
*cyclo*Pentanone	*cyclo*Pentanol	90
Diacetyl	Butan-2,3-diol	62
Mesityl oxide	4-Methyl-3-pentan-2-ol	77
Methyl ethyl ketone	*s*-Butanol	87
Levulinic acid	γ-Valerolactone	81
Reductions in Methanol Solution		
Anisaldehyde	Anisyl alcohol	96
Benzil	Hydrobenzoin	89
ω-Bromoacetophenone*	Styrene bromohydrin	71
Cinnamaldehyde	Cinnamyl alcohol	97
Di*cyclo*hexyl ketone	Di*cyclo*hexylcarbinol	88
p-Dimethylaminobenzaldehyde	*p*-Dimethylaminobenzyl alcohol	96
m-Hydroxybenzaldehyde	*m*-Hydroxybenzyl alcohol	93
m-Nitrobenzaldehyde	*m*-Nitrobenzyl alcohol	82

* Lithium aluminium hydride reduces this to 1-phenylethanol.

An alternative mechanism (Garrett *et al.*, 1953) for reduction has been suggested which is applicable to aqueous solutions and which does not postulate a large tetra molecular-borohydride-complex, nor does it necessitate borohydride adducts reacting with a carbonyl group at a faster rate than with borohydride alone. The mechanism involves the formation of the rate-determining 1 : 1 compound-borohydride complex followed by hydrolysis to the corresponding alcohol and a mono-substituted hydroxyborohydride which undergoes fast reaction with another molecule of the ketone. The reaction thus proceeds until four molecules of the ketone have been reduced.

$$R_2C{=}O + NaBH_4 \longrightarrow Na(R_2CHO)BH_3$$

$$Na(R_2CHO)BH_3 + H_2O \longrightarrow R_2CHOH + NaBH_3(OH)$$

$$R_2C{=}O + NaBH_3(OH) \longrightarrow Na(R_2CHO)BH_2(OH)$$

$$Na(R_2CHO)BH_2(OH) + H_2O \longrightarrow R_2CHOH + NaBH_2(OH)_2$$

$$R_2C{=}O + NaBH_2(OH)_2 \longrightarrow Na(R_2CHO)BH(OH)_2$$

$$Na(R_2CHO)B(OH)_3 + H_2O \longrightarrow R_2CHOH + NaBO_2 + 2H_2O$$

Table XIII shows the reduction products and percentage yields from aldehydes and ketones in aqueous and methanol solutions using sodium borohydride.

2. Selective Reductions of Keto-groups in the Presence of C=C and Acylamino Groups

Keto groups have been successfully reduced in the presence of, and without reacting with, associated unsaturated carbon double bonds and acylamino groups (Iliceto et al., 1956). The reduction of the keto group of 1-p-nitrophenyl-1-keto-2-acylamino-2-propenes is accomplished by sodium borohydride in aqueous ethanol/glycerol solution.

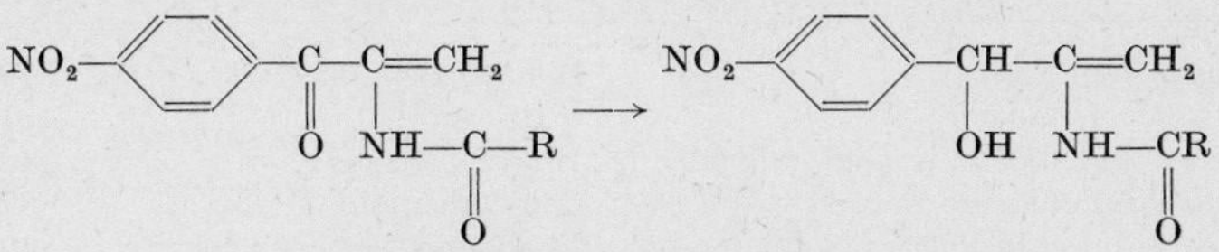

Where R is CH_3, $CHCl_2$, CCl_3 or C_6H_5.

A further example of selectivity is furnished by the reduction of the substituted santonin with keto groups at the 3 and 5 position (Tahara, 1956). Reduction is accomplished in methanol solution at room temperature when the 3-one is reduced, leaving the 5-one and the carboxylic groups untouched.

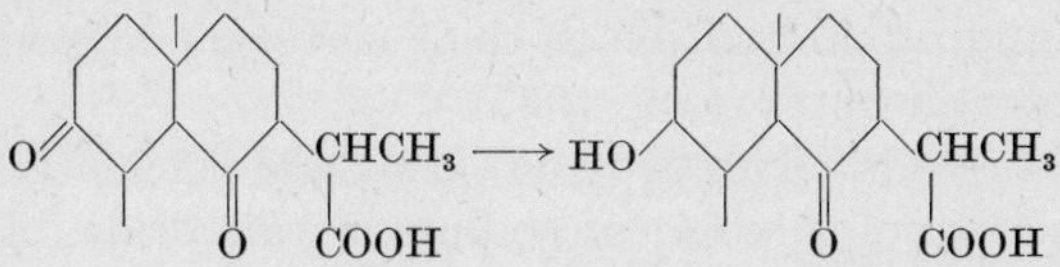

Lithium aluminium hydride reduces both keto groups along with the carbonyl of the carboxylic acid.

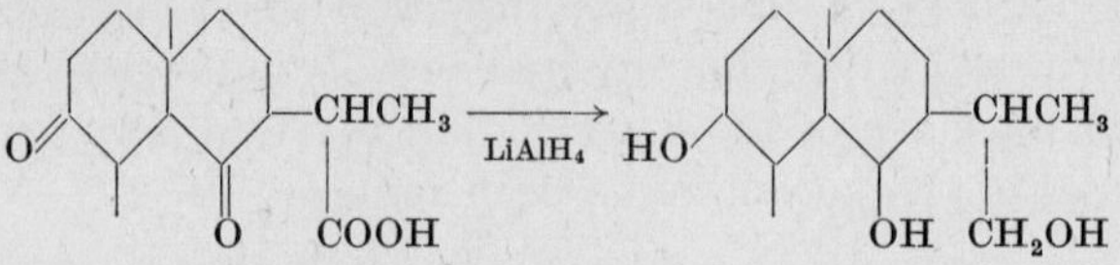

3. *Reduction of Acid Chlorides*

Reduction of both aliphatic and aromatic acid chlorides (Chaikin *et al.*, 1949) to the corresponding alcohols is conveniently accomplished by slow addition of the acid chloride dissolved in dioxan to a cooled suspension of sodium borohydride in excess in the same solvent. The mixture is heated on a steam bath for a short time and water then added to hydrolyse the intermediate complex (see Table XIV).

Some difficulties were encountered in the reduction of unsaturated and polyfunctional acid chlorides. Unsaturated alcohols are not obtained and it is assumed from analogy with lithium aluminium hydride reduction (Hochstein *et al.*, 1948) that borohydride adds to the double bond to form a C—B bond with the β-carbon atom. Unlike their aluminium analogue these compounds are resistant to hydrolytic fission.

TABLE XIV

Reduction of Acid Chlorides by Sodium Borohydride

Acid chloride	Product	Yield %
Benzoyl chloride	Benzyl alcohol	76
n-Butyryl chloride	*n*-Butanol	81
Monoethyl succinate acid chloride	Butyrolactone	40
Palmitoyl chloride	Cetyl alcohol	87
o-Phthalyl chloride	Phthalide	49
	Phthalyl alcohol	15
Crotonoyl chloride	not isolated	
Cinnamoyl chloride	Hydrocinnamyl alcohol	12

4. *Reduction of Nitroketones to Nitroalcohols*

Aliphatic primary, secondary and tertiary mononitro and primary and secondary gem-dinitro aldehydes and ketones have been selectively reduced by sodium borohydride to their corresponding mononitro and dinitro alcohols (Shechter *et al.*, 1952).

The selective reduction is effected rapidly and efficiently at 0–25°C by an aqueous methanol solution of sodium borohydride. The reduction is carried out preferably at pH 7–10.5, but in the case of alkali-sensitive nitro–carbonyl compounds, reduction is effected in acid solution and the pH is maintained at 3–4 by the addition of sulphuric acid as the reaction proceeds. The borohydride solution is added dropwise to the reaction mixture containing just sufficient methanol to keep the nitro compound in solution. Unreacted carbonyl compound is removed by washing ethereal solutions of the product with saturated sodium bisulphite solution. Aluminium *iso*propoxide is an alternative reducing reagent for this reaction, but sodium borohydride is to be preferred (see Table XV).

TABLE XV

Reduction of Nitroketones by Sodium Borohydride

Nitro ketone	Product	Yield % by $NaBH_4$	Yield % by $Al(Pr^iO)_3$
5-Methyl-5-nitro hexanone	-hexanol	98.7	93.3
5-Nitro-2-hexanone	-hexanol	50.7	48.0
5-Nitro-2-pentanone	-pentanol	61.2	24.0
4-Methyl-4-nitro-1-pentanone	-pentanol	54.8	69.3
4-Nitro-1-pentanone	-pentanol	56.3	26.7
4-Nitro-1-butanone	-butanol	33.4	—
5,5-Dinitro-2-hexanone	-hexanol	78.2	87.9
4,4-Dinitro-1-pentanone	-pentanol	67.6	62.8
5,5-Dinitro-2-pentanone	-pentanol	48.5	—

5 : 5-Dinitro-2-pentanone is only slowly reduced and not completely even with excess sodium borohydride. This difficulty is probably due to a ring-chain tautomerism involving the carbonyl group.

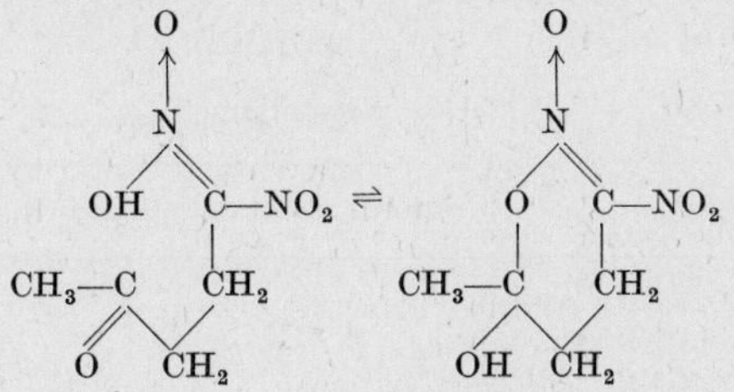

5. *Debromination Reduction. Synthesis of Nitrocycloalkanes*

α-Bromonitro*cyclo*alkanes are debrominated to nitro*cyclo*alkanes through reaction with sodium borohydride (Iffland *et al.*, 1953). The reaction occurs in aqueous methanol solution ($H_2O : MeOH$, 3 : 1). The method thus forms an attractive route for the conversion of a *cyclo*ketoxime to the corresponding nitro*cyclo*alkane. The borohydride debromination leads to a more favourable yield (see Table XVI) than the conventional method employing alcoholic potassium hydroxide solution.

TABLE XVI

Yields of Nitro*cyclo*alkanes

Product	Yield %
Nitro*cyclo*butane	15
Nitro*cyclo*pentane	76
Nitro*cyclo*hexane	80
Nitro*cyclo*heptane	76
2-Nitro-1-methyl*cyclo*hexane	63
4-Nitro-1-methyl*cyclo*hexane	53

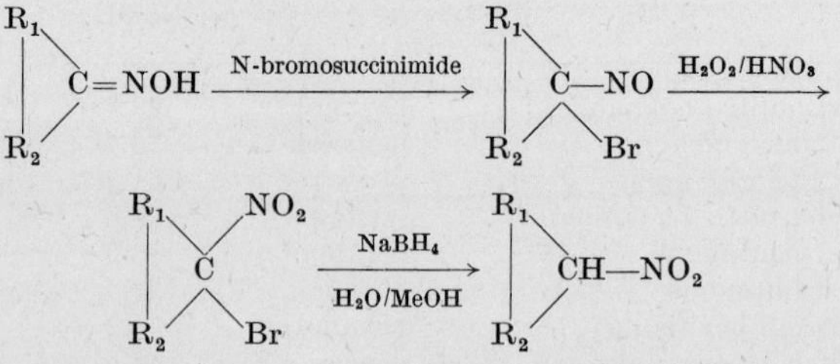

Further investigations by Iffland *et al.* (1954) showed that the debromination technique involving sodium borohydride could be successfully extended to include the preparation of secondary nitroalkanes from aliphatic ketoximes (see Table XVII). The debromination was accomplished by aqueous methanol solutions of sodium borohydride in a 5 : 1 mole excess.

TABLE XVII

Yields of Nitroalkanes

Product	Yield %
2-Nitropentane	38
3-Nitropentane	29
3-Nitro-2-methyl butane	48
2-Nitrohexane	19
3-Nitrohexane	25
2-Nitroheptane	16
3-Nitroheptane	29
4-Nitroheptane	28
3-Nitro-2,2-dimethyl butane	30
2-Nitro-octane	10

6. Reduction of Conjugated Nitroalkenes

Metal borohydrides are, in general, inert towards the nitro group and C=C double bonds. However, they react with C=C double bonds of conjugated nitroalkenes and afford an attractive route for the synthesis of primary and secondary nitroalkanes (Shechter et al., 1956). The mechanics of the reactions involve nucleophilic attack by the borohydride ion at the relatively positive end of the C=C of the nitroolefins. Addition probably results initially in the formation of a mono-alkanenitroborohydride ion, which may then undergo a series of reductive addition reactions with the conjugated nitro-olefin to give finally the tetraalkane nitroboron compound. Hydrolysis of this complex is accomplished by the addition of acid.

$$4R_2C=CRNO_2 + NaBH_4 \longrightarrow (R_2CHCR=NO_2)_4BNa$$
$$(R_2CHCR=NO_2)_4BNa + CH_3CO_2H + 3H_2O \longrightarrow 4R_2CHCHRNO_2 + H_3BO_3 + CH_3CO_2Na$$

7. Reduction of Esters

Brown et al. (1955, 1956) have shown that essentially quantitative reduction of esters is achieved by employing 25% excess reagent in 1 hour reaction time at 75°C, or in 3 hours at 25°C with 100% excess of sodium borohydride. The experimental procedure is simple and consists of adding the ester to a solution of sodium borohydride in diglyme, followed by the dropwise addition of aluminium chloride in diglyme (2.0 M) solution so that the temperature does not rise above 50°C.

Both aromatic and aliphatic esters are successfully reduced and give good yields of product. Reduction of unsaturated esters was, however, unsatisfactory, because it appears that the reagent reacts with the double bond to form the stable B—C bonds (see p. 152).

Substituted halogen and nitro esters are reduced to give good yields of the corresponding halogen and nitro alcohols.

8. Dechlorination of Substituted Halogeno Acetophenones

These reductions illustrate the selective nature of borohydride reactions and the degree of reduction dependent upon the metallic ion associated. Dehalogenation to the corresponding substituted styrene oxide is speedily obtained by employing aqueous dioxan solutions of the reagent, and good yields are obtained, as shown by Fuchs et al. (1954, 1956).

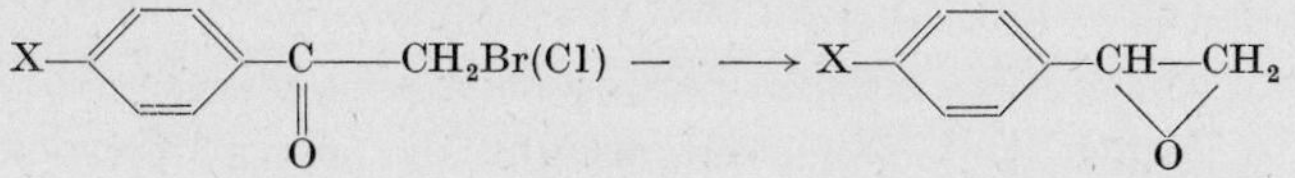

An ethereal solution of lithium borohydride, however, brings about reductive cleavage of the epoxide ring and gives good yields of the corresponding substituted primary and secondary phenylethanols.

9. *Reduction of Sugars and Sugar Derivatives*

Many papers on this subject have appeared. Sodium borohydride may be used to convert aldonic lactones into either an aldose or a glycitol in good yield, according to the conditions employed (Wolfrom *et al.*, 1951; Frush *et al.*, 1956). Lithium aluminium hydride cannot be employed as a reducing agent, and sodium amalgam and platinum catalysed hydrogenation generally give poor yields.

d-Gluco-*d*-gulo-heptoneo-γ-lactone was reduced to *d*-gluco-*d*-gulo-heptose by the dropwise addition of aqueous sodium borohydride solution. The temperature was maintained at 0–3°C and the pH adjusted to 3–4 by the continuous addition of 1 N sulphuric acid. Meso-gluco-gulo-heptanol was obtained from the same starting material by the reducing action of a slightly stronger aqueous sodium borohydride solution, and the temperature not allowed to exceed 50°C.

Other γ-lactones reduced to the corresponding aldose were:

d-galactono-γ-lactone	Aqueous methanol solution gave
d-ribono-γ-lactone	lower yield, absolute methanol solution gave
d-manno-*d*-gala-heptono-γ-lactone	very poor yield.

d-Arabitol was prepared from a mixture of *d*-araburonic acid and its lactone by reaction with aqueous sodium borohydride over a period of 18 hours (Gorin *et al.*, 1956).

Sodium borohydride is more effective for these particular reductions in aqueous solution than the corresponding lithium compound, according to Abdel-Akher *et al.* (1951). It has been previously shown that aqueous sodium borohydride solution reacted with reducing sugars, giving the boric acid complexes which are difficult to crystallize. This difficulty is overcome by completely acetylating the dehydrated residues from the reaction, wherefrom the acetylated sugar usually crystallizes with ease (see Table XVIII).

Other techniques developed involve ion exchange methods and the treatment of the dried reaction product with methanolic hydrogen chloride.

Reductions of other sugars have been reported (Bera *et al.*, 1956; Gorin *et al.*, 1956).

Attempts have been made by Abdel-Akher *et al.* (1951) and Hough *et al.* (1953) to employ sodium borohydride as a reagent to determine the chain length and molecular weight of polysaccharides. Other investigators have criticized this claim and have failed to obtain any evidence of reductive scission of polysaccharides with the reagent; furthermore, the reagent was regarded as converting only reducing sugars to the corresponding alcohols and no side reactions were observed (Whelan *et al.*, 1955).

TABLE XVIII

Reduction Products and Yields from Sugars

Sugar	Acetylated product	Yield %
d-Glucose	Sorbitol hexa-acetate	78
d-Mannose	*d*-Mannitol hexa-acetate	92
d-Galactose	Dulcitol hexa-acetate	87
l-Arabinose	*l*-Arabitol penta-acetate	87
d-Xylose	Xylitol penta-acetate	80
d-Fructose	Sorbitol and *d*-mannitol	75
Maltose	Malitol nona-acetate	70

More recent investigators working on the nature of the linkages and units of carbohydrates have utilized sodium borohydride reduction of the products of periodate oxidation (Hamilton *et al.*, 1956) of these sugars.

10. *Reduction of Steroidal Compounds*

Sodium borohydride is an excellent reducing agent for keto steroids and has been extensively used in this capacity. Its considerable success in this field has been enhanced by its selective nature. Investigations by Norymberski *et al.* (1954, 1955) have shown that keto steroids of the fully saturated pregnane series undergo selective reduction at C-3 keto position, keto groups in positions C-11, C-12, C-17, and C-20 not reacting except under more forcing conditions.

The introduction, however, of an unsaturated double bond in the position 4 : 5 sufficiently inhibits the reactivity of the 3-keto position, and favours reduction with respect to the oxo groups at C-17 and C-20.

The reactivity of steroidal ketones towards sodium borohydride decreases in the order 3-one > 17 and 20-one > 4-ene-3-one > 11-one.

11. *The Selective Reduction of Specific Keto Groups*

A technique has been evolved whereby the ketone groups which normally undergo preferential reduction with sodium borohydride are rendered inert to the reducing reagent, and the otherwise non-reactive ketone group is selectively reduced. A method which has enjoyed considerable success involves protection of the ketone by converting it to a dioxolane group. These dioxolane groups do not react with sodium borohydride and permit the selective reduction of the desired keto group. The keto groups are then regenerated by hydrolysis of the dioxolane group (Herzog et al., 1953). Several mono- and bis-ketals have been prepared and have enabled C-11 reductions to be accomplished without reducing associated C-20 and C-3 ketones (Oliveto et al., 1953) and C-17 (Herzog et al., 1953) and C-21.

12. Reduction of Carbonyl Compounds and Esters

Aldehydes and ketones are reduced rapidly by lithium borohydride in ether solution at room temperature in exothermic reactions, whereas esters react slowly and require long heating under reflux (Nystrom et al., 1949). Dibasic acid esters on reduction yield the corresponding glycol (see Table XIX). Nitrobenzene yields, after 18 hours' heating under reflux: aniline, 22%; nitrobenzene, 30%; dark oil, 30%.

TABLE XIX

Reduction Products and Yields from Carbonyl Compounds and Esters

Material	Product	Yield %
Carbonyl compound		
n-Heptaldehyde	n-Heptanol	83
Benzaldehyde	Benzyl alcohol	91
Methyl ethyl ketone	s-Butanol	77
Benzophenone	Benzhydrol	81
Esters		
n-Butyl palmitate	n-Hexadecanol	95
Ethyl benzoate	Benzyl alcohol	62*
Ethyl sebacate	Decamethylene glycol	60
β-Benzoyl propionic acid	γ-Phenylbutyrolactone	78
Ethyl levulinate	γ-Valerolactone	44

* Better yield with $NaBH_4/LiBr$ in diglyme.

13. Reduction of Substituted Styrene Oxides

Reduction is accomplished by lithium borohydride in ethereal solution and gives a mixture of primary and secondary alcohols, as shown by Fuchs et al. (1954, 1956) (see Table XX). The mechanism of the reduction involves bimolecular SN_2 epoxide ring opening; accordingly, the formation of a primary phenylethanol is favoured by the presence of electron-donating groups such as p-methoxy group; whilst electron-withdrawing groups facilitate the formation of a secondary phenylethanol.

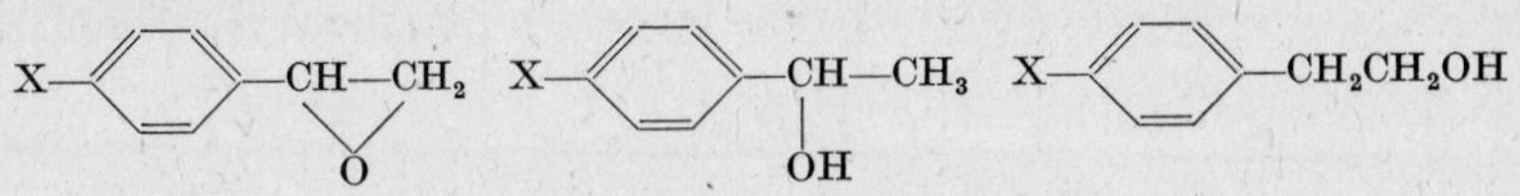

X—⟨benzene⟩—CH—CH₂   X—⟨benzene⟩—CH—CH₃   X—⟨benzene⟩—CH₂CH₂OH
                | |
                O   OH

TABLE XX

Yields of Primary and Secondary Alcohols

Substituted group	Secondary alcohol % yield	Primary alcohol % yield	Overall % yield
H	74	26	100
p-Methoxy	5	95	70
p-Methyl	32	68	—
p-Bromo	84	16	66
p-Nitro	38	62	64
m-Methoxy	77	23	—
3 : 4-Dichloro	95	5	—

The p-nitro is an exception to the general rule and reductive fissure favours the formation of the primary alcohol. The relative yields of the two alcohols depend to a large extent upon the electronic effect of the substituted group. Bromine in the p-position has a net electron-attracting effect and consequently the formation of the secondary alcohol is favoured. Investigations using other hydrides as reductive-cleavage agents have shown that steric factors are important in determining the position of attack and hence the direction of ring opening. Aliphatic epoxides have also been successfully reduced.

14. *Reduction of Amides*

Although amides are readily reduced by lithium aluminium hydride, both sodium and potassium borohydrides are without action. A mixture of potassium borohydride and lithium chloride in tetrahydrofuran as solvent has been investigated by Davis (1956) with regard to its action on various amides. The primary and secondary amides, benzamide and N-methyl benzamide, were recovered unchanged after 20 hours, but the tertiary N-N-dimethyl benzamide was completely reduced to N-N-dimethyl benzylamine (33%) and benzyl alcohol (58%).

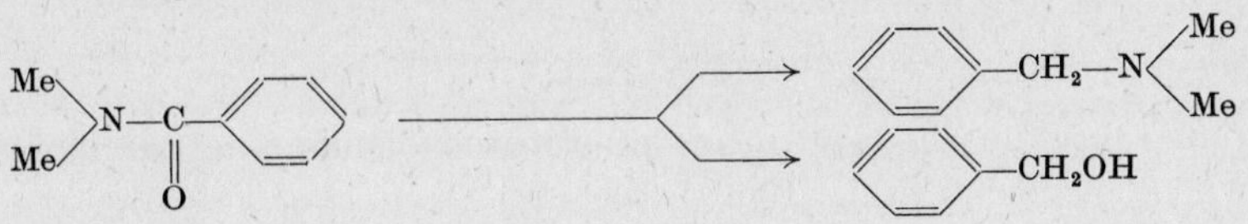

Me\
   N—C—⟨benzene⟩   ⟶   ⟨benzene⟩—CH₂—N⟨Me / Me⟩
Me/ ‖
     O   ⟨benzene⟩—CH₂OH

6

N-N-dimethyl p-nitrobenzamide yielded on reduction with lithium borohydride the corresponding azoxy compound in 41% yield, together with 52% of a neutral oil which was largely p-nitrobenzyl alcohol.

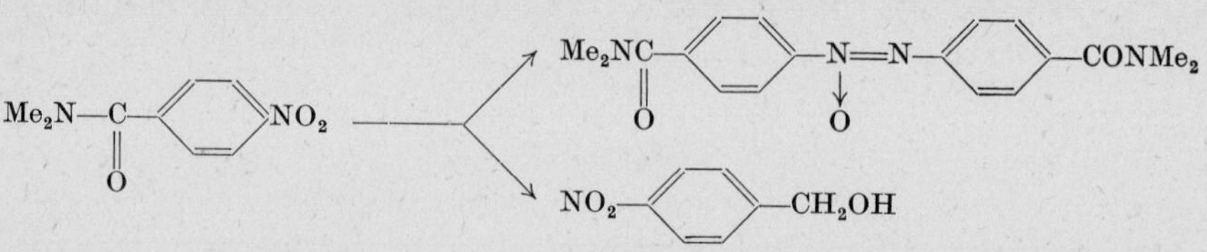

The reagent has been successfully employed in the selective reduction of certain substituted secondary amides, whereby carbonyl groups have been hydrogenated without any attack on the amide group (Pastour *et al.*, 1956). Various derivatives of acetyl acetanilide were investigated, and reduction occurred after heating under reflux for 3–8 hours.

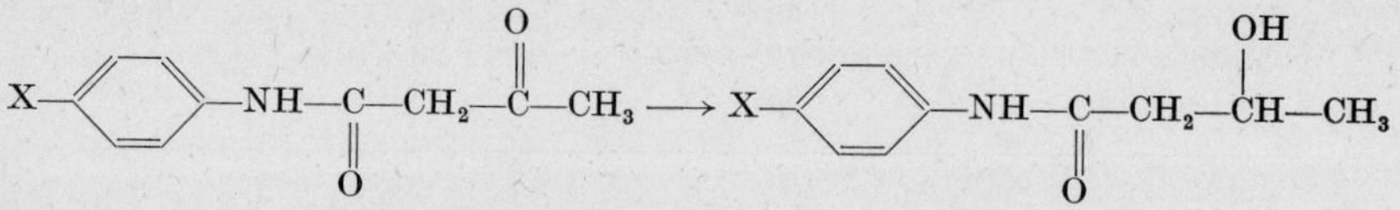

Derivative employed X =	Yield %
H	64
o-methyl	71
m-methyl	68
p-methyl	61

Crawhall *et al.* (1955) have shown that the cleavage of certain peptide bonds can be brought about by lithium borohydride, with the formation of the corresponding alcohol and amine.

D. *Hydroboration of Olefins*

The process designated hydroboration is rightly discussed in this chapter although it is one for the formation of boron–carbon bonds. Nevertheless, the reagents, the hydrido-compounds of boron, constitute an important system in themselves and are discussed earlier in this chapter.

When ethylene was heated with diborane in a sealed tube an unspecified amount of triethylboron was obtained (Hurd, 1948), and a study of the kinetics of this reaction has since been made by Whatley *et al.* (1954).

$$C_2H_4 + B_2H_6 \xrightarrow[\text{24 hours}]{100°C} (C_2H_5)_3B$$

Styrene, acrylonitrile, and other substituted olefins react similarly, as shown by Eméleus *et al.* (1950); (see also Stone *et al.*, 1955).

Sodium borohydride in the presence of aluminium chloride converts olefins into trialkylborons quickly at room temperature, and diborane itself adds to olefins in glycol ether solvents with ease at room

$$9RCH{=}CH_2 + 3NaBH_4 + AlCl_3 \longrightarrow 3(RCH_2CH_2)_3B + AlH_3 + 3NaCl$$

temperature (Brown *et al.*, 1956). In the absence of a catalyst, the reaction is slow, but in presence of even small amounts of the ether, the reaction is very fast. The operation can therefore be carried out by the addition of diborane to an ether solution of the olefin. The diborane may be made *in situ* by one of several reactions involving, on the one hand, the hydrides LiH, NaH, LiAlH$_4$, or the borohydrides LiBH$_4$, NaBH$_4$, KBH$_4$, or aminoborane, and on the other hand, boron trifluoride, boron trichloride, or aluminium trichloride (plus trimethyl borate) as acids.

With the borohydrides, hydrogen chloride or benzyl chloride may be used as acids, and as solvents, diethyl ether, T.H.F. and diglyme may be used. The following is a selection of the remarkable reactions which are available (Brown *et al.*, 1959).

$$3LiBH_4 + BF_3 \text{ (or } BCl_3) + 12RCH{=}CH_2 \longrightarrow 4(RCH_2CH_2)_3B + 3LiF \text{ (or } LiCl)$$

$$3LiBH_4 + AlCl_3 + 9RCH{=}CH_2 \longrightarrow 3(RCH_2CH_2)_3B + LiCl + AlH_3$$

$$3LiBH_4 + AlCl_3 + B(OMe)_3 + 12RCH{=}CH_2 \longrightarrow 4(RCH_2CH_2)_3B + 3LiCl + Al(OMe)_3$$

$$LiBH_4 + HCl + 3RCH{=}CH_2 \longrightarrow (RCH_2CH_2)_3B + LiCl + H_2$$

$$LiBH_4 + PhCH_2Cl + 3RCH{=}CH_2 \longrightarrow (RCH_2CH_2)_3B + LiCl + PhCH_3$$

Lithium borohydride is soluble in diethyl ether, T.H.F. and diglyme, whereas sodium borohydride is effectively soluble only in diglyme, and so for sodium borohydride this solvent is used. Potassium borohydride may be used in the presence of lithium chloride in T.H.F. (Paul *et al.*, 1953).

Types of olefin known to react are listed as follows:

1-hexene	2-octene	*trans*stilbene
1-octene	*cyclo*pentene	triphenylethylene
1-decene	*cyclo*hexene	allyl chloride
1-tetradecene	2,3-dimethyl-2-butene	allyl ethyl ether
3,3-dimethyl-1-butene	styrene	ethyl oleate
2-hexene	α-methylstyrene	ethyl cinnamate
3-hexene	1,1-diphenylethylene	

The system appears to be as general as the addition of hydrogen or bromine to a double bond, and at first glance appears to be similar to the addition of H—Al < (Ziegler *et al.*, 1954, 1956); but whereas H—B < adds rapidly at room temperature, H—Al < adds relatively slowly, and the addition proceeds satisfactorily only with terminal olefins at higher temperatures.

This system has remarkable potentialities, not only for the production of a wide variety of trialkylborons, but also for the compounds obtainable from the boron products. The trialkylborons undergo solvolysis to the corresponding hydrocarbons when heated with a carboxylic acid under reflux in diglyme.

$$3C_6H_5CH{=}CH_2 \xrightarrow{B_2H_6} (C_6H_5CH_2CH_2)_3B \xrightarrow[\text{in diglyme}]{RCOOH} C_6H_5CH_2CH_3$$

$$3C_6H_{13}CH{=}CH_2 \xrightarrow{B_2H_6} (C_6H_{13}CH_2CH_2)_3B \xrightarrow{RCOOH} C_8H_{18} \text{ in 90\% yield}$$

Thus is provided a new non-catalytic process for the hydrogenation of unsaturated molecules, even in the presence of certain atoms or groups which might interfere with the usual hydrogenation catalysts. Thus n-propylmethylsulphide was obtained in 80% yield from the corresponding unsaturated sulphide.

$$CH_3SCH_2CH{=}CH_2 \xrightarrow[RCOOH]{B_2H_6} CH_3SCH_2CH_2CH_3$$

Furthermore, as trialkylborons are readily oxidized by alkaline hydrogen peroxide, an anti-Markownikoff hydration of olefins is effected.

$$C_6H_{13}CH{=}CH_2 \xrightarrow{B_2H_6} (C_6H_{13}CH_2CH_2)_3B \xrightarrow[H_2O]{H_2O_2} C_6H_{13}CH_2CH_2OH$$

$$PhCH{=}CH_2 \xrightarrow{B_2H_6} (PhCH_2CH_2)_3B \xrightarrow[H_2O]{H_2O_2} PhCH_2CH_2OH$$

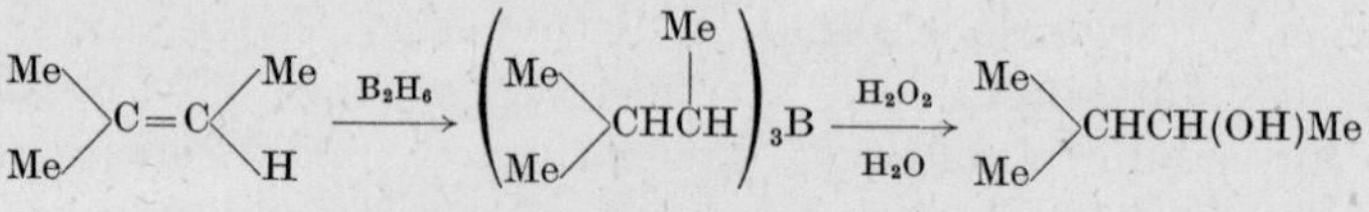

$$\underset{Me}{\overset{Me}{>}}C{=}C\underset{H}{\overset{Me}{<}} \xrightarrow{B_2H_6} \left(\underset{Me}{\overset{Me}{>}}CHCH\overset{Me}{|}\right)_3B \xrightarrow[H_2O]{H_2O_2} \underset{Me}{\overset{Me}{>}}CHCH(OH)Me$$

$$Me_3CCH{=}CH_2 \xrightarrow{B_2H_6} (Me_3CCH_2CH_2)_3B \xrightarrow[H_2O]{H_2O_2} Me_3C{.}CH_2CH_2OH$$

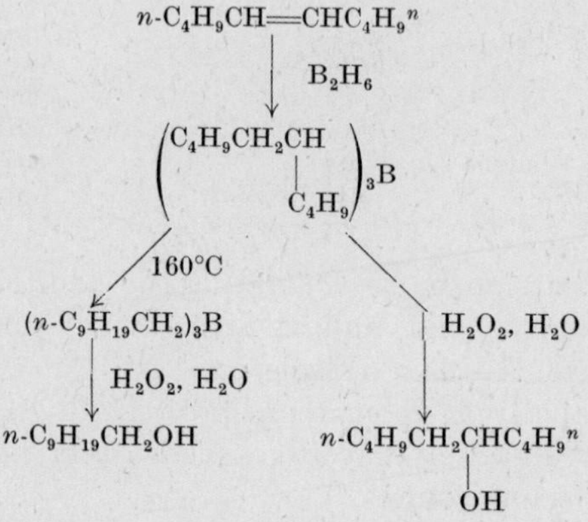

$$n\text{-}C_4H_9CH{=}CHC_4H_9{}^n$$
$$\downarrow B_2H_6$$
$$\left(\underset{C_4H_9}{\overset{C_4H_9CH_2CH}{|}}\right)_3B$$

$$\swarrow 160°C \qquad\qquad \searrow$$

$$(n\text{-}C_9H_{19}CH_2)_3B \qquad\qquad H_2O_2,\ H_2O$$

$$\downarrow H_2O_2,\ H_2O$$

$$n\text{-}C_9H_{19}CH_2OH \qquad\qquad n\text{-}C_4H_9CH_2CHC_4H_9{}^n$$
$$\overset{|}{OH}$$

A modified procedure for the hydroboronation of olefins is in the interaction of trialkylaminoboranes, R_3NBH_3, with olefins at 200°C (4 hours) (Hawthorne, 1958; Ashby, 1959). The yields are between 78–95%, no solvent is required, and the aminoboranes Me_3NBH_3, Et_3NBH_3, Bu_3NBH_3 and $C_5H_5NBH_3$ are readily prepared (see p. 160). Examples of olefins used are *iso*butylene, 1-hexene, 2-hexene (gives tri-*n*-hexylboron in 95% yield), *cyclo*hexene, and 1-octene.

It is believed that a sufficiently high temperature is required to weaken the B—N bond and effect partial dissociation. B—H addition then can proceed *via* the boron p-orbital interaction with the π-electrons of the olefin. As diborane reacts rapidly with an olefin in solution at room temperature, dissociation of the amineborane is probably the

$$Et_3NBH_3 \rightleftharpoons \tfrac{1}{2}B_2H_6 + NEt_3 \xrightarrow{\quad 3RCH=CH_2 \quad} (RCH_2CH_2)_3B + NEt_3$$

rate-determining step. It is possible that the olefin assists separation of the tertiary amine by the basic function of the π-electrons.

It is significant that both 2-hexene and 1-hexene with triethylaminoborane give tri-*n*-hexylboron. This appears to be in accordance with the isomerization of *s*- and *t*-alkylborons to corresponding primary alkylborons in absence of co-ordinating solvent (McCusker *et al.*, 1955). Although tri-*n*-octylboron was formed in the triethylamineborane-1-octene reaction, *trans*-2-octene was collected when the product was distilled at 169°/0.2 mm, and tetraoctyldiborane was formed, and showed an infra-red strong absorption band at 6.38 μ assigned to the B—H bridge absorption characteristic of a tetraalkyldiborane.

The extensive work of Brown *et al.*, as recently summarized (1959), points to such startling potentialities that a more detailed discussion is advisable. The important point to emphasize is the significance of the ether solvent, which can be conveniently diglyme or triglyme. Three experimental procedures are available.

A solution of sodium borohydride in diglyme can be added from a dropping funnel to a solution of boron trifluoride etherate in diglyme (Brown *et al.*, 1958). Diborane is evolved quantitatively as the borohydride is added, and may be passed into a solution of the olefin in diglyme or triglyme. The interaction of the diborane and olefin is quantitative under these conditions, and there are no other products to get rid of. Diborane is very soluble in T.H.F., and this ether is a good reaction medium for the diborane–olefin system.

Now when the boron trifluoride etherate is added *to* the sodium borohydride in diglyme, the diborane is not evolved at first, because it forms a complex, $NaBH_4, BH_3$ (Brown *et al.*, 1958). This accumulation of diborane can be used to advantage by adding the boron trifluoride etherate in diglyme to a mixture of sodium borohydride and the olefin in diglyme in the reaction flask, thus carrying out the hydroboration *in situ*.

$$12\ R.CH{=}CH_2 + 4BF_3 + 3NaBH_4 \longrightarrow 4(RCH_2{-}CH_2)_3B + 3NaBF_4$$

This procedure appears to be suitable for large-scale work in examples where the presence of sodium tetrafluoroborate is not a disadvantage.

Finally, the third procedure involves the sodium borohydride–aluminium chloride system mentioned, which does not give diborane, and appears useful for large-scale work, although it is rather slower in reaction than the other systems, and only about 75% of active hydrogen seems to be used.

After hydroboration, the trialkylboron may be oxidized *in situ*, without isolation, by addition of water and then hydrogen peroxide and alkali. As diglyme is more soluble in water than it is in diethyl ether, the latter may be used to extract the alcohol after the addition of water in excess. Modifications will be practised according to the nature of the alcohol, and in certain cases triglyme, b.p. 216°C, instead of diglyme, may be used to advantage. It is possible to isolate the alcohol by adding solid sodium hydroxide to the solution in order to get an upper layer of the alcohol in diglyme or triglyme from which the alcohol may be distilled.

The isomerization of trialkylborons has been discussed in connection with Hennion *et al.*'s work on redistribution in mixed trialkylborons (p. 102). Brown *et al.* (1957, 1959) show that the isomerization is more rapid when the boron compound is heated in diglyme. Tri-*s*-butylboron gave tri-*n*-butylboron in 2–4 hours in diglyme heated under reflux (160°C). Hydroboration of 2-pentene in diglyme yields tri-*s*-pentylboron, which on oxidation (without isolation) with hydrogen peroxide and water in the presence of alkali gives a 50 : 50% mixture of 2-pentanol and 3-pentanol, b.p. 116–119°C/745 mm. If the diglyme solution containing the tri-*s*-pentylboron is heated for 4 hours under reflux before oxidation, then on oxidation, 1-pentanol, b.p. 136–137°C/742 mm, is obtained. Similarly, 1-hexene gives 1-hexanol, and 2-hexene gives a 50 : 50% mixture of 2- and 3-hexanol, although if heated, before oxidation, 1-hexanol is isolated after oxidation. Yields of alcohols are about 75–90%.

Trialkylborons from 1-olefins did not undergo any significant change on being heated under reflux, or on being distilled at moderate pressures. Tri-n-hexylboron had b.p. 185–188°C/30 mm, and tri-n-octylboron had b.p. 144–145°C/2 mm. Köster (1958) reported a dismutation of tri-n-decylboron to 1-decene and di-n-decylborane, R_2BH, on attempted distillation. Rosenblum (1955) noticed a slow dismutation of tri-n-butylboron into n-butyldiboranes and butene on heating under reflux at 125–130°C for 10 days. It appears, therefore (Brown et al., 1959), that the 4-centre cis addition is reversible at elevated temperatures, and isomerization involves a movement of boron from an internal position to a terminal position. Preference for terminal

position may be because this involves a decrease in steric interactions. As ether catalyses the addition, so it may be assumed that it catalyses elimination. Therefore, when tri-n-pentylboron was heated with 1-hexene in a flask under a fractionating column, 1-pentene was distilled off, and tri-n-hexylboron remained in 85–93% yield. Therefore internal olefins may be isomerized to give primary alcohols and terminal olefins.

Two cases of probable steric hindrance are mentioned: 2-Me-2-butene and 2,4,4-trimethyl-2-pentene appear to give dialkylboranes, R_2BH, showing incomplete carbonation of boron.

The reagent resulting from mixing sodium borohydride and aluminium chloride in diglyme exhibits some of the characteristics of a solution of aluminium borohydride, $Al(BH_4)_3$, although the concentration must be small because the solution remains clear, thus showing the presence of only a small amount of sodium chloride, which is insoluble in diglyme. It is probable that the aluminium borohydride is generated as it is used up in the hydroboration reaction. It has been previously shown by Brokaw et al. (1950) that aluminium borohydride will hydroborate olefins, although higher temperatures are required than when the operation is performed with ether present. It is evident that ethers catalyse this reaction, and it is interesting to notice that, according to Schlesinger et al. (1940), aluminium borohydride forms a complex with ethers.

Graham *et al.* (1957) obtained tri-*n*-propylboron from diborane and *cyclo*propane.

E. Preparation of Borate Esters using Sodium Borohydride

Primary, secondary and tertiary borate esters can be prepared in good yield (see Table XXI) by means of sodium borohydride reacting with the appropriate alcohol in the presence of acetic acid (Brown *et al.*, 1956).

Primary and secondary alcohols react,

$$3ROH + NaBH_4 + HOAc \longrightarrow (RO)_3B + NaOAc + 4H_2$$

Under similar conditions, tertiary alcohols react to yield the corresponding di-alkyl hydrogen borate, but at elevated temperatures the tri-ester is obtained.

$$2ROH + NaBH_4 + HOAc \longrightarrow (RO)_2BH + NaOAc + 3H_2$$
$$ROH + (RO)_2BH \longrightarrow (RO)_3B + H_2$$

TABLE XXI

Yields of Borate Esters

Borate	Yield %
Ethyl	93·1
*iso*Propyl	88·5
t-Butyl	99·0
t-Amyl	85·6
Allyl	91·5

Quaternary ammonium borohydrides are precipitated from an alcoholic solution of sodium borohydride by the addition of the theoretical quantity of the appropriate tetra-alkyl ammonium hydroxide, as shown by Banus *et al.* (1952) and Bragdon (1956).

$$NaBH_4 + Me_4N.OH \longrightarrow Me_4NBH_4 + NaOH$$

V. Suggested Applications of Hydrido Compounds of Boron

Diborane itself is a good *fuel*, but it is an inconvenient one, being gaseous, toxic and relatively unstable at ordinary temperature, giving hydrogen and higher boranes. When it is heated, hydrogen and higher boranes are formed, and some measure of control can be exercised by varying the pressure of hydrogen. The stable pentaborane, B_5H_9, can be produced by circulating diborane with hydrogen through a reaction vessel at 200–250°C. It has been investigated as a rocket fuel; it is a relatively stable liquid, a highly reactive reducing agent, and its toxicity does not constitute a prohibitive factor. The formation of a slag of boric oxide and probably boranes higher than $B_{10}H_{14}$ has been a trouble, and amongst the procedures for avoiding this are the production of a sheath of helium, and the use of fluorine instead of oxygen.

$$B_5H_9 + 12F_2 \longrightarrow 5BF_3 + 9HF$$

It has been stated that pentaborane is an effective additive (10%) to conventional fuels, increasing flame velocity, and hindering "flame-outs" caused by fuel being blown away from the flame holder.

Result of spectroscopic observations on pentaborane–air flames and explosions have been reported (Berl et al., 1956), and the influence of diborane on flame speed of propane–air mixtures has been investigated (Kurz, 1956), and in a study of the effect of boron compounds on the combustion of gasoline by Hughes et al. (1956), it was found that 0.004% boron compound reduced surface ignition under severe operating conditions by 40%.

B-aminoborazoles and alkyl derivatives have been offered as rocket fuels (Gould, 1956), B-tris-aminoborazole (95% yield) was made by the low temperature ($-78°C$) reaction between trichloroborazole and liquid ammonia.

$$B_3N_3H_3Cl_3 + 6NH_3 \longrightarrow B_3N_3H_3(NH_2)_3 + 3NH_4Cl$$

Other examples are B-tris(dimethylamino)borazole, B-tris(diethylamino)borazole, and B-tris(methylamino)borazole obtained by analogous reactions.

Diborane reacts with a number of organic functional groups such as carbonyl groups and olefinic double bonds. Olefins form trialkylborons which can be caused to react as already described. It is suggested that waste olefin gases in the petroleum industry can be utilized for such reactions. There is no doubt that there is much investigation into the possible uses of diborane in connection with production of amines, polymers, transistors and rubber. Diborane reacts with ethylene oxide at low temperatures to form a solid wax-like polymer, and with propylene oxide a liquid polymer is produced. In non-aqueous systems diborane reduces carboxylic acids, and in contrast to the relative reducing rates with sodium borohydride, it reduces acid chlorides more slowly than it reduces nitriles or carboxylic acids, and it reacts with nitriles more rapidly than it does with esters. There is a strong move to develop the use of diborane as an industrial process chemical, and as already mentioned, transportation conditions have been worked out.

The reducing action of metal borohydrides has already been discussed, and it is clear that we are only in the beginning of this application in organic chemistry. Although the purpose of an organic chemist will often be to effect a certain change in his compound, and to this end the boron compound will be merely a reagent, nevertheless, during reduction organic compounds of boron are formed, and it is conceivable that there will be increasing interest in the study of such compounds apart from their transitory function during a reduction. A pertinent example (Wechter, 1959) is the almost quantitative formation of a dialkylboronic acid by the interaction of diborane and cholesterol, and the development of a study of steric factors involved in steroid hydro-boronylation, related to the stereo-specific overall *cis*-non-Markownikoff addition of the elements of water to simple cyclic olefins.

There is a hair-waving process in which sodium or potassium boro-hydride may be employed, thus avoiding the obnoxious odours of similar functioning agents (Bogaty *et al.*, 1956). During the production of cellular elastomeric articles, rubber foam is filled with hydrogen by decomposing sodium borohydride in the latex (Talalay *et al.*, 1956). The borohydride is stable at pH 10, but evolves hydrogen at pH 8–8.5. There is a dyeing and hair-tinting technique (Frohnsdorff *et al.*, 1956; Gillette Ind., 1957) which employs aqueous borohydride to reduce and solubilize certain quinone dyestuffs to form leuco-compounds. These stabilized leuco-hydroquinone derivatives are then applied, and acidification followed by aerial oxidation provides the dyeing process.

The pyridine complex of borine, $C_5H_5N.BH_3$, is prepared from the interaction of pyridine hydrochloride and the borohydride in pyridine, as shown by Taylor *et al.* (1955). Pyridine–borane is a pale yellow liquid, stable in dry air. Similar reactions have been used by Schaeffer *et al.* (1949) to prepare trimethylamine-borane and N-dimethylamine-borane. Pyridine borane and other amine-boranes are commercially available as reducing agents.

Chapter XI

BORON–NITROGEN COMPOUNDS

I. Co-ordination Compounds derived from Diborane

INTERACTION of ammonia and diborane was studied by Stock *et al.* (1926) and Wiberg *et al.* (1940) in connection with investigations into borazole, designated by Wiberg, "inorganic benzene". The white solid obtained by the interaction of ammonia and diborane under suitable conditions has been designated diammonia diborane, diborane diammoniate, and other names.

$$B_2H_6 + 2NH_3 \longrightarrow B_2H_6 \cdot 2NH_3$$

However, according to Schlesinger *et al.* (1938, 1942), the structure of the dimer is probably $NH_4^+(BH_3NH_2BH_3)^-$, and Schaeffer *et al.* (1956) have indicated the presence of a boron hydride ion. In a series of papers which has more recently appeared dealing with results of a programme of research under a U.S.A. Air Force contract sponsored by the Aeronautical Research Laboratory (Wright Air Development Center), Parry *et al.* (1958) discuss in detail the production of ammonium borohydride, NH_4BH_4, and diammonia diborane, formulated as $[H_2B(NH_3)_2]^+[BH_4]^-$, is described in detail. Ammonium borohydride, NH_4BH_4, is a white crystalline solid decomposing at room temperature to hydrogen + a solid $[BNH_6]_x$, and diammonia diborane, to which the authors assigned the structure:

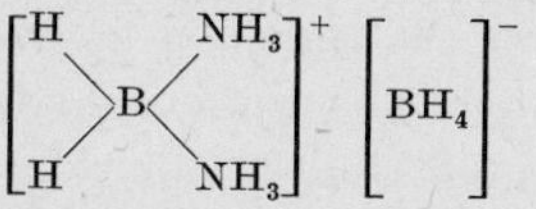

$$\left[\begin{array}{c} H \qquad NH_3 \\ {\scriptstyle>}B{\scriptstyle<} \\ H \qquad NH_3 \end{array}\right]^+ \left[\; BH_4 \;\right]^-$$

Diborane and ammonia interact by unsymmetrical cleavage, whereas an amine gives rise to symmetrical cleavage.

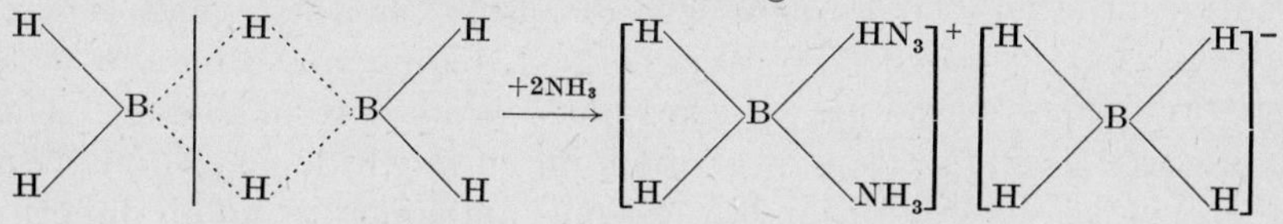

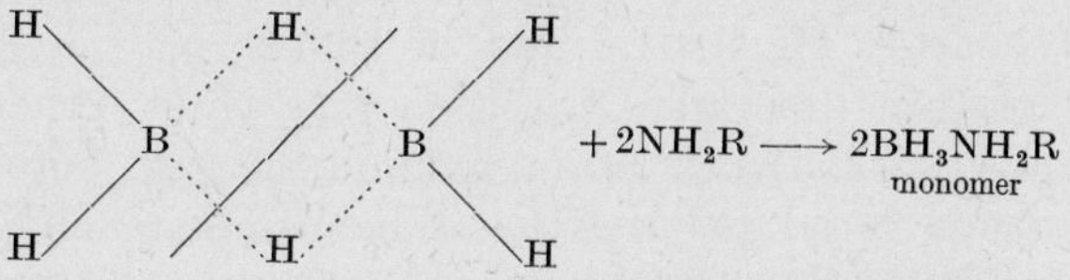

The reaction also takes place with NHR′R″ and NR′R″R‴.

The unstable nature of ammonium borohydride, and the lack of reaction between sodium borohydride and diammonia diborane, are pertinent to the structure of the diammonia diborane. Interaction with hydrogen chloride shows the common borohydride anionic structure:

$$2 \left[\begin{array}{c} H \\ \diagdown \\ H \diagup \end{array} B \begin{array}{c} NH_3 \\ \diagdown \\ \diagup NH_3 \end{array} \right]^+ \left[BH_4 \right]^- + 2HCl$$

$$\longrightarrow B_2H_6 + 2H_2 + 2[H_2B(NH_3)_2]^+ \, Cl^-$$

$$2M(BH_4) + 2HCl \longrightarrow B_2H_6 + 2H_2 + 2MCl$$

When an alternative route to the chloride of the diammoniate was sought by means of a heterogeneous reaction at room temperature between ammonium chloride and diammonia diborane in ethyl ether suspension, the unexpected monomeric ammonia borane, BH_3NH_3, was obtained.

$$[H_2B(NH_3)_2]^+ \, [BH_4]^- + NH_4Cl$$

$$\xrightarrow{\text{Et}_2O} [H_2B(NH_3)_2]^+ \, Cl^- + H_2 + H_3NBH_3$$

The crystalline ammonia borane was obtained in yields of 80% by filtering the ether slurry. Ammonia borane is monomeric in dioxane, in diethyl ether, in liquid ammonia, and also in the solid state. A polymeric form, $(H_3NBH_3)_n$ slowly precipitates from dry ether during several days at room temperature, whereas a trace of moisture causes rapid precipitation of a solid.

Although sodium borohydride with ammonium chloride in diethyl ether affords little hydrogen, and even with ammonium bromide, yield of ammonia borane is small, lithium borohydride reacts quickly enough with an ether slurry of ammonium chloride at room temperature and gives ammonia borane, H_3BNH_3 (cf. Schaeffer, G. W. *et al.*, 1949). Relevant points concerning the solubility of the alkali metal borohydride in ether and the presence of a solid phase in these reaction systems are discussed. Processes giving diammonia diborane rather than ammonia borane have a solid phase in absence of solvent. Thus, solid phase decomposition of ammonium borohydride gives diammonia diborane, as does addition of gaseous ammonia to solid dimethyletherate of borane, Me_2OBH_3; and in the normal procedure, *solid* ammonia picks up diborane. On the other hand, ether is present during the reaction between an ammonium salt and borohydride salt to give ammonia borane, BH_3NH_3. When ammonia in large excess was added to Me_2OBH_3 in liquid ether, Me_2O, at $-78°C$, the ammonia borane was obtained in 70% yield.

Conditions for the production of what is named *specification di-ammoniate* are as follows: A thin film of solid ammonia is condensed on bottom and lower wall of reaction tube. A band of solid diborane is condensed to several inches above the ammonia. The reaction vessel is warmed from $-140°$ to $-80°C$ (8 hours), whereupon the gaseous diborane at $-120°C$ adds to solid ammonia. The system is maintained at $-80°C$ whilst excess ammonia is sublimed. Only specification di-ammonia diborane will quickly give one equivalent of hydrogen when treated with sodium in liquid ammonia.

This series of papers is of considerable interest, and the details are not easily summarized further. Their significance in this monograph relates to techniques, constitution of fundamental compounds, and to the foundations of polymerization-condensation in the boron–nitrogen systems.

A number of what may be called substituted borazanes of the formula R_3NBR_3, in which R=H or hydrocarbon group, have been made from diborane or alkylboranes, and amines or ammonia. Some of these first appeared in connection with researches of Stock et al. (1926) and Wiberg et al. (1940, 1947, 1948) into the so-called inorganic benzene, and during the early work of Schlesinger et al. (1936, 1937, 1938, 1942) and Burg et al. (1949). Hewitt et al. (1953), Taylor et al. (1955), Hough et al. (1955), Mikheeva et al. (1956, 1957) and Köster (1957) have also contributed to this section of organic chemistry of boron. The compounds have considerable heat stability when there is no hydrogen on nitrogen; thus trimethylamine borane, Me_3NBH_3, m.p. 94°C, is stable to prolonged heating at 125°C. Trimethylamine methyl-borane, Me_3NBH_2Me, does not change at 100°C, and the dimethyl-borane compound, Me_3NBHMe_2, is stable up to 70°C. At higher temperatures redistribution occurs when the amine is tertiary and there is hydrogen attached to boron.

$$3Me_3NBH_2Me \xrightarrow{200°C} 2Me_3NBH_3 + BMe_3 + NMe_3$$

$$2Me_3NBHMe_2 \xrightarrow{70°C} Me_3NBH_2Me + BMe_3 + NMe_3$$

When there is hydrogen on nitrogen, then it is easily split off, and the first stage of dehydrogenation gives a "borazene", R_2NBH_2, R_2NBHMe, in which the monomeric form boron is trivalent.

$$R_2N—B\bigg\langle {}^H_H \qquad R_2N—B\bigg\langle {}^H_{Me}$$

Although earlier work was done by means of high vacuum technique already mentioned, and diborane or its methyl derivatives were mixed with ammonia or amine at $-40°$ to $-90°C$, the thermally stable borazanes, such as Me_3NBH_3 and $C_5H_5NBH_3$, may be prepared by ordinary bench procedures, as shown by Schaeffer et al. (1949) and Burg et al. (1951).

$$MBH_4 + Me_3NHCl \longrightarrow MCl + Me_3NBH_3 + H_2$$

M = Li for example   In ether, or pyridine at room
                          temperature. Yield 80–90%.

When the borazane thus prepared contains hydrogen on nitrogen, removal of the ether, and heating to the required temperature affords the corresponding "borazene", e.g. $Me_2HNBH_3 \rightarrow Me_2NBH_2 + H_2$ in 90% yield.

The interaction of N-methylhydroxylamine and diborane (Campbell et al., 1958) shows an opening to an unlimited field of investigation. Campbell (1957) has studied the interaction of tetramethyldiborane and ammonia in presence of calcium.

Of special interest are such amine–boranes as Me_2HNBH_3, m.p. $36°C$, Me_3NBH_3, m.p. $94°C$, and $C_5H_5NBH_3$, m.p. $10–11°C$, b.p. $65°C/1$ mm. Emerging from development of high-energy boron fuels, these compounds are found to have many interesting properties. They have a special function as reducing agents. Dimethylamine borane, Me_2HNBH_3, is stable, is hydrolysed very slowly in neutral or alkaline aqueous solutions. In benzene it reduces acetone to isopropanol, and in dioxan, citronellal to citronellol. Trimethylamine borane, Me_3NBH_3, can be heated for hours without change, is slowly hydrolysed by long boiling in water, and in acetic acid reduces ketones to alcohols. Pyridine borane, being a liquid, may be used without solvent, or it may be used in pyridine or in ether. For aldehydes and ketones which are insoluble or unstable in water or alcohol, or for substances sensitive to alkaline solutions, such amine boranes appear to have a special function as selective reducing agents. They have potentialities as polymerizing catalysts, as inhibitors for acrylates and vinyl compounds, as antioxidants and stabilizing agents, and as biological agents of one function or another. They are now commercially available.

Reference may here be made to certain physico-chemical studies related to the N—B bond: strain effects, Burg et al. (1956); vibrational frequency, Stewart (1955), Taylor et al. (1958); spectra, Rice et al. (1957); thermodynamics, Levitin et al. (1959); vapour pressure, Alton et al. (1959); and dipole moments, Bax et al. (1958).

II. Co-ordination of Trialkyl and Triarylborons with Nitrogen Bases

A considerable number of 1 : 1 co-ordination compounds of trialkyl- and triarylborons with ammonia and amines are known, and are formed usually by simple admixture at low temperatures; melting-points and vapour pressure at 0°C are accepted as simple indication of complex formation and stability (see Table XXII). These have been investigated by the following workers: Stock *et al.* (1921), Krause *et al.* (1924, 1928, 1930, 1931), Schlesinger *et al.* (1939), Fowler *et al.* (1940), Brown *et al.* (1944 and onwards), Wiberg *et al.* (1947, 1948), Goubeau *et al.* (1951), Wittig *et al.* (1951), Brown, C. A., *et al.* (1952), Razuvaev *et al.* (1953), and Sujishi *et al.* (1954). Some of these compounds have been of special interest in the B- and F-strain theory (Brown *et al.*, 1942 and onwards) to account for disorderliness in basic and Lewis acid strength sequences. Whereas trimethylboron formed a complex with ammonia, methylamine, dimethylamine, trimethylamine, ethylamine, diethylamine, and certain other amines, it did not do so with di-*iso*propylamine, the vapour pressure of the mixture being more than 165 mm at 0°C, compare results of Venkataramaraj Urs *et al.* (1952) on trialkyl borates.

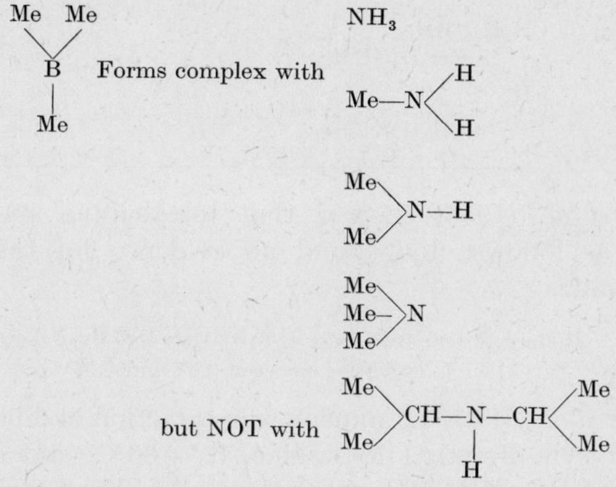

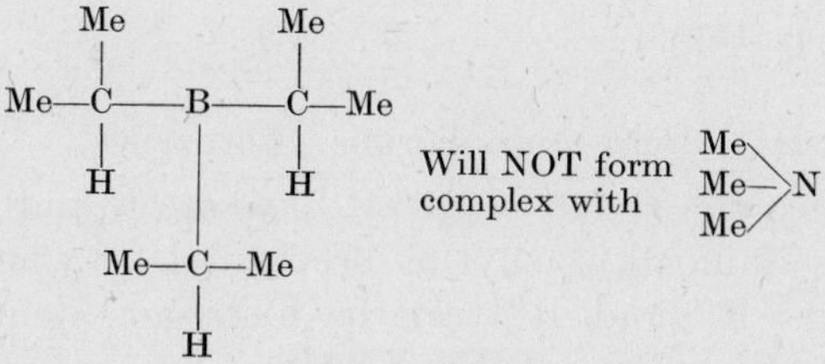

Tri-*iso*propylboron and trimethylamine did not form a complex; and there are indeed several other such examples of steric hindrance. Trimethylboron complexes with each of the amino- groups in ethylenediamine, and 1,3-diaminopropane, as shown by Goubeau *et al.* (1955).

It is of interest to mention that the occurrence of this type of compound was noticed by Frankland in 1862.

TABLE XXII

Examples of Complexes and their Properties

Complex	Melting point °C	Vapour pressure at 0°C/mm
H_3NBMe_3	56	1.0
MeH_2NBMe_3	27	0.16
Et_2HNBMe_3	26–28	0.3
Et_3NBMe_3	-18 to -14	238
$C_5H_5NBMe_3$	54–55	0.03
$2\text{-}CH_3C_5H_4NBMe_3$	28–33	3.2
$MeH_2NBBu^t_3$	—	1.5
$Me_2HNBBu^t_3$	44.5	—
$MeH_2NB(\alpha\text{-}C_{10}H_7)_3$	192–193	—
$\begin{array}{c} CH_2H_2NBMe_3 \\ \vert \\ CH_2H_2NBMe_3 \end{array}$	139	—
$\begin{array}{c} \diagup CH_2H_2NBMe_3 \\ CH_2 \\ \diagdown CH_2H_2NBMe_3 \end{array}$	104	—

Burg *et al.* (1953) showed that tetraborane and trimethylamine reacted as follows, and found no evidence for the formation of a B—B bond.

$$B_4H_{10} + (3+x)Me_3N \longrightarrow 3Me_3NBH_3 + BH(Me_3N)_x + \text{polymers}$$
$$(x \text{ varies between } 0.37 \text{ and } 0.54)$$

Steindler *et al.* (1953) examined the interaction of diborane with hydrazine and sym-dimethyl hydrazine at $-80°C$, and obtained a white solid $H_3BNHR \cdot NHRBH_3$ (R = Me or H). The behaviour on pyrolysis is described (see p. 169).

III. Mono-aminoborons "Borazenes"

The general formula is $(RR'N)BR''R'''$, where R and R' are hydrogen, alkyl, aryl, aminoalkyl, silyl or bromosilyl, or where $RR'N$ is a heterocyclic ring. R'' and R''' can be hydrogen, deuterium, alkyl, aryl, halogen, or alkoxyl (see Table XXIII).

The simplest borazene, H_2NBH_2, does not appear to have been isolated; for when borazane, H_3NBH_3 (as dimer) is heated at 90°C, a solid non-volatile polymer $(H_2NBH_2)_x$, insoluble in many solvents, is formed, see Wiberg *et al.* (1948) and Schaeffer *et al.* (1955).

Aminoboranes, in which R'' and R''' are both hydrogen, or one is hydrogen, are represented by dimethylaminoborane, Me_2NBH_2, prepared by the reaction shown (Burg *et al.*, 1949; Wiberg, 1948); penta-

$$2Me_2NH + B_2H_6 \xrightarrow{-42°C} 2Me_2NH,B_2H_6 \xrightarrow{130-160°C} 2Me_2NBH_2 + H_2$$

borane may be used in place of diborane (Burg, 1957). A general procedure for establishing a B—N bond is by the agency of lithium borohydride (Schaeffer *et al.*, 1949), and other metal borohydrides.

$$Me_2NH_2^+Cl^- + LiBH_4 \longrightarrow Me_2NBH_2 + 2H_2 + LiCl$$

"Borazenes", e.g. $MeHNBH_2$, result by the loss of hydrogen from substituted "borazanes", and several such compounds have been described.

$$H_3NBH_2Me \longrightarrow H_2NBHMe + H_2$$

TABLE XXIII

Properties of Borazenes

Formula	Form at room temperature	M.p. °C	B.p. °C	References
Me_2NBH_2	Dimer	73.5	—	{ Wiberg *et al.* (1948) Burg *et al.* (1951)
Me_2NBMe_2	Monomer	−92.2	65	{ Coates (1950) Wiberg (1949)
$MeHNBMe_2$	Monomer	—	38.3	Wiberg *et al.* (1947)
$C_6H_5HNBMe_2$	Colorless liquid	—	—	Wiberg *et al.* (1948)

Molecular weight determination (Bissot *et al.*, 1955) shows a trimeric form for $MeHNBH_2$, obtained by heating the borazane, MeH_2NBH_3, to 100°C, and it is plausible that the dimeric and polymeric forms are

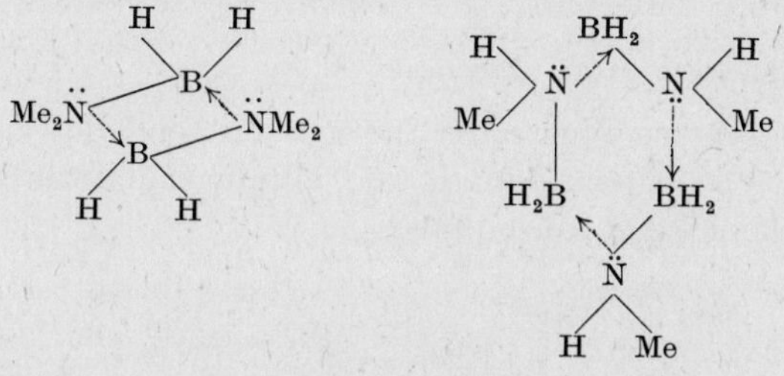

due to co-ordination. The dimers and polymers are normally relatively unreactive with water, hydrogen halides, and diborane; but the monomeric forms, resulting from depolymerization under suitable conditions, are open to addition, and a number of such diborane

$$2Me_2NBH_2 + B_2H_6 \xrightarrow{135°C} 2Me_2NB_2H_5$$

derivatives have been described by Schlesinger *et al.* (1938), Burg *et al.* (1949, 1950), and Wiberg (1955). The parent aminodiborane, $H_2NB_2H_5$, m.p. $-66.5°C$, b.p. $76.2°C$, decomposes fairly quickly at room temperature, decomposition in gas phase giving diborane and non-volatile polymers, $(H_2NBH_2)_x$. Methyl on nitrogen increases stability, and the vapour of dimethylaminodiborane is stable for hours at 90°C. When the borazene Me_2NBH_2 is heated for some time at about 100°C, N-dimethylaminodiborane, $Me_2NB_2H_5$, and bis-dimethylamine borane, $(Me_2N)_2BH$, result, probably due to redistribution, and of course borane

$$2Me_2NBH_2 \longrightarrow (Me_2N)_2BH + [BH_3]$$

then adds to a third molecule of the borazene to give $Me_2NB_2H_5$. It is, however, an equilibrium system; for crystals of the borazene, Me_2NBH_2, form when the two final products are heated together. Further redistribution in the bis-compound $(Me_2N)_2BH$ gives the tris-dimethylaminoboron $(Me_2N)_3B$. With dimethylamine the borazene, Me_2NBH_2, affords hydrogen and firstly the bis-compound and then the tris-dimethylaminoboron.

Interactions of hydrogen halides and borazenes have been described by Wiberg *et al.* (1947, 1948).

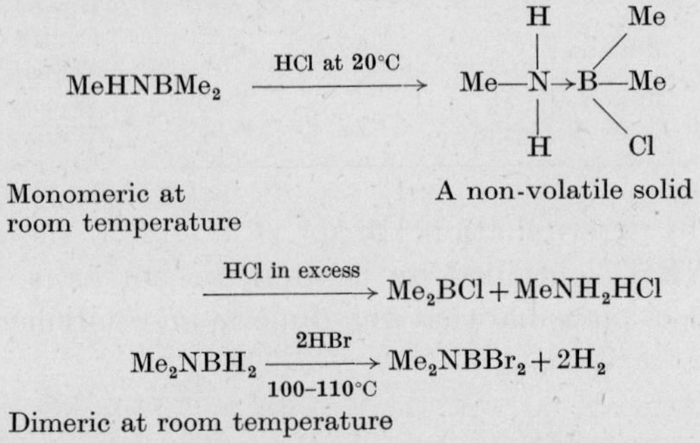

The following scheme illustrates the general tendency of addition to the borazene type of compound, and elimination from the borazane type to give a borazene compound.

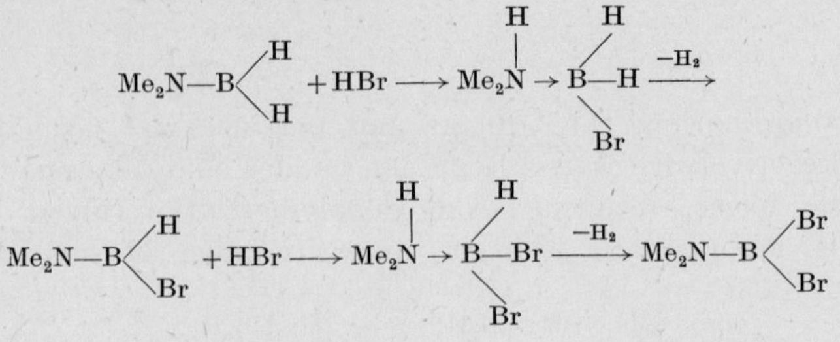

Dimethylaminoborane is a dimeric solid at room temperature, but dissociation occurs at 105°C/10 mm, 90% then existing as the monomer (Wiberg *et al.*, 1948). The dimer does not react with hydrogen bromide or water; but the monomer does.

$$Me_2NBH_2 + 2HBr \longrightarrow Me_2NBBr_2 + 2H_2$$

$$Me_2NBH_2 + 3H_2O \longrightarrow B(OH)_3 + Me_2NH + 2H_2$$

The deuterium analogue has been prepared by Burg (1952).

Diethylaminoborane, prepared by heating diethylamine with diborane at 200–230°C, is a colourless crystalline dimer at low temperatures, but dissociates on being heated. Steindler *et al.* (1953) obtained the 1 : 1 addition complex of symmetrical dimethylhydrazine and diborane at low temperature, and showed that on heating it gave hydrogen and the borane $H_2BN(Me)N(Me)BH_2$, which at higher temperature polymerized to a solid.

A number of bis-aminoborons, $(RR'N)_2BR''$, where R and R' are hydrogen or alkyl, alike or different (or where $(RR'N)_2$ is an aliphatic diamino group), and R'' is hydrogen, hydroxyl, alkyl, or halogen, have been prepared. For example, bis-dimethylaminoborane, $(Me_2N)_2BH$, m.p. $-45°C$, b.p. 109°C, has been prepared as shown.

$$Me_2NH + Me_2NBH_2 \xrightarrow[\text{10 hours}]{200°C} (Me_2N)_2BH + H_2$$

$$(Me_2N)_2BCl + LiAlH_4 \longrightarrow (Me_2N)_2BH + LiCl + AlH_3$$

It is a colourless liquid, monomeric, which does not redistribute when pure, and with a limited amount of water gives bis-dimethylaminoborinic acid, $(Me_2N)_2BOH$.

Burg (1957) has prepared bis(dimethylamino)borane by the interaction of dimethylamine and the compound $(Me_2N)_3B_3H_4$, showing that interest in this field of work is being maintained. Stock *et al.* (1901) and Wiberg *et al.* (1931, 1933) prepared the bis(dimethylamino)boron halides (Cl, Br) by the interaction of the dialkylamine and boron trihalide. Johnson (1912), in his isolated, but historically very interesting work, prepared bis(monomethylamine)boron bromide from the amine and boron tribromide, and Burg *et al.* (1954) added another procedure involving trimethylamine and dimethylamineboron difluoride.

The following reactions leading to a bis-aminoboron halide have been reported.

$$4Me_2NH + BX_3 \longrightarrow (Me_2N)_2BX + 2Me_2NH_2X \ (X = Cl \text{ or } Br)$$

$$Me_3N + 2Me_2NBF_2 \longrightarrow (Me_2N)_2BF + Me_3NBF_3$$

$$2MeNH_2 + BBr_3 \longrightarrow (MeNH)_2BBr + 2HBr$$

The fluoride reacts with boron trifluoride thus:

$$(Me_2N)_2BF + BF_3 \longrightarrow 2Me_2NBF_2$$

Ethylene diamine and trimethylboron leads to a three-stage reaction, the elimination of methane being completed at 475°C (Goubeau et al., 1955).

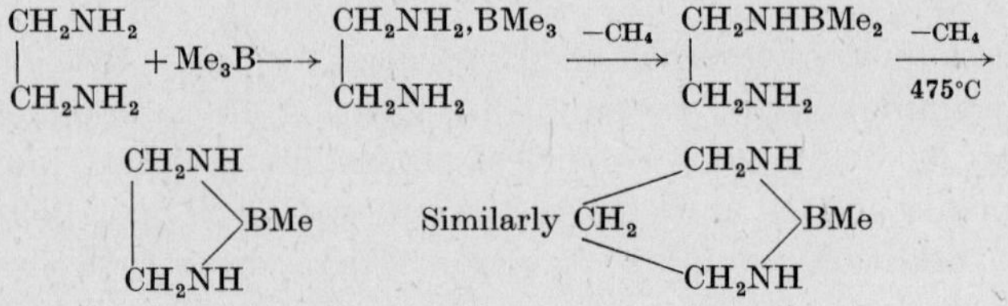

Each of these cyclic compounds adds 2 moles of hydrogen chloride rapidly at room temperature.

Tris-aminoborons are of the type $B(NRR')_3$, where R and R' are alike or different and can be hydrogen, alkyl, or aryl, or can be part of a heterocyclic system. They have been prepared by a number of reactions, and it will be obvious that a number of these are involved in other systems.

1. $BX_3 + 6RR'NH \longrightarrow (RR'N)_3B + 3RR'NH_2X$ (X = Cl, Br)

2. $RR'HNBCl_3 + 5RR'HN \longrightarrow (RR'N)_3B + 3RR'NH_2X$
   Isolated 1 : 1 complex

3. $3R_2NBCl_2NMe_3 \xrightarrow{120°C} (R_2N)_3B + Me_3N + 2Me_3NBCl_3$ (R = Me)

4. $3(R_2NBCl_2)_2 + 4NMe_3 \xrightarrow{140°C} 2(R_2N)_3B + 4Me_3NBCl_3$ (R = Me) (Dimer)

5. $2RNH_2 + RH_2NBF_3 + 3Li \longrightarrow (RHN)_3B + 3LiF + 3/2H_2$ (R = Et)

6. $6R_2NH + B_2H_6 \longrightarrow 2(R_2N)_3B + 6H_2$ (R = Me)

7. $3(R_2N)_2BH + Me_3N \underset{}{\overset{140°C}{\rightleftharpoons}} 2(Me_2N)_3B + Me_3N,BH_3$ (R = Me)

8. $3KNH_2 + H_3NBF_3 \xrightarrow{\text{Liquid } NH_3} (NH_2)_3B + 3KF + NH_3$

The simplest tris-aminoboron, $(NH_2)_3B$, has been prepared by methods 1 and 8, Joannis (1902); Stock et al. (1908); Keenan et al. (1954); tris(dimethylamino)boron by methods 1, 3, 4, 5, 6, Burg et al. (1951); Brown, C. A., et al. (1952); Wiberg et al. (1955); tris-ethylaminoboron by method 5, Kraus et al. (1930); tris-phenylaminoboron, m.p. 166–169°C, by method 2, Jones et al. (1939); tris-p-tolylaminoboron, $(p\text{-}CH_3C_6H_4NH)_3B$, m.p. 165–166°C, by method 2, Kinney et al. (1942); and tris-p-methoxyphenylaminoboron, $(p\text{-}CH_3OC_6H_4NH)_3B$, by method 1, Kinney et al. (1943).

A general Grignard procedure has recently been devised by Dornow *et al.* (1956) and an interesting variety of tris-aminoborons, from such as $RR'NH = N$-methylstearylamine, 4-chloroaniline, pyrrolidine, and carbazole, have been prepared. A number of these have high m.p.'s, that from carbazole has m.p. 348°C (see Table XXIV).

Tris-aminoborons are easily hydrolysed, and with alcohols give the trialkylborate; there is a certain dependence on the nature of the amino group, as Dornow *et al.* (1956) found, and some of the compounds just mentioned show a certain steric hindrance to hydrolysis.

TABLE XXIV

Tris-aminoborons Prepared by the Grignard Reaction

$$3RMgX + 3R'R''NH + BF_3, Et_2O \longrightarrow (R'R''N)_3B + 3RH + 3MgXF + Et_2O$$

R'R"HN	Yield %	B.p.°	M.p.°
Me$_2$NH	67	147	—
Et$_2$NH	62	220	—
Bu$^n_2$NH	50	—	11
N-methyl stearylamine	65	—	155
N-methylaniline	80	—	210
N-ethylaniline	25	—	164
aniline	62	—	170
4-chloroaniline	60	—	210
2 : 4 : 6-tribromoaniline	59	—	225
α-naphthylamine	80	—	unsharp
pyrrolidine	45	—	49
carbazole	91	—	348
1 : 2 : 3 : 4-tetrahydrocarbazole	94	—	338
1 : 2 : 3 : 4 : 5 : 6 : 7 : 8-octahydrocarbazole	92	—	325

Interaction of tris-aminoboron with hydrogen chloride is of considerable interest. Those derived from dimethylamine (Wiberg *et al.*, 1933), *p*-anisidine and *p*-toluidine (Kinney *et al.*, 1942, 1943) react with hydrogen chloride in such a way that chlorine becomes attached to boron.

$$(RR'N)_3B + 5HCl \longrightarrow RR'NBCl_2, HCl + 2RR'NH_2Cl$$

Of intense current interest is the observation of Jones *et al.* (1939) that tris-phenylaminoboron gave the chloroborazole.

$$3(PhNH)_3B + 9HCl \longrightarrow B_3Cl_3N_3Ph_3 + 6PhNH_3Cl$$

Becher's work (1956) on infra-red and Raman spectra of tris-dimethylaminoboron indicates a planar arrangement for the boron and three atoms of nitrogen. The B—N bond length is 1.46 Å, and B—N—C angles are 120°. See Becher's work on spectroscopic and other features of B—N bonds (1956, 1957), and p. 229.

The mention of polymers in connection with boron compounds has always caused immediate response of interest during these last few years. Burg *et al.* (1951) refer to a complex reaction between diborane and tris(methylamino)boron, whereupon dimethylaminoborane, dimethylaminodiborane, and a copolymer were formed.

The first example of an alkoxyboron compound with amino-groups also attached to boron is *n*-butyl-bis-diethylaminoborinate, $Bu^nOB(NEt_2)_2$, which was described by Gerrard *et al.* (1957). This is a colourless mobile liquid which is monomeric in *cyclo*hexane, used in the molecular weight determination. It is prepared by the following reactions.

$$Bu^nOBCl_2 + 4Et_2NH \longrightarrow Bu^nOB(NEt_2)_2 + 2Et_2NH,HCl$$
$$Bu^nOH + B(NEt_2)_3 \longrightarrow Bu^nOB(NEt_2)_2 + Et_2NH$$
$$Bu^nOH + (Et_2N)_2BCl + Et_3N \longrightarrow Bu^nOB(NEt_2)_2 + Et_3NHCl$$

Interactions with water, *n*-butanol (typical of ordinary alcohols), di-*n*-butylamine, and hydrogen chloride are illustrated by the following equations.

$$Bu^nOB(NEt_2)_2 + 3H_2O \longrightarrow B(OH)_3 + 2Et_2NH + Bu^nOH$$
$$Bu^nOB(NEt_2)_2 + 2Bu^nOH \longrightarrow B(OBu^n)_3 + 2Et_2NH$$
$$Bu^nOB(NEt_2)_2 + 2Bu^n_2NH \longrightarrow Bu^nOB(NBu^n_2)_2 + 2Et_2NH$$
$$2Bu^nOB(NEt_2)_2 + 7HCl \longrightarrow (Bu^nO)_2BCl + Et_2NBCl_2HCl + 3Et_2NHHCl$$

The borinate showed no tendency to rearrange at 145–155°C during 12 hours. Similarly to trialkyl borates (see p. 12), it did not co-ordinate with pyridine, due most likely to the reduction of electrophilic power of boron by back–co-ordination from adjacent oxygen and nitrogen atoms. Relevant to this are the facts (see equations) that whereas the diethylamino group was replaced by the di-*n*-butylamino group, and the *n*-butoxy group replaces each amino-group, the latter will not replace the alkoxyl group. Likewise diethylamine will not effect the replacement of an alkoxyl group in tri-*n*-butyl borate. An explanation is not at all clear at present, and further work is needed. If a three-centre "end-on" mechanism is postulated for the reaction, we are faced with the formation of the R_2N^- anion.

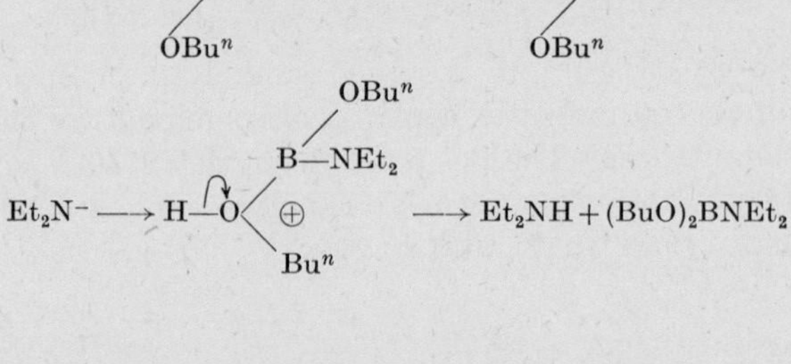

On the other hand, a four-centre broadside reaction would not involve this objection, and might lead to a satisfactory explanation.

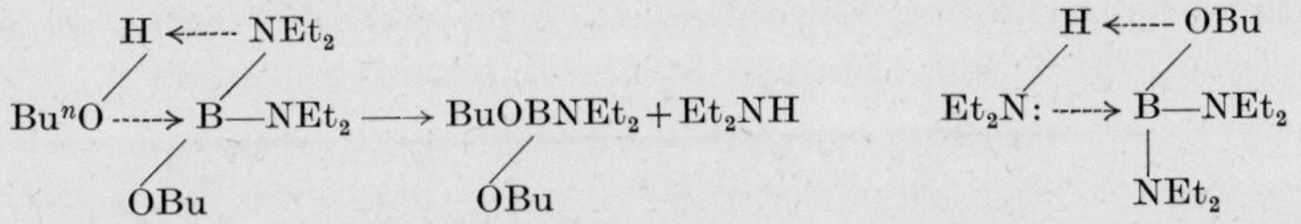

$$\begin{array}{ccc}
\text{H} \longleftarrow \text{NEt}_2 & & \text{H} \longleftarrow \text{OBu} \\
\diagup \quad \diagup & & \diagup \quad \diagup \\
\text{Bu}^n\text{O} \dashrightarrow \text{B}\!-\!\text{NEt}_2 \longrightarrow \text{BuOBNEt}_2 + \text{Et}_2\text{NH} & & \text{Et}_2\text{N}\!: \dashrightarrow \text{B}\!-\!\text{NEt}_2 \\
\diagup \quad \diagup & & | \\
\text{OBu} \qquad \text{OBu} & & \text{NEt}_2
\end{array}$$

The corresponding structure for the transition state involving replacement of alkoxyl by amino group might be unattainable because of some requirement of the N—H and B—O bonds, although a steric factor might be intruding. If reactivity depends only on the electrophilic function of boron (see p. 103), then back–co-ordination from oxygen to boron in the trialkyl borate would appear to be greater than back–co-ordination from nitrogen to boron in aminoborons.

Related to this point is the fact that attempts to prepare a tris-(diethylamino)boron by interaction with boric acid were unsuccessful according to Gerrard et al. (1957).

$$\text{B(OH)}_3 + 3\text{EtOH} \longrightarrow \text{(EtO)}_3\text{B} + 3\text{H}_2\text{O}$$

$$\text{B(OH)}_3 + 3\text{Et}_2\text{NH} \xrightarrow{\text{no reaction}} \text{B(NEt}_2)_3 + 3\text{H}_2\text{O}$$

General rules relating to redistribution in boron compounds have still to be elucidated; but there are discernible tendencies for such mutual replacement to show a drift in one direction such as has just been mentioned with reference to alkoxyl and amino groups.

The amino-groups were readily replaced by halogen even at $-80°\text{C}$.

$$\text{Bu}^n\text{OB(NEt}_2)_2 + 2\text{BCl}_3 \longrightarrow \text{Bu}^n\text{OBCl}_2 + 2\text{Et}_2\text{NBCl}_2$$

It remains to mention two other related prototypes—dialkyl amino-boronates, $\text{(RO)}_2\text{BNRR}'$, and alkyl aminohalogenoborinates, ROB(X)NRR'—described by Gerrard et al. (1958). Di-n-butyl diethylamino-boronate, $\text{(Bu}^n\text{O)}_2\text{BNEt}_2$, b.p. 65–68°C/0.45 mm, was prepared in 81%

$$\text{(Bu}^n\text{O)}_2\text{BCl} + 2\text{Et}_2\text{NH} \longrightarrow \text{(Bu}^n\text{O)}_2\text{BNEt}_2 + \text{Et}_2\text{NH}_2\text{Cl}$$

yield. It is a colourless monomeric liquid, which will not co-ordinate with pyridine, thus showing strong back–co-ordination from oxygen and nitrogen, but nevertheless it is easily hydrolysed. The following reactions readily occurred.

$$\text{(Bu}^n\text{O)}_2\text{BNEt}_2 + \left\{\begin{array}{l} \text{Bu}^n\text{OH} \longrightarrow \text{(Bu}^n\text{O)}_3\text{B} + \text{Et}_2\text{NH} \\ \text{Bu}^n_2\text{NH} \longrightarrow \text{(Bu}^n\text{O)}_2\text{BNBu}^n_2 + \text{Et}_2\text{NH} \\ \text{HCl} \longrightarrow \text{(Bu}^n\text{O)}_2\text{BCl} + \text{Et}_2\text{NH}_2\text{Cl} \end{array}\right.$$

It must be emphasized that the n-butoxyl group was not replaced by a dialkylamino group. The compound, $\text{(Bu}^n\text{O)}_2\text{BNBu}^n_2$, b.p. 107–110°C/0.2 mm, underwent redistribution on being heated, and thereby was different from the di-n-butyl-diethylaminoboronate, which was stable.

n-Butyl diethylaminochloroborinate, b.p. 76–78°C/14 mm, an easily hydrolysed liquid, was obtained by the following reactions.

$$Bu^nOBCl_2 + 2Et_2NH \longrightarrow Bu^nOB(Cl)NEt_2 + Et_2NH_2Cl$$
$$(Et_2N)_2BCl + (Bu^nO)_2BCl \longrightarrow 2Bu^nOB(Cl)NEt_2$$

IV. Borazines (XBNY), Borazole Derivatives

Dehydrogenation or dealkylation of a borazene would give a mono-meric borazine.

$$MeHNBH_2 \longrightarrow MeN{=}BH + H_2$$
$$MeNHBHMe \longrightarrow MeN{=}BMe + H_2$$
$$H_2NBMe_2 \longrightarrow HN{=}BMe + CH_4$$

However, there does not appear to be any definitely established example of a compound containing a B=N double bond; trimerization occurs, the simplest case, borazole itself, being mentioned in 1926 by Stock *et al.*, and recognized as a six-membered ring of alternating boron and nitrogen atoms. Wiberg *et al.* (1940) referred to it as "inorganic benzene", and a series of papers under this title was published in 1940, 1947, and 1948. Wiberg also reviewed the subject in 1948.

The first method of preparation entailed heating diammonia diborane at 200°C in a sealed tube.

$$3B_2H_6, 2NH_3 \longrightarrow$$

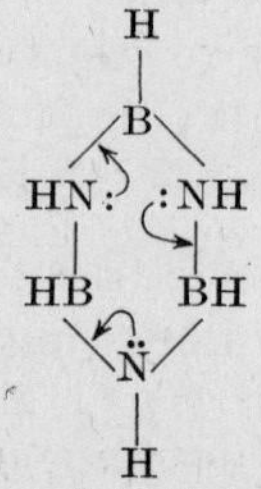

Colourless liquid,
b.p. 55°C
m.p. −58°C
d_4^0 0.86

Such early preparations involved the preparation of diborane attended by all the rigid discipline of high-vacuum technique and complete absence of oxygen and water. For the study of the properties of borazole, 0.6 c.c of liquid was laboriously prepared, but in later improvements in high-vacuum technique, and the use of a carrier inert gas enabled larger amounts of such materials to be handled.

This so-called "inorganic benzene" has received much attention, and original papers have been reviewed extensively by Schlesinger *et al.* (1942), Bauer (1942), Bell *et al.* (1948) and Stone (1955). Electron diffraction shows a planar ring, comparable with benzene (see Bauer, 1942); the bond angle is 120°, the B—N bond length 1.44 Å, in con-trast with an accepted single-bond length of 1.54 Å for B—N, and a hypothetical length of 1.36 Å for B═N. A resonance structure involving an internal co-ordination of N with B is indicated by Raman and infra-red spectra. Borazole is fairly stable at room temperature in sealed containers, but a white solid is slowly deposited and hydrogen is formed, whereas at 300°C there is a rapid evolution of hydrogen, and formation of a non-volatile solid which is probably boron nitride.

Derivatives result from substitution directly or indirectly of hydrogen on boron (B) or on nitrogen (N) or on both. At 0–20°C, 1 mole of borazole adds 3 moles of water, hydrogen chloride or methanol, and thermal decomposition of these addition compounds results in formation of substitution derivatives.

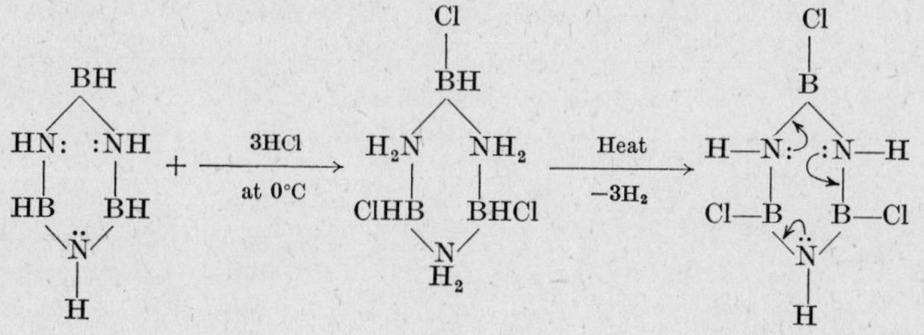

B-trichloroborazole

B-methyl derivatives add hydrogen chloride, but split off methane and form *B*-trichloroborazole. At low temperature water gives a trihydrate which on heating undergoes ring splitting or a substitution reaction according to whether the alkyl group is on boron or nitrogen.

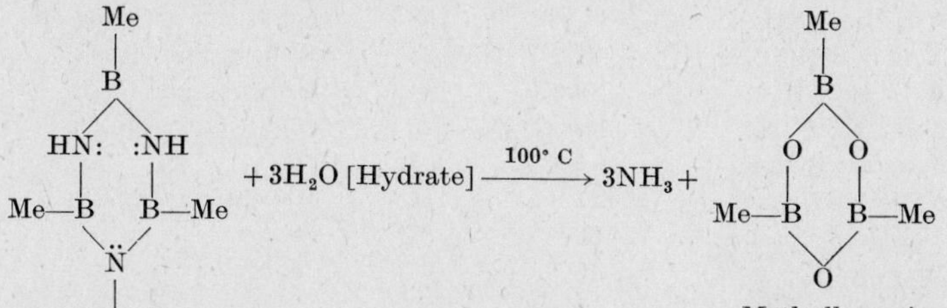

(Methyl on boron)

Methylboronic, anhydride trimer.

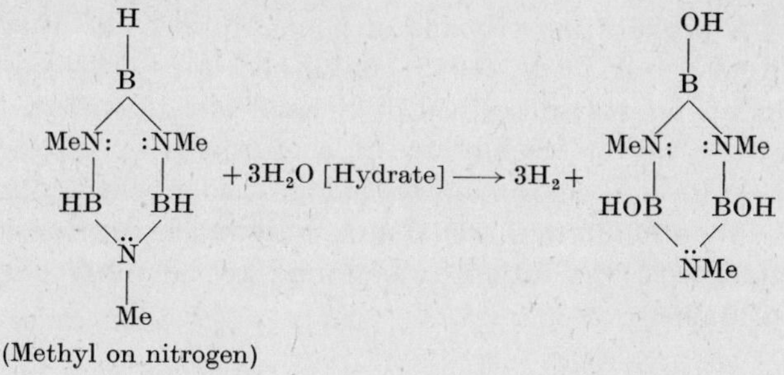

(Methyl on nitrogen)

Hydrolytic breakdown occurs in boiling aqueous acid solutions, and
B-methyl borazole, $Me_3B_3N_3H_3$, thereby gives methylboronic acid,

$$H_6B_3N_3 + (H_2O + HCl) \longrightarrow 3B(OH)_3 + 3H_2 + 3NH_4Cl$$

$MeB(OH)_2$, instead of boric acid, and of course, a *N*-alkyl derivative
gives the salt of the alkylamine.

At 100°C borazole and alcohol are stated (Wiberg *et al.*, 1948) to
give ROB=NH, which does not trimerize, but which at 200°C gives
alcohol and boron nitride.

TABLE XXV

Borazole and its Alkyl Derivatives

Formula	Melting point °C	Boiling point (calc.)°C	References
$B_3N_3H_6$	−56.1	53	Stock *et al.* (1926, 1930) Wiberg *et al.* (1940) Eddy *et al.* (1955)
N-$MeB_3N_3H_5$	—	84	Schlesinger *et al.* (1938)
B-$NMe_2B_3N_3H_4$	—	124	*Ibid.*
N,B,B-$Me_3B_3N_3H_3$	—	139	*Ibid.*
N,N-$Me_2B_3N_3H_4$	—	108	*Ibid.*
N,N,N-$Me_3B_3N_3H_3$	—	134	*Ibid.* and Wiberg *et al.* (1947, 1948); Hough *et al.* (1955)
N,B,B,B-$Me_4B_3N_3H_2$	—	158	Schlesinger *et al.* (1938)
N,N,N,B,B,B-$Me_6B_3N_3$	97.1	221	Wiberg *et al.* (1947)
N,N,N-$Et_3B_3N_3H_3$	−49.6	184	Hough *et al.* (1955)
N,N,N-n-$Pr_3B_3N_3H_3$	glass	225	*Ibid.*
N,N,N-iso-$Pr_3B_3N_3H_3$	− 6.5	203	*Ibid.*
B,N,N-$Me_3B_3N_3H_3$	31.8	127	Wiberg *et al.* (1947)
B-$MeB_3N_3H_5$	−59	87	*Ibid.*
B,B-$Me_2B_3N_3H_4$	−48	107	*Ibid.*

Because of its synthetic potentialities, trichloroborazole should be
mentioned. A preparation capable of being stepped up was first outlined by Brown, C. A., *et al.* (1949), published later (1955), but in the
meantime independently described by Schaeffer *et al.* (1954). It entails
the delivery of boron trichloride in a carrier gas (nitrogen) into
ammonium chloride in chlorobenzene heated under reflux. Purification
can be difficult; sublimation at 0.3 mm, followed by extraction of the
chloroborazole with dry ether, was found to be most effective by
Gerrard *et al.* (1960).

Schaeffer *et al.* (1954, see also 1951) have reduced trichloroborazole to borazole by adding the former to a mixture of lithium borohydride, dibutyl ether, and powdered glass in an apparatus flushed with nitrogen at room temperature. The yield was stated to be 65%, but the formation of diborane is a little troublesome, and means that it has to be

$$6LiBH_4 + 2H_3B_3N_3Cl_3 \longrightarrow 2H_6B_3N_3 + 3B_2H_6 + 6LiCl$$

absorbed in such a compound as sodium trimethoxyborohydride, $NaBH(OMe)_3$. 0.2 Molar quantities were used; but there appears to be a good chance of being able to use this technique on a larger scale.

Pyrolysis of the diammonia compounds of methylated diborane gives mono-, di-, or trimethylborazoles (Schlesinger *et al.*, 1936), the alkyl group being, of course, attached to boron (see Table XXV). *B*-triethylborazole was obtained in 70% yield by pyrolysis of the borazane, H_3NBEt_3 (Zhigach *et al.*, 1956). It is unaffected by cold water, but is slowly hydrolysed by hot hydrochloric acid. Pyrolysis at 450°C/20 atm of the borazane, MeH_2NBMe_3, gives *B*,*N*-hexamethylborazole and methane, as shown by Wiberg *et al.* (1947, 1948). Hydrolysis of this compound is slow, and ultimately gives methylboronic acid, and methylamine.

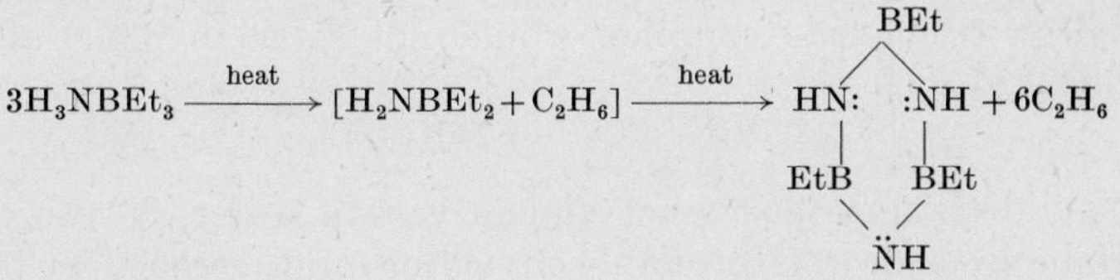

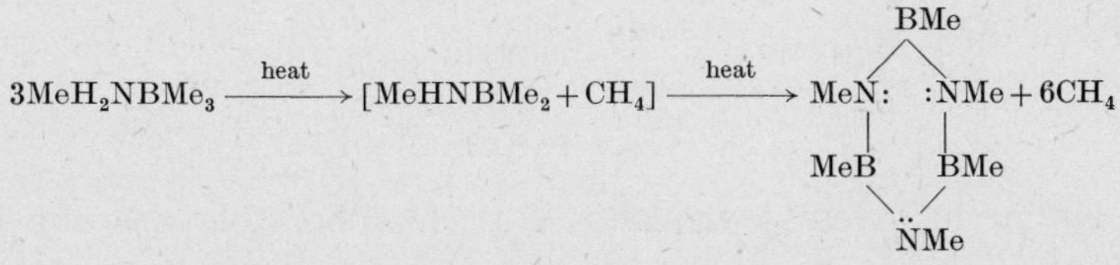

$$3H_3NBEt_3 \xrightarrow{\text{heat}} [H_2NBEt_2 + C_2H_6] \xrightarrow{\text{heat}} \begin{array}{c} BEt \\ HN: \quad :NH + 6C_2H_6 \\ EtB \quad BEt \\ \ddot{N}H \end{array}$$

$$3MeH_2NBMe_3 \xrightarrow{\text{heat}} [MeHNBMe_2 + CH_4] \xrightarrow{\text{heat}} \begin{array}{c} BMe \\ MeN: \quad :NMe + 6CH_4 \\ MeB \quad BMe \\ \ddot{N}Me \end{array}$$

An entirely different approach to the whole of this field of chemistry has been introduced by the discovery of the metal borohydrides. Schaeffer *et al.* (1949) prepared *N*-alkyl derivatives by interaction of methylamine hydrochloride and lithium borohydride in ether, and this procedure was later improved by Hough *et al.* (1955). In an *N*-filled dry box, 0.1 mole quantities of the two solids are mixed in a flask which has a long neck. Ether (40 c.c) is added, whereupon hydrogen is evolved. Ether is removed, and the solid is pyrolysed to give the substituted borazole in 84–94% yield, a heating unit (400°C) being used to cover the whole surface of the apparatus. N-Trimethyl-, N-tripropyl-, and N-tri-isopropylborazole were thus prepared.

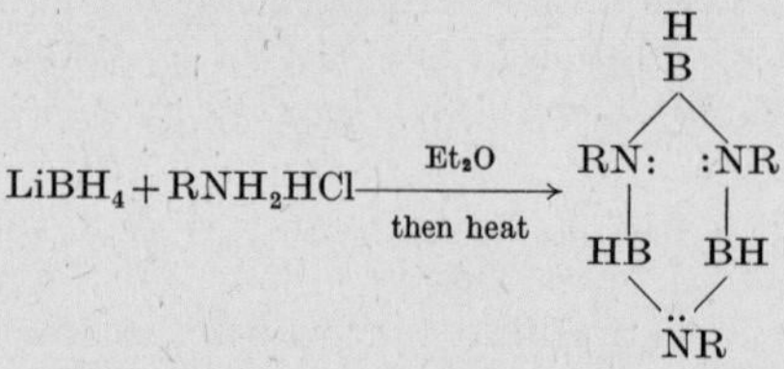

$$LiBH_4 + RNH_2HCl \xrightarrow[\text{then heat}]{\text{Et}_2\text{O}}$$

By the interaction of boron trichloride and methylamine at 250°C, Wiberg *et al.* (1951) prepared N-trimethyl-B-trichloroborazole.

The chemistry of derivatives of borazole is merely in its infancy, and the whole subject needs detailed scrutiny.

Interaction of boron halides and alkylated or arylated boron halides with amines has become a topic of considerable interest and potentialities.

In early work, Rideal (1889) obtained a borazine (formulated PhN=BCl) by the interaction of boron trichloride and aniline. In a later report (Jones *et al.*, 1939) the reaction is described as violent, and in an attempt to form the aniline–boron trichloride complex, benzene was used as a solvent after it was found that the dimethylamine–boron trichloride complex would not serve in place of boron trichloride.

$$PhNH_2 + BCl_3 \longrightarrow PhH_2NBCl_3$$

Not only did the complex react violently with water, it also reacted with aniline by a process probably involving prior conversion into the borazene.

$$PhH_2NBCl_3 \xrightarrow{PhNH_2} PhHNBCl_2 + PhNH_3Cl$$
$$\downarrow PhNH_2$$
$$PhHNBCl_2 . NH_2Ph$$

With aniline in excess N-triphenyl-B-trichloroborazole was obtained.

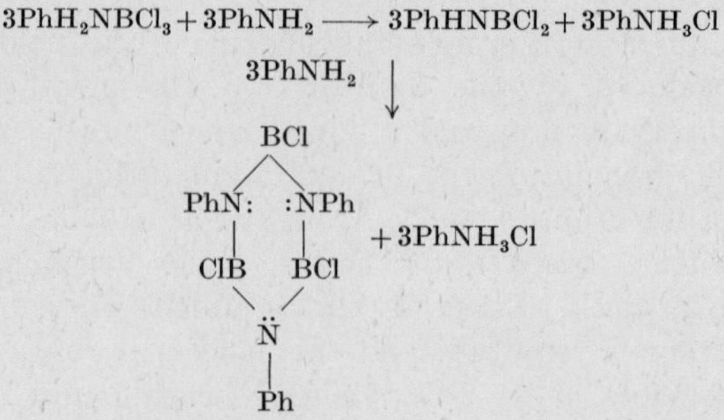

$$3PhH_2NBCl_3 + 3PhNH_2 \longrightarrow 3PhHNBCl_2 + 3PhNH_3Cl$$

$$3PhNH_2 \downarrow$$

With aniline in still greater excess, triphenylaminoboron was formed, and this with hydrogen chloride afforded the chloroborazole.

$$PhH_2NBCl_3 + 5PhNH_2 \longrightarrow B(NHPh)_3 + 3PhNH_3Cl$$

$$3B(NHPh)_3 + 6HCl \longrightarrow (PhNBCl)_3 + 3PhNH_3Cl + 3PhNH_2$$

Whereas similar observations were made with p-toluidine (Kinney *et al.*, 1942), p-anisidine–boron trichloride complex prepared in this way readily lost hydrogen chloride and gave p-anisylaminoboron dichloride, as shown by Kinney *et al.* (1943).

$$MeOC_6H_4 . NH_2,BCl_3 \longrightarrow MeOC_6H_4NHBCl_2 + HCl$$

The 1 : 1 complex was obtained by treating a cold suspension of p-anisidine hydrochloride in benzene with boron trichloride, and on being heated, the complex gave the chloroborazole.

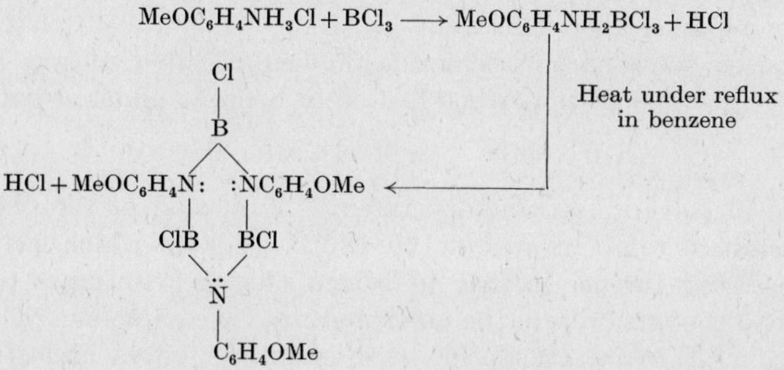

$$MeOC_6H_4NH_3Cl + BCl_3 \longrightarrow MeOC_6H_4NH_2BCl_3 + HCl$$

Contrary to a previous comment by Kinney *et al.* (1943), B-trichloro-N-trialkylborazoles were prepared from aliphatic primary amines and boron trichloride by Wiberg *et al.* (1947).

$$3MeNH_2 + 3BCl_3 \text{ at } 250°C \longrightarrow (MeNBCl)_3 + 6HCl$$

$$3MeHNBClMe \xrightarrow{\text{Heat}} (MeNBCl)_3 + 3CH_4$$

Where R is methyl, *cyclo*hexyl, phenyl, p-tolyl or p-anisyl, but not in certain other examples, the trichloroborazole, $R_3N_3B_3Cl_3$, was obtained from the amine hydrochloride and boron trichloride. Methylammonium chloride and boron trichloride first gave the 1 : 1 complex, purified by recrystallization from benzene–petroleum ether; the chloroborazole was then obtained by treatment in toluene suspension with triethylamine (Turner *et al.*, 1958).

$$3MeH_2NBCl_3 + 6Et_3N \longrightarrow (MeNBCl)_3 + 6Et_3NHCl$$
$$\text{Sublimed} \quad \text{Filtered off}$$

Analogous N-ethyl-and N-n-butylborazoles were thus made.

In recent work by Gerrard *et al.* (1958, 1959, 1960), interaction of aniline, *p*-toluidine or *p*-bromoaniline in methylene dichloride with boron trichloride gave the insoluble 1 : 1 complex, and from the filtrate impure arylaminoboron dichloride was obtained.

$$ArNH_2 + BCl_3 \longrightarrow ArH_2NBCl_3 + ArHNBCl_2$$

The 1 : 1 complexes are white solids, insoluble in benzene, toluene, *n*-pentane or chloroform, and are violently decomposed by ethanol. In accordance with a previous observation, the *p*-anisidine complex was difficult to purify, because of its tendency to give the aminoboron dichloride.

It is remarkable that although nitrobenzene forms a 1 : 1 complex with boron trichloride, *p*-nitroaniline forms neither a 1 : 1 nor a 1 : 2 complex. Infra-red spectra still showed bands at 2.88 μ (3,472 cm^{-1}) and 3.0 μ (3,333 cm^{-1}) due to —NH$_2$ stretching modes.

Contrary to an earlier statement by Jones *et al.* (1939) that the 1 : 1 complexes gave boron trichloride on being heated, it was later found that they reacted thus when heated in benzene under reflux.

$$3ArH_2NBCl_3 \longrightarrow (ArNBCl)_3 + 6HCl$$

In absence of solvent, satisfactory rates of formation of the chloroborazole required temperatures of 160–180°C. A point which has still to be considered further is that hydrogen chloride continues to be evolved after conversion into the chloroborazole is complete.

Interaction of boron trichloride with diamines offers almost unlimited potentialities; but little is yet known (Schupp *et al.*, 1955). An adduct of *p*-phenylenediamine with boron trichloride underwent dehydrohalogenation at 400°C, giving a white solid approximately (C$_6$H$_5$N$_2$B), which was stable in a vacuum up to at least 500°C.

There has been a recent quickening of interest in the detailed disposition of electron density in the 1 : 1 complexes and related compounds. The complex Me$_2$HNBCl$_3$ and pyridine underwent base exchange (Brown, J. F., 1952), but triethylamine caused dehydrochlorination despite the fact that triethylamine–boron trichloride complex can be

$$Me_2HNBCl_3 + C_5H_5N \longrightarrow Me_2NH + C_5H_5NBCl_3$$
$$Me_2HNBCl_3 + Et_3N \longrightarrow Me_2NBCl_2 + Et_3NHCl$$

made from the constituents, and that triethylamine is a stronger proton base than pyridine. It is probably a matter of F-strain (see Brown *et al.*, 1944, and other papers, for discussion on F-strain).

Both dimethylaminoboron dichloride and diethylaminoboron dichloride form with hydrogen chloride a 1 : 1 complex, which could have an ionic structure (Gerrard *et al.*, 1957; Brown, C. A., *et al.*, 1952; Goubeau *et al.*, 1954).

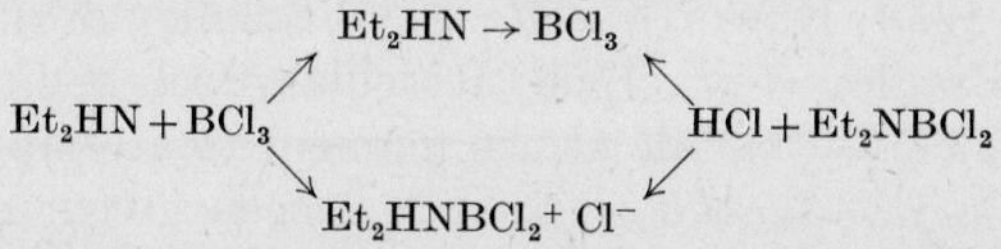

$$\begin{array}{ccc}
 & Et_2HN \to BCl_3 & \\
Et_2HN + BCl_3 & & HCl + Et_2NBCl_2 \\
 & Et_2HNBCl_2{}^+ \ Cl^- &
\end{array}$$

The facts that the product by each reaction was only slowly hydrolysed by water, and had the same low electrical conductivity in nitrobenzene, rather point to an externally non-ionic structure $Et_2NH \to BCl_3$.

Most detailed infra-red study by Gerrard *et al.* (1958, 1959, 1960) of the 1 : 1 complex of a primary arylamine (aniline, *p*-toluidine or *p*-bromoaniline) point neither to $ArH_2NBCl_2{}^+Cl^-$ nor to ArH_2NBCl_3, and some other structure involving hydrogen bonding, or a π-complex, must be admitted as a possibility. *cyclo*Hexylamine in methylene chloride added to boron trichloride in the same solvent at room temperatures gave a white precipitate of *cyclo*hexylammonium tetra-chloroborate, and a solution of *cyclo*hexylaminoboron dichloride.

$$2RH_2N + BCl_3 \longrightarrow [RNH_3]^+ [BCl_4]^- + RHNBCl_2$$
$$R = cyclohexyl \text{ or benzyl}$$

The 1 : 1 complex is probably not an intermediate because diethyl-ammonium tetrachloroborate is converted into the 1 : 1 complex on being heated in benzene under reflux, and there was no redistribution of the complex.

$$[Et_2H_2N]^+[BCl_4]^- \xrightarrow[\text{benzene}]{\text{Heat in}} Et_2HNBCl_3 + HCl$$

Similar results were obtained with other aliphatic primary amines such as *n*-butylamine. Whereas the tetrafluoroborate anion is well known, the corresponding tetrachloroborate anion is of recent recognition. The first reported examples of preparation from an amine have just been mentioned. A pyridinium halide and the corresponding boron trihalide (Cl, Br) have afforded the tetrahaloborate (Lappert, 1957), and whereas the interaction of alkylammonium chloride with boron trichloride in boiling toluene usually gives the 1 : 1 complex, RH_2NBCl_3 (Turner *et al.*, 1958), in boiling chloroform the reaction affords the tetrachloroborate (Kynaston *et al.*, 1958).

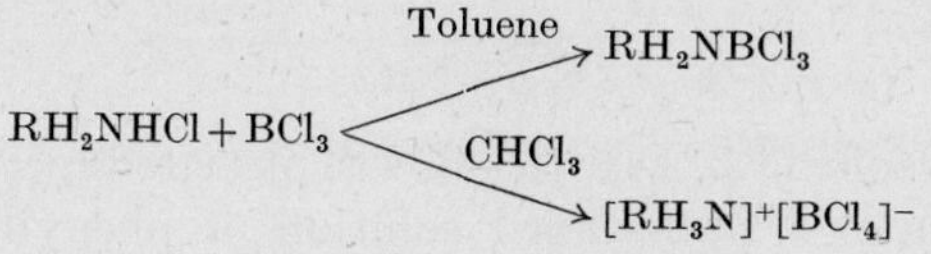

$$RH_2NHCl + BCl_3 \begin{array}{c} \xrightarrow{\text{Toluene}} RH_2NBCl_3 \\ \\ \xrightarrow[\text{CHCl}_3]{} [RH_3N]^+[BCl_4]^- \end{array}$$

There is intense activity in the investigation of chemistry related to borazole (Gerrard *et al.*, 1958, 1959, 1960; Niedenzu *et al.*, 1959; Turner *et al.*, 1958; Haworth *et al.*, 1959; Smalley *et al.*, 1959; Groszos *et al.*, 1958; Bradley *et al.*, 1959; Ryschkewitsch *et al.*, 1958) and the situation is much too mobile at the moment to attempt an assessment of the position in advance of what has been given in the preceding paragraphs.

V. Borazole Polymers

Aimed at the production of applicable polymers to withstand high temperatures and other testing conditions, a research project labelled "Research on Boron Polymers" sponsored by the United States Government was initiated by Ruigh *et al.* (1955). One of the starting-points of the project was a previous observation by Booth *et al.* (1952) that an oily polymer named "*n*-butylboronimine" of attractive properties was formed by the interaction of *n*-butylboron dichloride, *n*-BuBCl$_2$, and liquid ammonia in the presence of sodium. It was later shown that this material was probably *B*-tri-*n*-butylborazole, which could be produced without the agency of sodium. Entertaining strong belief in the potentialities of suitable substituted borazole compounds, a programme of work was started with a view to the technical production of *B*-tributylborazole and *B*-triphenylborazole. The starting-points for this production were the alkyl and arylboron dichloride, RBCl$_2$, and the convenient preparation of these was a problem in itself, as has been mentioned by Gerrard (1957, 1959). *n*-Butylboron dichloride and ammonia were mixed at $-80°$C, and the mixture allowed to warm to room temperature. The product, *B*-tributylborazole, was unaffected by water, 6N NaOH, and 50% KOH, and only slowly attacked by 6N HCl. By using *n*-butylamine in place of ammonia, the fully substituted *B,N*-hexa-*n*-butylborazole was prepared. Similarly, phenylboron dichloride and ammonia gave the *B*-triphenylborazole. Thermal stability appeared great, but no definite information as to strict evaluation of these materials was given.

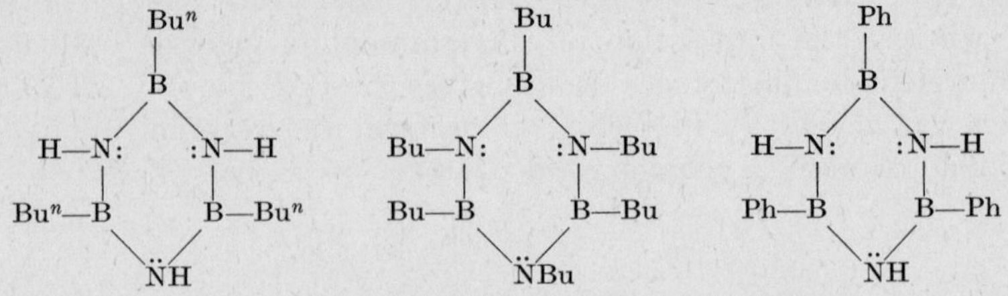

B-triphenylborazole, m.p. 184–185°C, was prepared by passing ammonia onto the surface of a well-stirred solution of phenylboron dichloride in benzene at 10–25°C. The solvent was removed, and the product was recrystallized from a mixture of chloroform and hexane.

The potentialities of borazole systems are at present boundless; for one can consider the use of diamine, and other means of linking a wide variety of substituted borazole rings together (Gerrard, 1959).

VI. Interaction of Boron Halides and Nitriles

A number of 1 : 1 complexes of nitriles and boron halides, RCN,BX_3, where $R=CH_3$, C_2H_5, C_3H_7, C_6H_5, $o\text{-}CH_3C_6H_4$, $m\text{-}CH_3C_6H_4$, $p\text{-}CH_3C_6H_4$ or $p\text{-}MeOC_6H_4$, and $X=F$, Cl or Br, have been prepared, and certain physico-chemical properties have been recorded by Gautier (1866), Patein (1891), Johnson (1912), Bowlus et al. (1931), Nespital (1932), Laubengayer et al. (1945), Brown et al. (1950), Hoard (1950), Geller et al. (1951), and Coerver et al. (1958). The halide has been added to the nitrile, sometimes in presence of a solvent, and the solid complex separated and purified.

$$\begin{array}{ccc} H\ \ H & & Cl \\ \diagdown\ | & & | \\ C{-}C{\equiv}N\ :\ B{-}Cl \\ \diagup & & \diagdown \\ H & & Cl \end{array}$$

For example, CH_3CN,BCl_3 has m.p. 180°C (decomp.), $C_2H_5CNBCl_3$ has m.p. 140°C, and $C_3H_7CNBCl_3$ has m.p. 92°C. Previous studies of the chemical reactions of these complexes have been limited, being mainly concerned with ease of hydrolysis and their relative thermal stability.

Recently, Gerrard et al. (1960) have made a much more detailed examination of $RCNBCl_3$ complexes. These are still prepared by the simple direct interactions of the components, and interesting additional members of the type $RCNBCl_3$ are $R=C_4H_9$ (m.p. 34–38°C, b.p. 61°C/12 mm), $n\text{-}C_5H_{11}$ (b.p. 52–55°C/0.5 mm), and $CH_2{=}CH$, $C_6H_5CH_2$, $p\text{-}CH_3C_6H_4$, $p\text{-}Cl.C_6H_4$, $p\text{-}CH_3OC_6H_4$, $p\text{-}NO_2C_6H_4$ are solids. In contrast to the tertiary amine adducts with boron trichloride they are readily hydrolysed by cold water. All the reactions studied show that the boron trichloride structure remains intact, and the products are those expected from boron trichloride itself.

$$RCNBCl_3 + H_2O\ (\text{excess}) \longrightarrow 3HCl + B(OH)_3 + RCN$$

$$MeCNBCl_3 + 2n\text{-}BuOH \longrightarrow 2HCl + (n\text{-}BuO)_2BCl + MeCN$$

$$RCNBCl_3 + 3n\text{-}BuOH \longrightarrow 3HCl + B(OBu)_3 + RCN$$

$$MeCNBCl_3 + 3t\text{-}BuOH \longrightarrow 3t\text{-}BuCl + B(OH)_3 + MeCN$$

Acetonitrile effected a redistribution with n-butyl dichloroborinate.

$$2n\text{-}BuOBCl_2 \longrightarrow (n\text{-}BuO)_2BCl + BCl_3$$

The resulting boron trichloride remained complexed with the nitrile.

$$2Bu^nO—B{\overset{Cl}{\underset{Cl}{\diagdown}}} + CH_3CN \longrightarrow (Bu^nO)_2BCl + CH_3C{\equiv}N : BCl_3$$

o-Nitrophenol gave only the dichloroborinate with the complex because of the stabilization by chelation (see p. 35).

$$MeCNBCl_3 + o\text{-}NO_2C_6H_4OH \longrightarrow HCl + o\text{-}NO_2C_6H_4OBCl_2 + MeCN$$

Members of the following series are placed in order of acceptor power; only the trichloride will form an adduct with nitriles.

$$BCl_3 > ROBCl_2 > (RO)_2BCl > (RO)_3B$$

Some indication of the position of nitriles amongst ligands (L) with reference to the acid BCl_3 may be given by a consideration of ligand exchange.

$$MeCNBCl_3 + L \rightleftharpoons MeCN + LBCl_3$$

Pyridine, tetrahydrofuran and di-n-butyl sulphide displace acetonitrile, whereas diethyl ether and 2,2'-dichlorodiethyl ether will not, and indeed the latter is displaced from its adduct with boron trichloride by acetonitrile. These and other considerations place the ligands in the following order with respect to boron trichloride as acceptor.

$$C_5H_5N > T.H.F. > (n\text{-}Bu)_2S > MeCN > acyclic ethers$$

Spectroscopic evidence shows that in the 1 : 1 complex of p-methoxy- and of p-nitrobenzonitrile, the nitrile group is the donor.

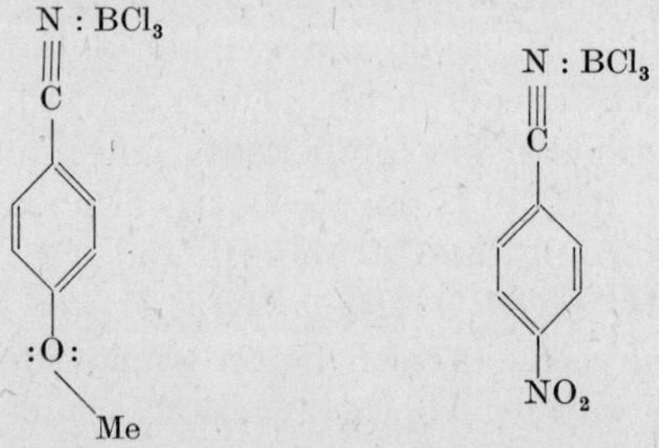

Not at the oxygen atom   Not at the nitro group

Attempts to follow replacement reactions of boron trihalides (BF_3, BCl_3, BBr_3) with respect to acetonitrile were inconclusive.

It had been hoped that pyrolysis of the nitrile complexes would lead to interesting polymeric structures. However, infra-red studies show that the complexes are $R—C{\equiv}N : BCl_3$ (: is the co-ordinate link) and not $R—C(Cl){=}N—BCl_2$, and results up to the present simply show that there are no definite signs of the possibility of producing boron-containing polymers this way.

VII. Interaction of Boron Trihalides and Amides

A number of boron trifluoride–amide $1 : 1$ complexes have been examined by Muetterties *et al.* (1953), Bowlus *et al.* (1931) and Sugden *et al.* (1932). The complexes with formamide, acetamide and N,N-dimethylformamide are immediately hydrolysed by water. Whereas the N,N-dimethylformamide complex can be distilled (b.p. $100°C/0.1$ mm), the formamide complex gives carbon monoxide and hydrogen cyanide, and the acetamide one gives acetic acid on being heated. The acetamide complex has been shown (Sowa *et al.*, 1933) to give the corresponding acetate when heated with a number of alcohols.

$$CH_3CONH_2BF_3 + ROH \longrightarrow CH_3CO_2R + H_3NBF_3$$

Heated in acetic acid, acetamide (2 moles) and boron trifluoride (1 mole) gives acetonitrile in 97% yield, and propionitrile is similarly obtained in 95% yield (Sowa *et al.*, 1937). Heated with an amine, however, the acetamide complex gives the corresponding N-substituted

$$CH_3CONH_2BF_3 + CH_3CONH_2 \longrightarrow CH_3CN + CH_3COOH + H_3NBF_3$$

acetamide. Furthermore, the benzamide–boron trifluoride complex gives benzanilide in 95% yield, and the propionamide complex gives propionanilide in 97% yield, when heated with aniline.

$$CH_3CONH_2BF_3 + RNH_2 \longrightarrow CH_3CONHR + H_3NBF_3$$

More recently, Gerrard *et al.* (1960) have examined in detail the interaction of boron trichloride and boron tribromide with amides. A selection of amides and N-substituted amides were shown to form $1 : 1$ complexes with boron halides (Cl, Br) by simple mixing of the reactants. Again in contrast with amine complexes, but in resemblance with nitrile complexes, the amide ones are readily hydrolysed by cold water. Alcoholysis affords the trialkyl borate and amide and not the

$$C_6H_5CONH_2BCl_3 + H_2O \text{ (excess)} \longrightarrow 3HCl + B(OH)_3 + C_6H_5CONH_2$$

acetate as with the boron trifluoride complex just mentioned, thus

$$C_6H_5CONH_2BCl_3 + 3n\text{-BuOH} \longrightarrow 2HCl + (n\text{-BuO})_3B + C_6H_5CONH_2HCl$$

providing still another example of the differences between boron trifluoride and boron trichloride systems, consequent on the greater reactivity of the B—Cl bond.

Pyrolysis appears to incur the following interesting reactions.

$$R.CONH_2BCl_3 \xrightarrow{-HCl} RCONHBCl_2 \xrightarrow{-HCl} RCONBCl \longrightarrow$$
$$\text{(I)} \qquad\qquad \text{(II)} \qquad\qquad \text{(III)}$$

$$RCN + [\tfrac{1}{3}(BOCl)_3 \longrightarrow \tfrac{1}{3}B_2O_3 + \tfrac{1}{3}BCl_3] \longrightarrow RCNBCl_3$$
$$\text{(IV)}$$

When $R = CH_3$, (II) was obtained as a glassy solid, and when $R' = CH_3$ or C_6H_5, the corresponding compound $RCONR'BCl_2$ was obtained, from N-methylformamide (as a yellow solid), and from acetanilide (as a yellow glass). The complex $CH_3CONMe_2BCl_3$ remains unchanged after being heated at $120°C/15$ mm for several hours.

An outstanding structural question relates to the point of co-ordination, whether "O" or "N".

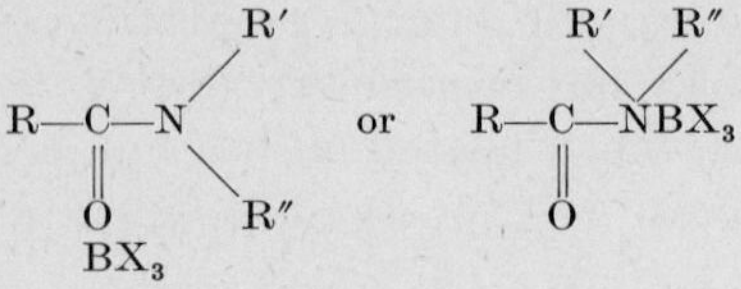

$$
\begin{array}{ccc}
& R' & \\
& \diagup & \\
R\text{—}C\text{—}N & & \text{or} \\
\| & \diagdown & \\
O & R'' & \\
BX_3 & &
\end{array}
\qquad
\begin{array}{c}
R' \quad R'' \\
\diagup \quad \diagup \\
R\text{—}C\text{—}NBX_3 \\
\| \\
O
\end{array}
$$

$$\text{``}O\text{''} \qquad\qquad\qquad \text{``}N\text{''}$$

For the complex $CH_3CONMe(Ph)BF_3$ the parachor, 453.2, was compared with the calculated value 452.5, based on $B = 16.4$; $F = 25.0$, and the co-ordinate bond $= -1.6$ units (Sugden *et al.*, 1932); but the acetanilide complex (BF_3) decomposed at the m.p. The infra-red band at $1,695$ cm^{-1} for the carbonyl group in dimethylformamide did not shift in the complex (BF_3) (Muetterties *et al.*, 1953). Very recent infra-red spectroscopic studies by Gerrard *et al.* (1960) have shown considerable difficulties of interpretation because many absorptions are not localized within particular atomic pairs. However, the "O" complex, with boron trichloride attached to oxygen, is the favoured one, and of course this then brings in the possibility of a cis or trans disposition.

Evidence points generally to boron tribromide being a better acceptor than boron trichloride in these systems.

It is not out of place to mention that nuclear magnetic resonance technique has been brought to bear on the problem of point of attachment, and again "O" structure is the favoured one.

Chapter XII

BORON–PHOSPHORUS COMPOUNDS

Burg *et al.* (1953) gave the first report on a systematic investigation into the preparation of compounds having phosphorus–boron bonds, which were most attractive from the viewpoint of production of new polymeric materials having great hydrolytic and thermal resistance. Later work, however, has done little towards the realization of earlier promise.

The chemistry is based on borane, the simplest example of a phosphinoborane being the compound resulting from the interaction of phosphine and diborane. This is an operation not easily performed, and of more significance in this monograph are the compounds derived

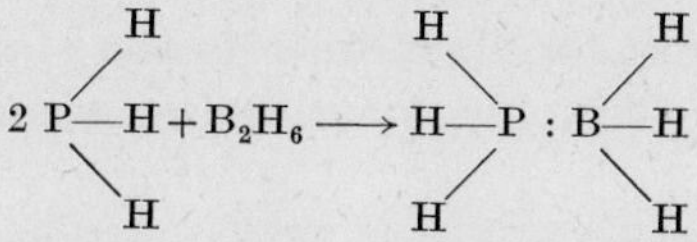

$$2\ P\!\!-\!\!H + B_2H_6 \longrightarrow H\!\!-\!\!P : B\!\!-\!\!H$$

from substituted phosphine. Thus dimethylphosphine and diborane, mixed at $-150°$ to $-80°C$, gave a liquid, dimethylphosphinoborane.

$$2Me_2PH + B_2H_6 \longrightarrow 2Me_2HPBH_3 \text{ (m.p. } -22.6°C, \text{ b.p. } 174°C)$$

Methylphosphine reacts similarly, and an example of interaction with a substituted diborane is also mentioned.

$$MePH_2 + B_2H_6 \longrightarrow MeH_2PBH_3 \text{ (m.p. } -49°C, \text{ b.p. } 150°C)$$

$$Me_2PH + Me_4B_2H_2 \longrightarrow Me_2HPBHMe_2$$

These compounds correspond with the borazanes, boron and phosphorus being linked by co-ordination $Me_2HP : \rightarrow BH_3$. Pyrolysis, however, usually results in the elimination of volatile matter, hydrogen, for example, and the establishment at first of a covalent B—P linkage, with the attendant possibility of polymerization.

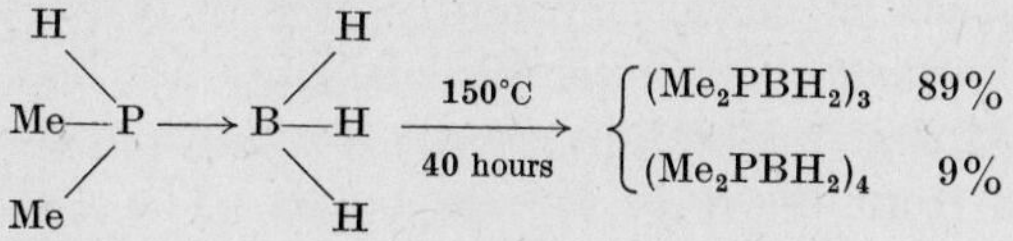

$$Me\!\!-\!\!P \longrightarrow B\!\!-\!\!H \xrightarrow[\text{40 hours}]{150°C} \begin{cases} (Me_2PBH_2)_3 & 89\% \\ (Me_2PBH_2)_4 & 9\% \end{cases}$$

187

These are white solids, crystallizable from benzene and methanol, stable in air, not attacked by water, acids or alkalis, except on prolonged heating. Hydrolysis with hydrochloric acid in a pyrex bomb at 300°C gives a dimethyl phosphorus acid, hydrogen and boric acid, showing that the carbon–phosphorus bond is much more stable than the phosphorus–boron bond under these conditions.

$$(Me_2PBH_2)_3 \xrightarrow{\text{Hydrolysis}} Me_2P\overset{\displaystyle O}{\underset{\displaystyle OH}{\big\langle}} + H_2 + B(OH)_3$$

Other examples of pyrolytic effects are shown as follows.

$$MePH_2BH_3 \xrightarrow[\text{20 hours}]{100°C} (MeHPBH_2)_n$$

A viscous oil, not hydrolysed by 4N—HCl

$$Me_2HPBMe_3 \xrightarrow{39°C/78\,mm} Me_2PH + BMe_3$$

92% dissociation

$$Me_2HPBBrMe_2 \xrightarrow[\text{in } C_6H_6]{+Et_3N} (Me_2PBMe_2)_3 \quad \text{m.p. } 334°C$$

A probable structure for the trimer is the standard six-membered ring with alternating phosphorus and boron atoms.

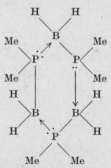

It is argued that this structure would not account for the slow rate of hydrolysis, since hydrochloric acid readily attacks B—H bonds. The fully methylated compound, $(Me_2PBMe_2)_3$, is as stable, and the acceptor power of boron should be reduced by the two methyl groups attached to it. As an aside, however, it might be doubted if the inductive effect of the methyl groups would be sufficient to decrease the acceptor power of boron sufficiently here to make any difference. An explanation based on non-classical valency has been propounded. It was supposed that the stability is due to multiple B—P π-bonding, using electrons from B—H or B—Me bonds. The methyl groups on phosphorus should increase the electron density on P and increase the strength of the $P \rightarrow B$ σ-bonding. Methyl groups, as compared with hydrogen on boron, will decrease the P—B σ-bonding by reducing acceptor power of boron; but it is believed the more important strengthening influence will be the π-bonding. CH_3HPBH_2 is a viscous oil of low volatility, suggesting a much higher polymer than corresponding to the trimer-ring, the monomer, H_2PBMe_2, is easily attacked by methyl alcohol to give phosphine and methyl dimethylborinate, Me_2BOMe, due to lack of polymeric condition.

From an occasional mention, it is apparent that further work has been supported by United States Government contracts, in the hope of producing polymers for specialized high-temperature performance. Such compounds as $[(CF_3)_2PBH_2]$ (Burg et al., 1958; see also work on stibinoborine, Burg et al., 1959) have been looked at, but there does not appear to have been any clear progress towards the production of applicable polymers based on this chemistry.

Other interests in phosphorus–boron bonds are evident from the kinetic study of the interaction of diborane with phosphine (Brumberger et al., 1956), the study of the interaction of phosphorus trifluoride with diborane (Parry et al., 1956), and the observation that dimethyl-phosphinoborohydride trimer, $(Me_2PBH_2)_3$, gives hydrogen and a small amount of methane at 360°C, and elementary phosphorus and methyl-boranes at 510°C (Florin et al., 1954). Applicable plastics would appear to be still a way off.

Relating to this chemistry and purpose, an entirely different approach has been made by Gerrard et al. (1960). A trialkyl phosphite such as triethyl phosphite contains a phosphorus lone pair electrons which are nucleophilically reactive judging from the ready interaction with hydrogen halides and halogens (Gerrard, 1940).

It was surprising to find that even at low temperatures ($-70°C$), boron trichloride and tri-n-butyl phosphite reacted to give a mixture

$$\begin{array}{c} RO \\ \diagdown \\ RO—P: \longrightarrow H——Cl \longrightarrow Cl^- \left[\begin{array}{c} RO \\ \diagdown \\ RO—PH \\ \diagup \\ RO \end{array}\right]^+ \longrightarrow RCl+ \begin{array}{c} RO \\ \diagdown \\ PH \\ \diagup \parallel \\ RO\ \ O \end{array} \\ \diagup \\ RO \end{array}$$

of chloro-esters without sign of P—B complexing. Boron trichloride and trialkyl phosphate such as triethyl, however, gave a complex at

$$(RO)_3P + BCl_3 \longrightarrow (RO)_2PCl + ROBCl_2$$

$-80°C$, presumably by attachment to the double-bonded oxygen atom, $(RO)_3P=OBCl_3$, as shown by Gerrard et al. (1959). Between 0°C to

$300°C$ alkyl chloride is evolved stepwise, leaving "boron phosphate" which in this condition is easily hydrolysed, but after it has been heated at $1,000°C$ the product is then hydrolysed with great difficulty. It is probable that the elimination of alkyl chloride is by more than

$$(RO)_3POBCl_3 \xrightarrow{\text{heat}} 3RCl + [PBO_4]_x$$

one mechanism, and involves the gradual assembly of $-\overset{\overset{\parallel}{}}{P}-O-B-O$ linkages as a three-dimensional network.

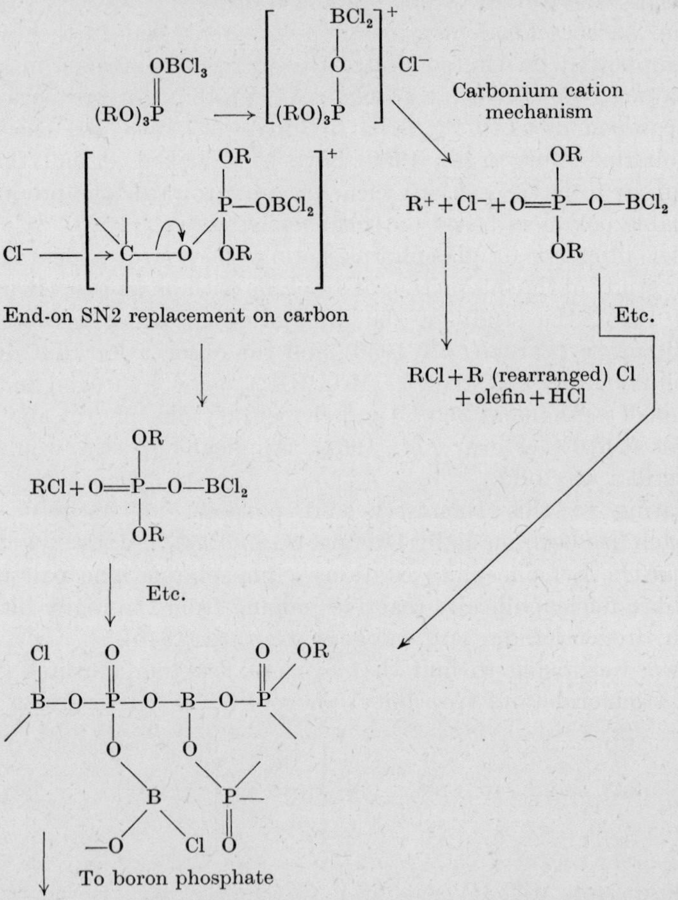

$$
\begin{array}{c}
OBCl_3 \\
\parallel \\
(RO)_3P
\end{array}
\longrightarrow
\left[
\begin{array}{c}
BCl_2 \\
| \\
O \\
| \\
(RO)_3P
\end{array}
\right]^{+} Cl^{-}
$$

Carbonium cation mechanism

$$
\left[
\begin{array}{c}
OR \\
| \\
P-OBCl_2 \\
\nearrow \\
Cl^- \longrightarrow C-O\ OR \\
|
\end{array}
\right]^{+}
$$

End-on SN2 replacement on carbon

$$R^+ + Cl^- + O{=}\overset{\overset{OR}{|}}{\underset{\underset{OR}{|}}{P}}-O-BCl_2$$

Etc.

$$RCl + O{=}\overset{\overset{OR}{|}}{\underset{\underset{OR}{|}}{P}}-O-BCl_2$$

$$RCl + R \text{ (rearranged) Cl} + \text{olefin} + HCl$$

Etc.

$$
\begin{array}{c}
Cl\quad O\qquad\quad O\ \ OR \\
| \quad\ \parallel \qquad\ \parallel\ / \\
B-O-P-O-B-O-P \\
/ \qquad\quad\ \ | \qquad | \qquad\ \ \backslash \\
O \qquad O \\
\backslash \qquad / \\
B \qquad P- \\
/\ \backslash \qquad \parallel \\
-O \quad Cl\ O
\end{array}
$$

To boron phosphate

A number of trialkyl phosphates, such as isomeric butyl propyl and others, have now been investigated in this way (Gerrard et al., 1960).

It has been shown by Frazer et al. (1959) that triphenyl phosphite behaves similarly to an ordinary trialkyl phosphite, in that there is a mutual replacement of RO and Cl; but there are differences in rate of reaction, and due to the different chemistry of the phenyl boron ester

$$(PhO)_3P + BCl_3 \longrightarrow (PhO)_2PCl + PhOBCl_2$$

system (p. 34), bearing especially on the non-formation of phenyl chloride. This means that the whole system can be examined under forcing conditions without losing the phenyl group as phenyl chloride. In other words, the chlorine can be kept in play, whereas in the alkyl system, analysis of the stepwise reaction system is seriously hampered by incidental decomposition of the chloro-boron ester.

When triphenyl phosphite (1 mole) and boron trichloride (1 mole) were mixed at $-78°C$, and the mixture allowed to warm to $-10°C$, boron trichloride was almost quantitatively recovered at $-10°C/20$ mm, and the residue was the phosphite, thus showing that complexing $(PhO)_3P:\rightarrow BCl_3(?)$ or $(PhO)_2PO:\rightarrow BCl_3$ must at best have been very

Ph

weak. At $25°C$ replacement reactions occurred, there being a distinct one-way drift towards triphenyl borate and phosphorus trichloride.

PhO   O—Ph   PhO
    $\diagdown$ P   B—Cl $\longrightarrow$ $\diagdown$ PCl $\xrightarrow{\text{Etc.}}$ PhOPCl$_2$ $\longrightarrow$ PCl$_3$
PhO   Cl   Cl   PhO   dichloridite

                      chloridite

$$+ \text{PhOBCl}_2 \longrightarrow (PhO)_2BCl \longrightarrow (PhO)_3B$$
dichloroborinate   chloroboronate

It is believed that in all such systems the driving force is the nucleophilic function of the oxygen atom, depending on the electron density thereon. An exchange of one phenoxyl group on phosphorus by chlorine should, by the inductive effect of the latter, reduce electron density on the remaining two atoms of oxygen, and hence reduce their ability to undergo further replacement reaction with boron trichloride. Furthermore, the electrophilic power of boron is reduced in the alkyl or aryl chloroboronate, $ROBCl_2$, because of back–co-ordination from oxygen to boron. It is therefore understandable that in the approach to phosphorus trichloride and to the triphenyl borate, the successive steps will be of slower and slower rate of reaction. It is remarkable that triphenyl borate appears to be completely inert to phosphorus trichloride.

Triphenyl phosphite and boron tribromide have been similarly investigated, and the results fall into a similar pattern as for the boron trichloride, and this system has been described by Frazer et al. (1959).

Similarly to the trialkyl phosphates, the aryl phosphates form 1 : 1 complexes with boron trihalides; but in sharp distinction, the aryl phosphate complexes are very much more thermally stable and do not give aryl chloride on heating. Some detailed work has been done by Frazer *et al.* (1960) on this system, and infra-red spectroscopic studies have been described. The complexes examined are $(RO)_3P=OBX_3$ where $R=C_6H_5$, m-MeC_6H_4, p-MeC_6H_4, p-ClC_6H_4 or $2,4$-$Cl_2C_6H_3$, when $X=Cl$ or Br; but not the last two when $X=F$. Substituents in the ring do not appear to have a very marked influence on the stability of the complex, an observation relating to transmission of electronic effects through phosphorus. It is of further interest to state that facts do support the following order of acceptor power: $BBr_3 > BCl_3 > BF_3$.

Chemical reactions showed the preservation of the BX_3 entity. The complexes were easily hydrolysed and alcoholized, the fate of the BX_3 part being what it would be if free. The phenyl phosphate–boron

$$(PhO)_3P=OBCl_3 \xrightarrow{\text{H}_2\text{O}} (PhO)_3P=O + 3HCl + B(OH)_3$$

$$(PhO)_3P=OBCl_3 \xrightarrow{n\text{-BuOH}} (PhO)_3P=O + 3HCl + B(OBu^n)_3$$

trichloride complex could be heated at 100°C for 5 hours without change; at 200°C during 3 hours there was a slight loss in weight, but the easily hydrolysed chlorine almost entirely remained, and at higher temperatures, decomposition, to as yet unidentified products, occurred as distinct from dissociation.

Some detailed infra-red studies point to a structure

$$(RO)_3P \overset{\curvearrowright}{=} O - BX_3$$

The complexes, $(RO)_2P(O)Cl,BCl_3$ and $ROP(O)Cl_2,BCl_3$ ($R=$ aryl), were much less stable than those of the triaryl phosphates, and infra-red data point rather to a chlorine boron trichloride co-ordination, $(RO)_2P(O)Cl \rightarrow BCl_3$, i.e. $[(RO)_2P=O]^+BCl_4^-$. It has been suggested that the complexes of phosphoryl halides and boron trihalides are $[POX_2]^+[BX_4]^-$.

The complex, $PhOP(O)Cl_2,BCl_3$, for example, had m.p. 35–42°C, and on being slowly heated to 300°C, afforded the phosphoryl chloride–boron trichloride complex, boron trichloride, and diphenyl phosphoro-chloridate. The complex dissociated when heated under reduced pressure.

There are interesting potentialities in the extension of this work. Instead of trialkyl phosphates, dialkyl alkanephosphonates $(RO)_2PR'(O)$ may be used. This would reduce the functionality of phosphorus and lead to polymeric structures such as:

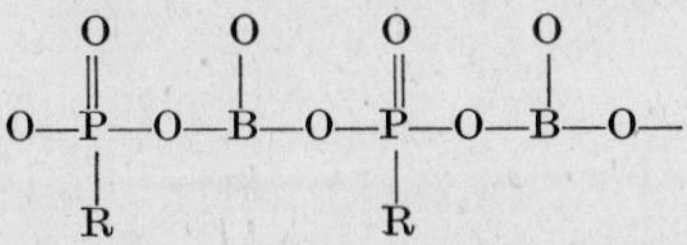

$$
\begin{array}{cccc}
O & O & O & O \\
\parallel & | & \parallel & | \\
O—P—O—B—O—P—O—B—O— \\
| & & | \\
R & & R
\end{array}
$$

Furthermore, by using also an alkyl or aryl boron dichloride, $RBCl_2$, in place of boron trichloride, further properties may be introduced:

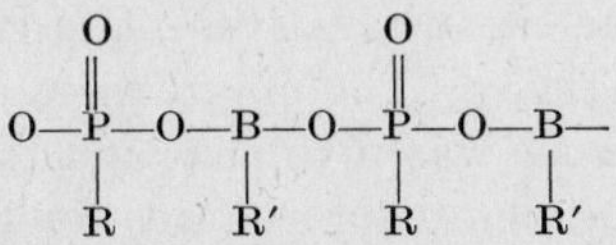

$$
\begin{array}{cc}
O & O \\
\parallel & \parallel \\
O—P—O—B—O—P—O—B— \\
| \quad | & | \quad | \\
R \quad R' & R \quad R'
\end{array}
$$

A beginning has been made (Chainani $et\ al.$, 1960). Thus diethyl ethylphosphonate forms a liquid 1 : 1 complex with boron trichloride

$$
\begin{array}{c}
(EtO)_2PEt. \\
\parallel \\
OBCl_3
\end{array}
$$

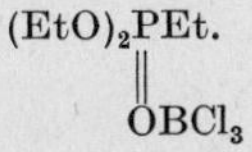

The complex decomposes very slowly at room temperature, more quickly at 80°C/18 mm, giving off ethyl chloride, and eventually leaving a solid approximating to $(EtPO_3BCl)_n$, which was insoluble in a number of non-hydrolytic solvents. At 370°C the easily hydrolysed chlorine decreased to 4%, and on ignition boron phosphate resulted. A material, $[(EtPO)_3B_2O_6)]_n$, which did not melt at 350°C, and was insoluble in common organic solvents, was formed from the interaction of the phosphonate $(EtO)_2PEt(O)$ (3 moles) with boron trichloride (1 mole).

Some attractive exploratory work has been done with such phosphonates as diethyl trichloromethylphosphonate,

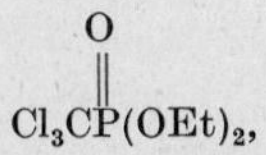

$$
\begin{array}{c}
O \\
\parallel \\
Cl_3CP(OEt)_2,
\end{array}
$$

and of course phenylboron dichloride has been used in place of boron trichloride. There is no doubt a fruitful field for research here.

Chapter XIII

CERTAIN BORON–SULPHUR COMPOUNDS

A SERIES of papers on the derivatives of $B_3H_3S_3$ was published by Wiberg *et al.* (1953, 1955). The impression appears to be that a "borole" type of structure was involved, and the papers contain brief statements about experimental procedures for ringing the changes on the groups attached to the atoms in the heterocycle.

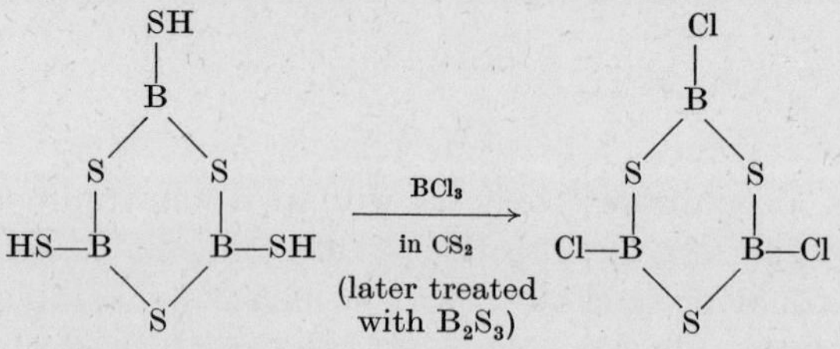

The trimeric chloroborosulphole, $(BClS)_3$, is very sensitive to water, and readily undergoes redistribution to boron sulphide, B_2S_3, and boron trichloride. The corresponding bromo-compound, which has analogous properties, was prepared by means of boron tribromide.

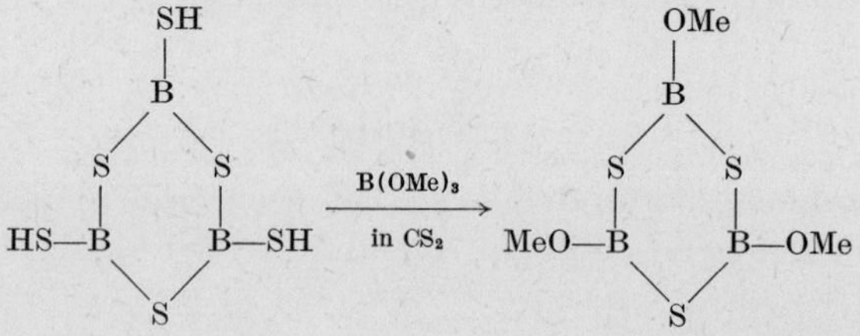

Trimethoxyborosulphole
m.p. 27.5°C

The reagents were heated under reflux, and the product was eventually obtained as a white solid, which fumed in air, was soluble in benzene, and was rapidly hydrolysed with water. Tris-dimethylaminoborosulphole, $[B(NR_2)S]_3$, was likewise prepared by means of a mixture of bis-dimethylaminoboron bromide, and tris-dimethylaminoboron.

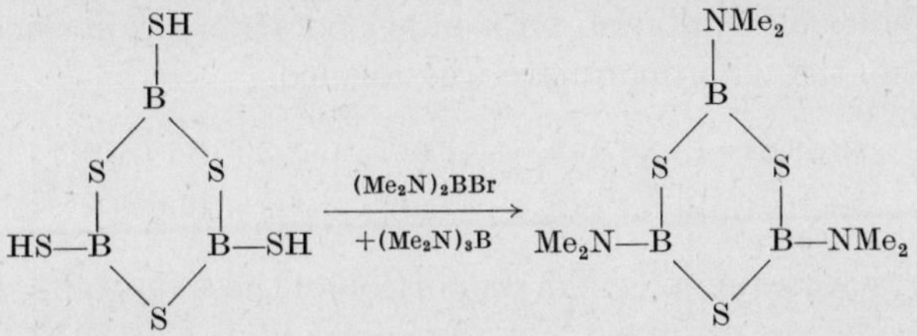

It is easily hydrolysed in moist air, and the amino-groups form salts with hydrogen bromide. By the interaction of trimethylboron, trimethylborosulphole was obtained, and the corresponding triphenylborosulphole was prepared.

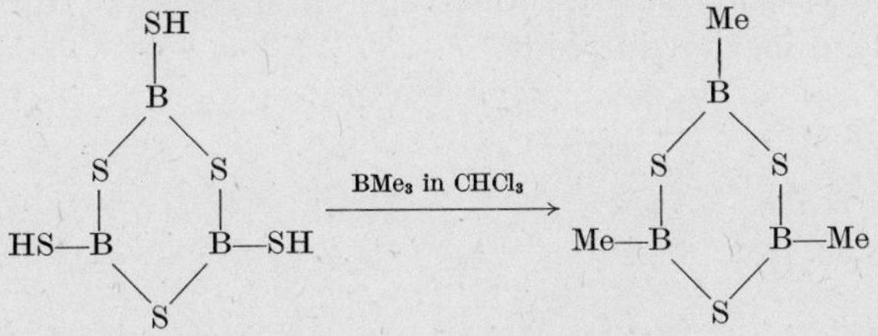

$$3H_2S + 3PhBBr_2 \longrightarrow (PhBS)_3 + 6HBr$$

white needles,
m.p. 232C°

Ethyl metathioborate, SB(SEt), is referred to as the trimer, and is obtained from the trimeric metathioboric acid, $[B(SH)S]_3$, and triethyl orthothioborate, $B(SEt)_3$.

Zvonkova (1958) has reported on the crystal structure of $Br_3B_3S_3$.

A. Extension to Dialkyl Sulphides

Investigations of Gerrard et al. into ethers was extended by Lappert (1953) to di-n-butyl sulphide. In strong distinction, di-n-butyl sulphide formed a stable complex, $Bu^n_2S.BCl_3$, which was a colourless stable liquid, 75% being recovered after the compound had been heated at 215°C for 32 hours. Decomposition appeared to involve no more than dissociation into the constituent boron trichloride and thio-ether. Pyridine removed the boron trichloride as the complex, $C_5H_5N.BCl_3$, and water reacted with the boron trichloride part, as if it were intact as a group.

$$Bu^n_2S.BCl_3 + C_5H_5N \longrightarrow C_5H_5N.BCl_3 + Bu^n_2S$$

$$Bu^n_2S.BCl_3 + 3H_2O \longrightarrow Bu^n_2S + H_3BO_3 + 3HCl$$

7

With 3 moles of n-butanol, tri-n-butyl borate was obtained, whereas with 2 moles, the chloroboronate was formed.

$$2Bu^nOH + Bu^n_2S.BCl_3 \longrightarrow 2HCl + Bu^n_2S + (Bu^nO)_2BCl$$

$$3Bu^nOH + Bu^n_2S.BCl_3 \longrightarrow 3HCl + Bu^n_2S + (Bu^nO)_3B$$

However, with equimolecular proportion of these reagents, an entirely different result was obtained; for the compound, $Bu^n_3S.BCl_4$, possibly $[Bu^n_3S]^+[BCl_4]^-$, was obtained as crystals.

Burg et al. (1956) have investigated compounds such as $MeSBMe_2$ and $[MeS]_2BMe$, and Coyle et al. (1959) include the following in the comprehensive study of molecular addition compounds of boron (see p. 161): $(CH_2)_4SBH_3$, $Et_2S.BH_3$, $Me_2S.BH_3$, $(CH_2)_4S.BF_3$, $Et_2S.BF_3$. Wartik et al. (1958) have studied the interaction of diboron tetrachloride with dimethyl sulphide.

CHAPTER XIV

MISCELLANEOUS RING SYSTEMS

RELATED to the ortho-phenylene ring systems described on p. 35, a number of similar possible ring systems may be imagined, and some of these have been realized. Dewar *et al.* (1958) have prepared the following compounds, and there is a suggestion that compounds such as these might be of beneficial biological activity. The series provides an exercise in nomenclature, and it will be noticed that phenylboron dichloride is the basic boron reagent.

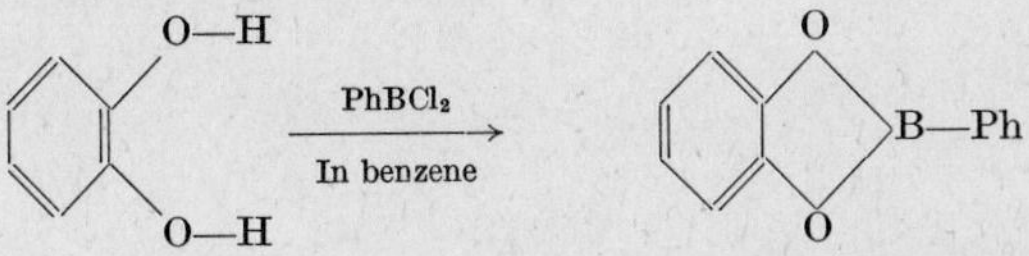

m.p. 94°C
2-phenyl-2,3-boroxocoumarone

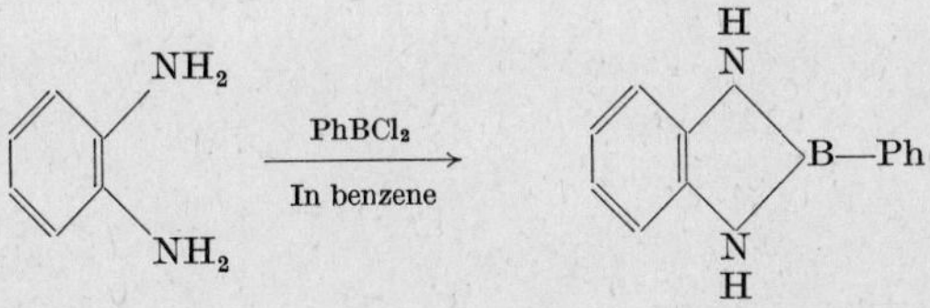

2-phenyl-2,3-borazaindole
m.p. 201°C

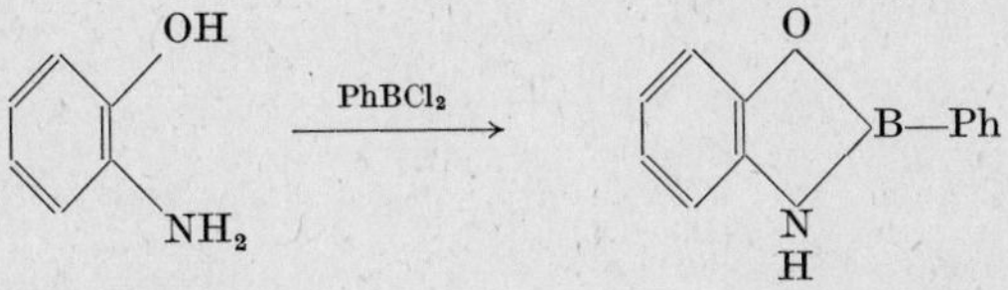

2-phenyl-2,3-boroxindole
m.p. 105°C

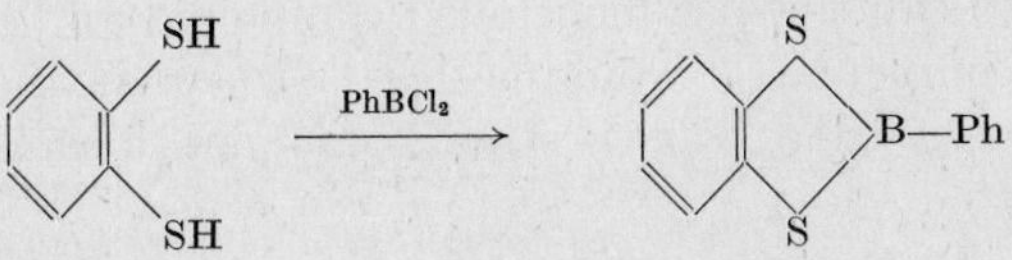

Dithiocatechol

m.p. 130°C
2-phenyl-2,3-borothianaphthene

197

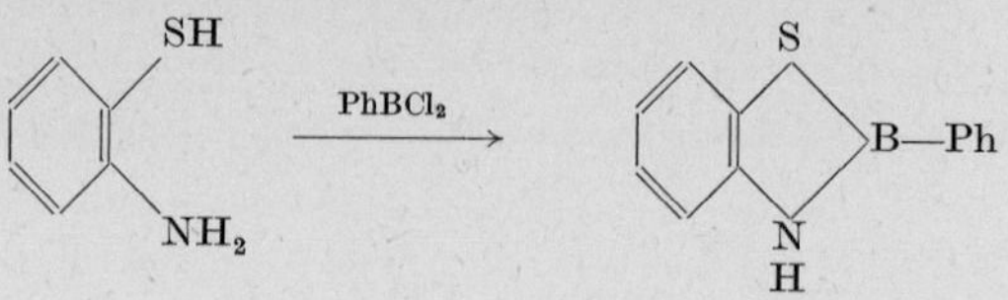
SH
NH2
PhBCl2
S
B—Ph
N
H

m.p. 152–154°C
2-phenyl-2, 3-borothiaindole

In extension of this work, Dewar *et al.* (1959) have carried out the following reactions, and have studied the isoconjugation aspect by spectroscopy.

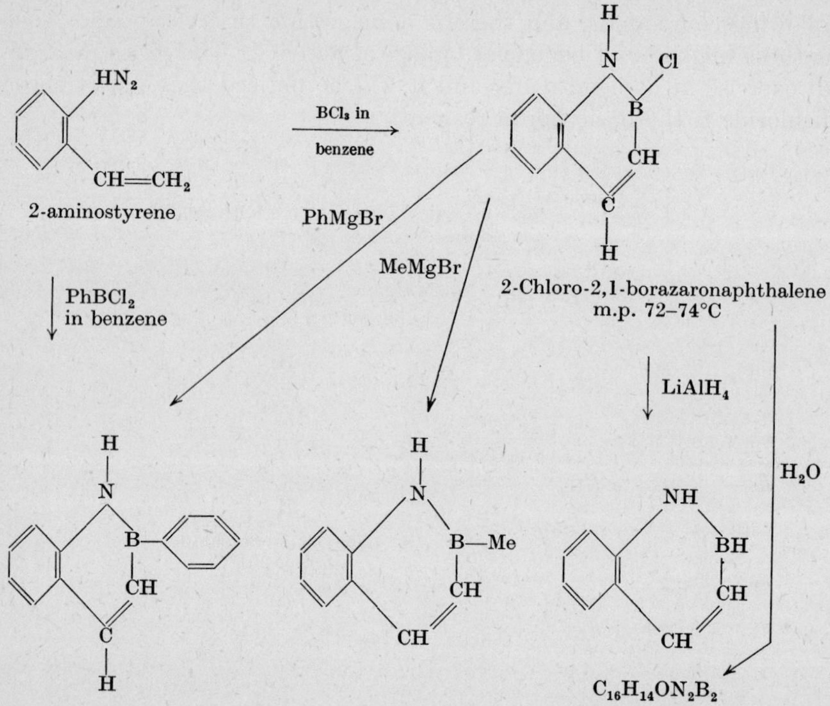
HN2
CH=CH2
2-aminostyrene
BCl3 in
benzene
H
N
Cl
B
CH
C
H
2-Chloro-2,1-borazaronaphthalene
m.p. 72–74°C
PhMgBr
MeMgBr
PhBCl2
in benzene
LiAlH4
H2O
H
N
B
CH
C
H
H
N
B—Me
CH
CH
NH
BH
CH
CH
C16H14ON2B2

2-phenyl-2,1-borazaro-naphthalene m.p.137.5–139°C

Bis-2,1-borazaro-naphthyl ether

It is strange to call the last-mentioned compound an "ether" if it is a $>$B—O—B$<$ compound; it would be better to retain the established name, borinic anhydride. The "ether" was not formulated in the paper quoted.

The question of nomenclature is taken up, and it is suggested that *boraz* is followed by *aro* and the name of the cyclic hydrocarbon with which the compound is isoconjugate. The compound named 9-aza-10-boraphenanthrene in the previous paper should therefore have been named an azobora-derivative of dihydrophenanthrene. It is deemed wise to leave the situation thus for the time being, especially as American workers are adopting different ways of naming these compounds.

Developments in the formation of heterocyclic boron rings are further illustrated by the results of Letsinger *et al.* (1959), who also provide new phenylboron systems, especially those with two boron atoms.

2,2′-Tolandiboronic acid was produced from 2,2′-dibromobibenzyl by a series of reactions, the last two of which are shown as follows:

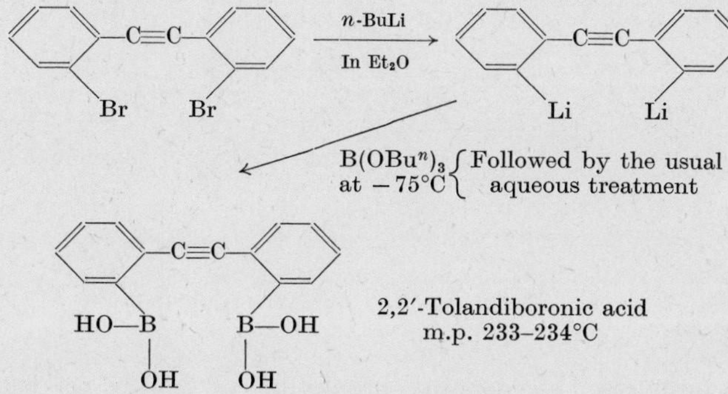

2,2′-Tolandiboronic acid
m.p. 233–234°C

The acid is less stable in alkaline solutions than phenylboronic acid. With *o*-phenylene diamine, the acid gives bis-dihydrobenzoboradiazole, in accordance with the work of Letsinger *et al.* (1958).

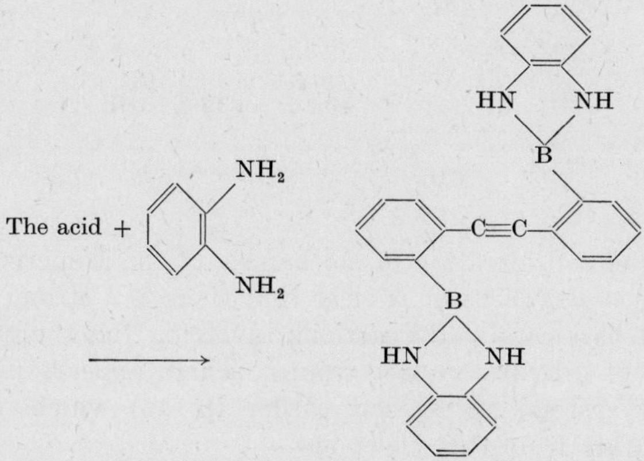

Isomerization occurred when the acid was in mild alkaline solution, and another acid was obtained on acidification.

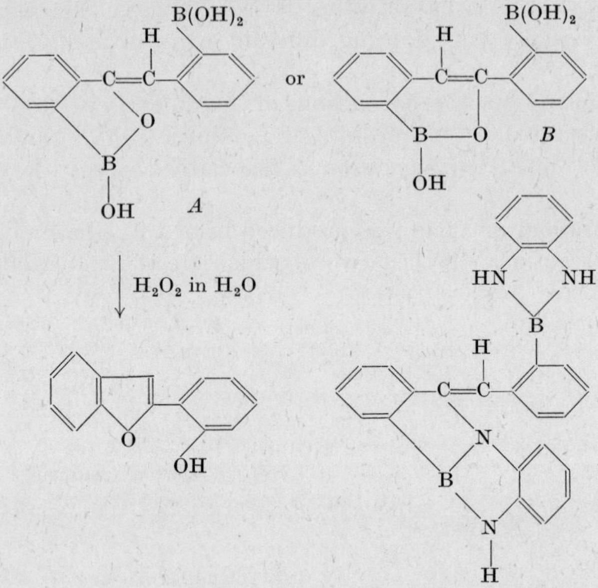

Evidence supports the general proclivity of boron to form five-membered cyclic structures. Furthermore, the relative hydrolytic stability of the B—O—C linkage has correlation with the reputed stability of the compound referred to by Snyder *et al.* (1958); see also Torssell (1957).

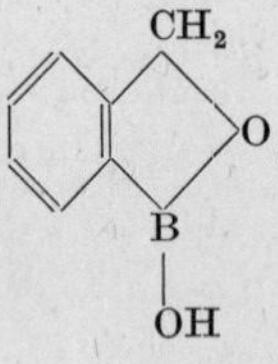

Stable in 10% HCl
and in hot 15% KOH

From a brief consideration of mechanism of the isomerization given by Letsinger *et al.* (1958), it is clear that there is a strong correlation with certain base-catalysed reactions involving functional groups in the vicinity of dihydroxyboryl groups. There appears much in the polyhydroxy systems mentioned earlier (p. 16) which would bear detailed scrutiny from this viewpoint.

Related work (Letsinger *et al.*, 1959) deals with 8-quinolineboronic acid, prepared for study in connection with selective catalysis in base-catalysed reactions. The following reactions were formulated.

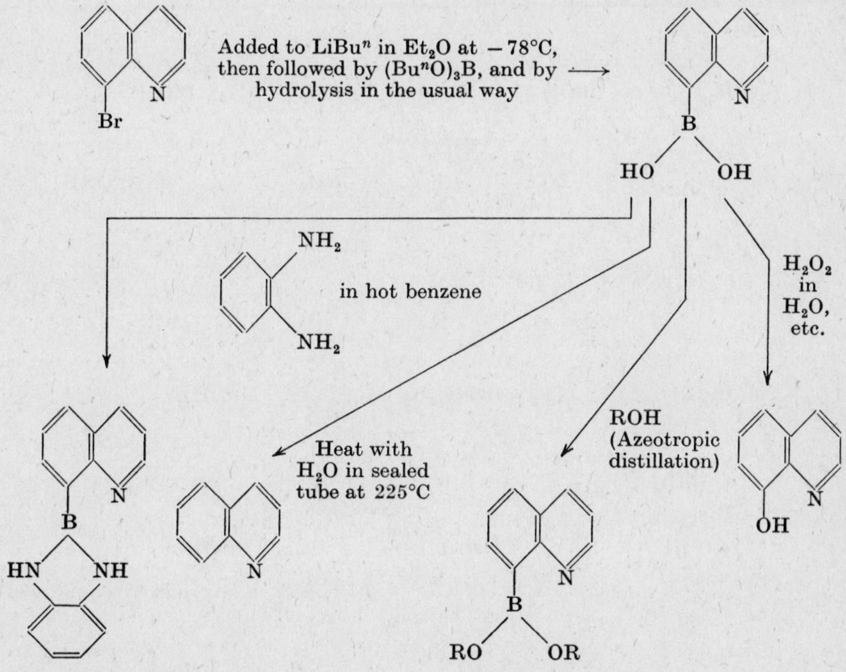

The liberation of chloride ion from a chloroalcohol (2-chloroethanol, 3-chloropropanol and 4-chlorobutanol) by interaction with 8-quinoline-boronic acid was examined.

Some detailed work by Soloway (1958) on the stability of phenyl-boronic acids shows clearly the decided interest in the search for biologically active boron compounds. In preliminary experiments the *m*- or *p*-carboxylic group, or a meta-ureido function, offered the most promise with regard to toxicity and tumour-to-brain boron ratio in neutron-capture irradiation.

Soloway's synthetic studies provide examples further to those already discussed (p. 68) for the preparation of substituted phenylboronic acids, particularly those involving a carboxylic group, which seems to stabilize the C—B link against fission reactions. *p*-Carboxyphenyl-boronic acid is readily nitrated at room temperature by a mixture of fuming nitric acid and sulphuric acid to give the 2-nitro-4-carboxy-phenylboronic acid, as shown by Torssell (1957), who also showed that *m*-carboxyphenylboronic acid gives 3-nitro-5-carboxyphenylboronic acid. The following scheme should be correlated to that already given on p. 68. The preparation of *p*-carboxyphenylboronic acid was described by Michaelis (1901), and that of the *m*-isomer by Bettman *et al.* (1934).

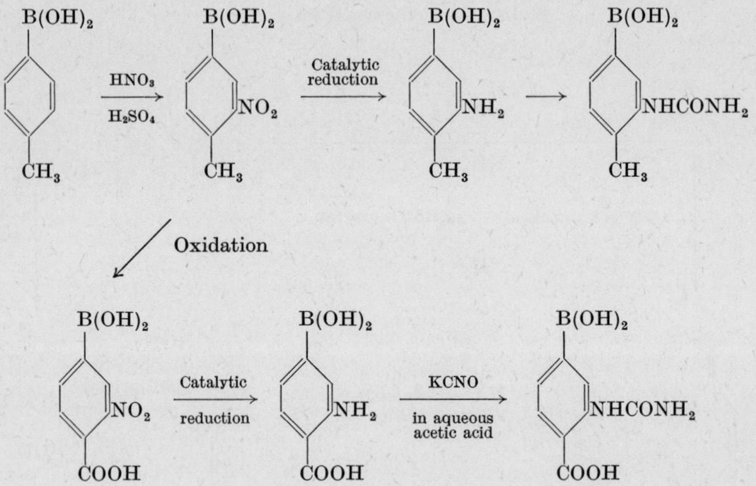

$B(OH)_2$ / CH_3 — $\xrightarrow[H_2SO_4]{HNO_3}$ — $B(OH)_2$, NO_2 / CH_3 — $\xrightarrow{\text{Catalytic reduction}}$ — $B(OH)_2$, NH_2 / CH_3 — $\longrightarrow$ — $B(OH)_2$, $NHCONH_2$ / CH_3

Oxidation

$B(OH)_2$, NO_2 / $COOH$ — $\xrightarrow[\text{reduction}]{\text{Catalytic}}$ — $B(OH)_2$, NH_2 / $COOH$ — $\xrightarrow[\substack{\text{in aqueous}\\\text{acetic acid}}]{KCNO}$ — $B(OH)_2$, $NHCONH_2$ / $COOH$

3-Ureido-4-carboxyphenylboronic acid

$B(OH)_2$ —COOH — $\xrightarrow[H_2SO_4]{HNO_3}$ — $B(OH)_2$, O_2N — COOH

$B(OH)_2$, NO_2 — $\longrightarrow$ — $B(OH)_2$, NH_2 — $\xrightarrow{}$ — $B(OH)_2$, NHCO

with $\overset{O}{\underset{Cl}{C}}$ (pyridine-carbonyl chloride)

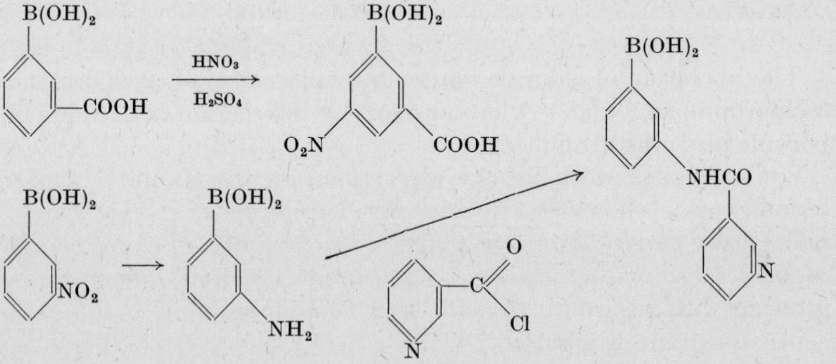

Gilman *et al.* (1958) have studied the preparation of arylboronic acids by means of mercury, zinc, and cadmium aryls; and they have also examined some C—B cleavage reactions with phenylboronic acid and with the anhydride of *o*-hydroxyphenylboronic acid. The C—B bond in phenylboronic acid was cleaved with lithium *n*-butyl, and in the *o*-hydroxy-compound, with bromine (under certain conditions) to give 2,4,6-tribromophenol, with hydrogen peroxide to give catechol, and with acetyl chloride in pyridine, to give phenol. Related work with hydroxyphenylboronic acid and anhydride has also been reported by Gilman *et al.* (1957). *o*-Hydroxyphenylboronic anhydride was prepared by the following reaction scheme,

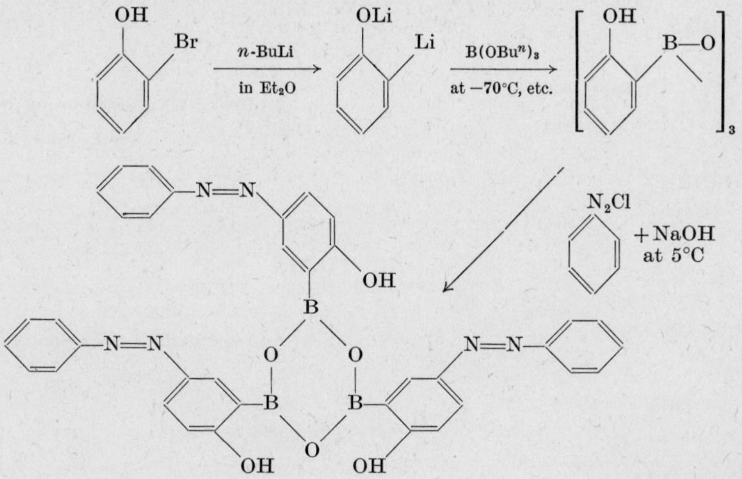

OH — Br → *n*-BuLi in Et$_2$O → OLi — Li → B(OBu^n)$_3$ at −70°C, etc. → [OH — B—O]$_3$

N$_2$Cl + NaOH at 5°C

and by interaction with benzene diazonium chloride afforded the azo-compound shown.

It is now clear that an unlimited field of investigation in organic chemistry is before us. All the reactions which have constituted the study of organic chemistry may now be re-examined for systems containing the boron atom. To what extent reactions may be applied to boron systems will be determined firstly by the influence the boron atom has on the electronic requirements of the reactions, according to the environment of the boron atom, and secondly, by the fission susceptibility of boron–carbon, boron–hydrogen, boron–oxygen, boron–nitrogen, and boron–halogen bonds under conditions required for the particular reaction.

Snyder *et al.* (1958) have given some interesting examples of this development. By bromination with *N*-bromosuccinimide (N.Br.S.) under suitable conditions (in the presence of benzoyl peroxide and carbon tetrachloride) *p*-tolueneboronic acid was converted into ω-bromo-*p*-tolueneboronic acid, and its anhydride. Even the ω,ω-dibromo-compounds could be obtained by using a larger proportion of N.Br.S. The scheme shows the reactions which were carried out on the organic functional group.

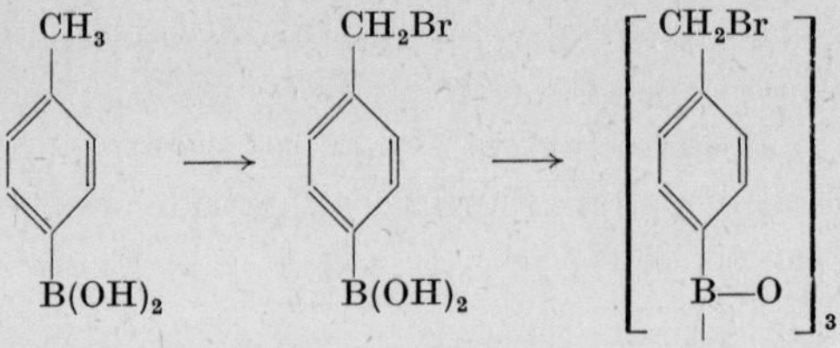

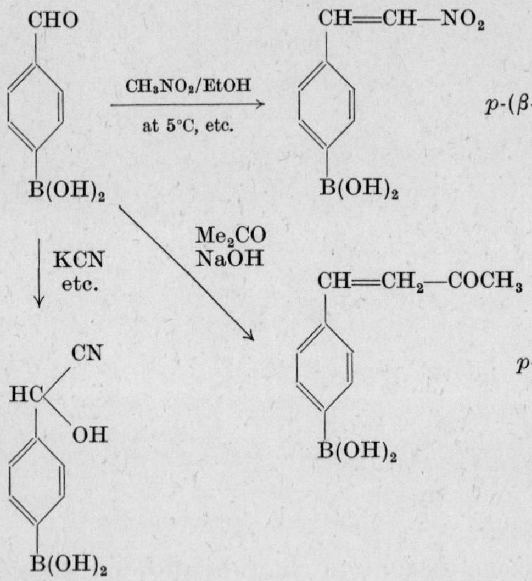

CH$_3$ → CH$_2$Br → [CH$_2$Br]

B(OH)$_2$ B(OH)$_2$ [B—O]$_3$

CHO

$\xrightarrow[\text{at 5°C, etc.}]{\text{CH}_3\text{NO}_2/\text{EtOH}}$ CH=CH—NO$_2$ p-(β-nitrovinyl)-phenylboronic acid

B(OH)$_2$ B(OH)$_2$

| KCN etc.

Me$_2$CO NaOH

CH=CH$_2$—COCH$_3$ p-(β-acetovinyl)-phenylboronic acid

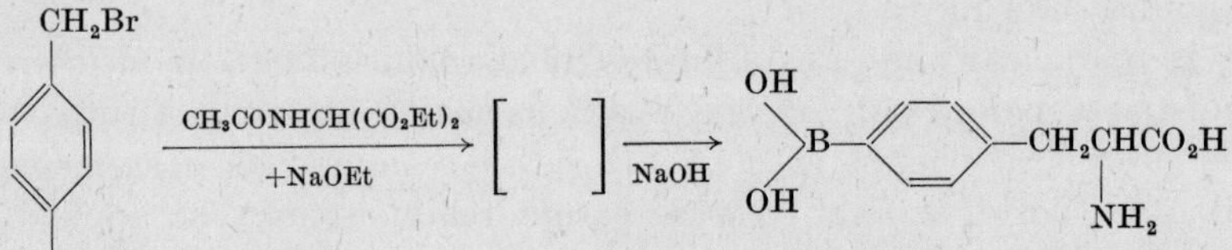

B(OH)$_2$

CN HC< OH

B(OH)$_2$

CH$_2$Br

$\xrightarrow[\text{+NaOEt}]{\text{CH}_3\text{CONHCH(CO}_2\text{Et)}_2}$ [ ] $\xrightarrow{\text{NaOH}}$ OH \B—〈 〉—CH$_2$CHCO$_2$H / OH NH$_2$

B(OH)$_2$ p-(2-carboxy-2-aminoethyl)-phenylboronic acid

Certain reactions with the corresponding o-bromomethyl compound were described, one of special interest being formulated.

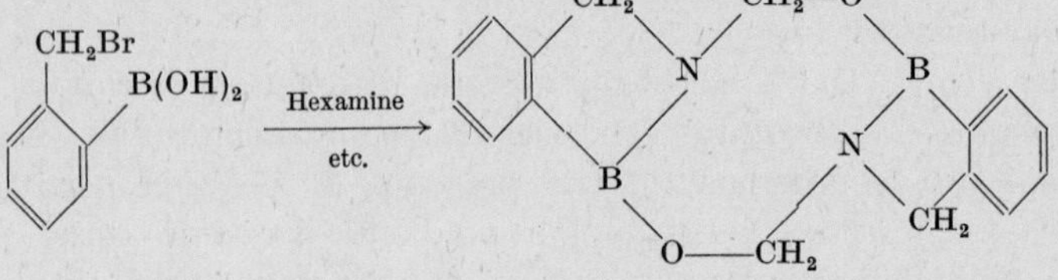

CH$_2$Br B(OH)$_2$

$\xrightarrow[\text{etc.}]{\text{Hexamine}}$ CH$_2$ CH$_2$—O N B B N O—CH$_2$ CH$_2$

In a related paper points about complex formation between phenylboronic anhydride and certain bases such as pyridine, morpholine, and acridine were discussed.

It is not out of place to mention the preparation of mesitylene-boronic acid and its anhydride by Hawthorne (1958) as an example of the application of standard technique.

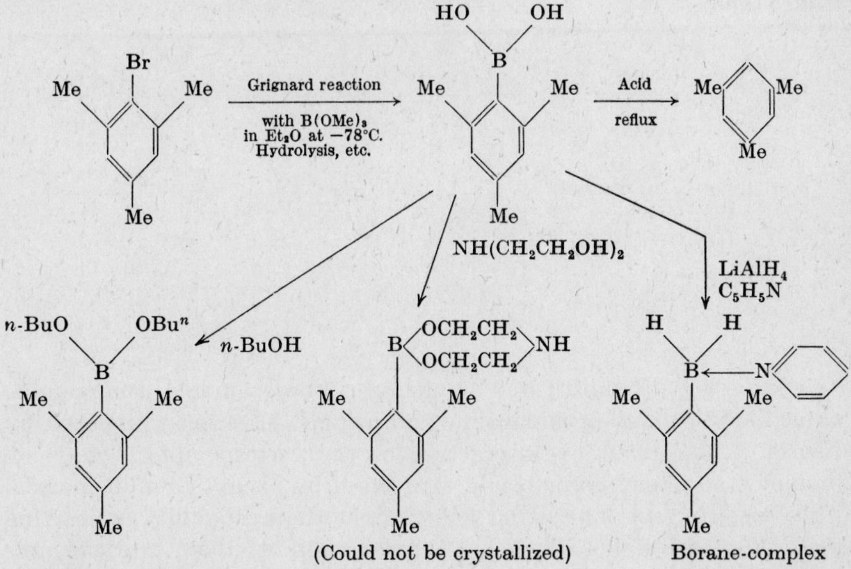

Grignard reaction
with B(OMe)$_3$
in Et$_2$O at $-78°$C.
Hydrolysis, etc.

Acid
reflux

NH(CH$_2$CH$_2$OH)$_2$

LiAlH$_4$
C$_5$H$_5$N

n-BuOH

(Could not be crystallized)   Borane-complex

Another example of an alkyl side-chain reaction is afforded by Hoffmann *et al.* (1959), who apply the N-bromosuccinimide technique to convert p-ethylphenylboronic anhydride into the α-bromoethyl-phenyl compound, and from this the interesting p-vinylphenylboronic acid and esters were obtained. Letsinger *et al.* (1959) have opened a

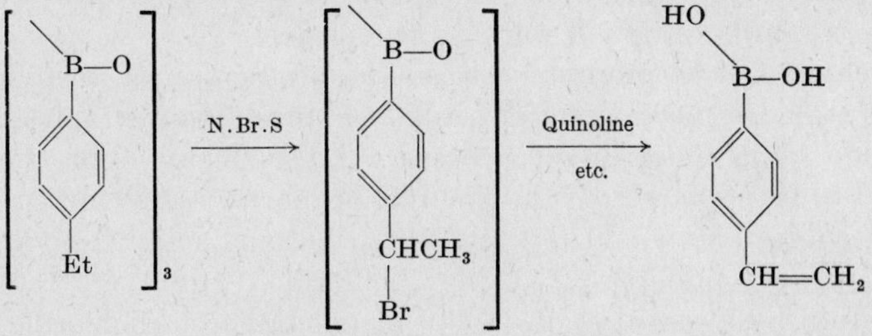

N.Br.S

Quinoline
etc.

new approach to the polymerization of styrene by the production of so-called popcorn polymers, using the vinylphenylboronic acids,

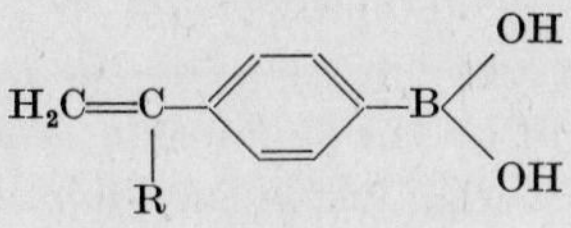

where R=H$_1$ or CH$_3$.

Further examples of the interaction of phenylboronic acid with vicinal hydroxyl structures, and the possible applicability of such esters are afforded by the work of Wolfrom *et al.* (1956) and Sugihara *et al.* (1958).

```
      CHO                         CHO                    CH₂O
       |                           |                        \
  H────C──OH                  H────C──O                      B──Ph
       |                           |    \                 CHO/
 HO────C──H                   H────C──O   B──Ph            |
       |           PhB(OH)₂        |    /                 CHO
 HO────C──H        ─────────►  O───C──H                    \
       |                   Ph─B      \                       B──Ph
     CH₂OH                      \  O──CH₂               CHO/
                                                          |
   l-Arabinose                                          CHO
                                                          \
                                                           B──Ph
                                                    CH₂O/

                                             d-Mannitol tris-
                                             (phenylboronate)
```

A number of substituted 2-phenyl-2,3-borazaindole compounds, named as 2-phenylborimidazoline compounds, have been prepared by Nyilas *et al.* (1959) by interaction of the corresponding phenylene diamine and phenylboronic acid, instead of by phenylboron dichloride. This raises a very interesting point about the availability of starting material. In Ruigh's research (W.A.D.C. report, 1956) into the production of boron polymers based on the borazole structure (p. 182), he realized the importance of developing a direct synthesis of phenylboron dichloride which could be operated commercially. He and his co-workers spent much time in attempts to bring Pace's (1929) method, the catalysed interaction of boron trichloride and benzene, into technically satisfactory operation. The situation was rather uncertain when Ruigh's team discontinued work on the project.

The choice lies at present between the production of phenylboronic acid by the Grignard reaction, and the production of phenylboron dichloride by a more direct process. If the phenylboron dichloride still had to be prepared *via* the phenylboronic, as was so until recently, then of course one would try to use the boronic acid, as Nyilas *et al.* have done, in preference to the dichloride. However, the recent preparation by Gerrard *et al.* (1960) of phenylboron dichloride by the interaction of the commercially available tetraphenyltin and boron trichloride (p. 207), makes the dichloride the more economic starting reagent. On the other hand, modification in structure obviously may be achieved by ringing the changes in the aryl group attached to boron in the boronic acid or the dichloride, and in development work at present, the arylboron dichloride would probably come only from the arylboronic acid.

PhMgBr
+B(OR)₃

| Followed
↓ by hydrolysis

PhB(OH)₂

| −H₂O
↓

(PhBO)₃

| BCl₃ or PCl₅
↓

PhBCl₂

Ph₄Sn

|
| BCl₃
| Direct to
| dichloride
|
↓

PhBCl₂

Substituted phenyl

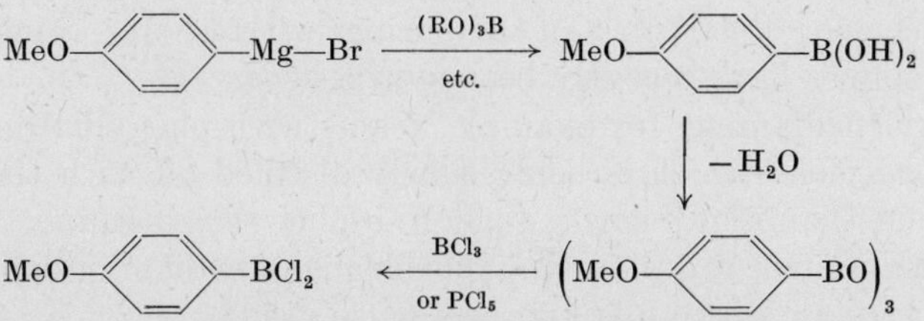

$$MeO\!-\!\langle\ \rangle\!-\!Mg\!-\!Br \xrightarrow[\text{etc.}]{(RO)_3B} MeO\!-\!\langle\ \rangle\!-\!B(OH)_2$$

$$\downarrow -H_2O$$

$$MeO\!-\!\langle\ \rangle\!-\!BCl_2 \xleftarrow[\text{or PCl}_5]{BCl_3} \left(MeO\!-\!\langle\ \rangle\!-\!BO\right)_3$$

Such substituted phenylboron reagents may be expected to react as follows and give a variety of compounds with variations in the group having a carbon atom attached to boron, superimposed on variations in the other part of the molecule.

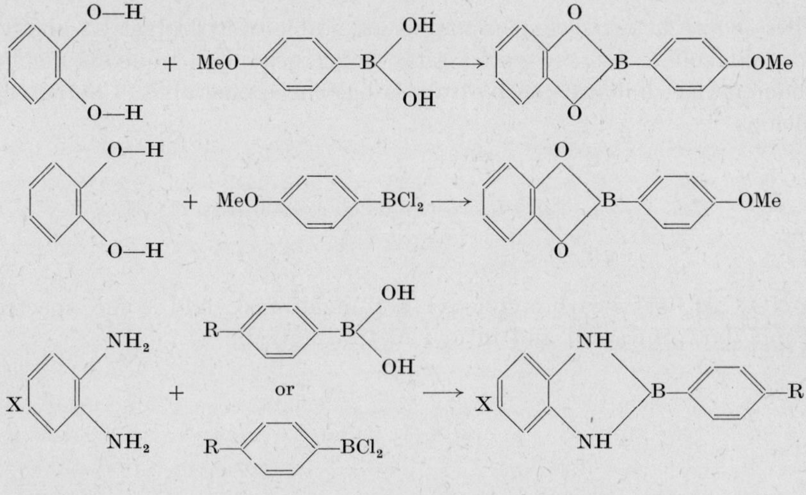

$$\text{(catechol)} \begin{matrix} O\!-\!H \\ O\!-\!H \end{matrix} + MeO\!-\!\langle\ \rangle\!-\!B\!\!\begin{matrix} OH \\ OH \end{matrix} \longrightarrow \begin{matrix} O \\ O \end{matrix}\!\!B\!-\!\langle\ \rangle\!-\!OMe$$

$$\begin{matrix} O\!-\!H \\ O\!-\!H \end{matrix} + MeO\!-\!\langle\ \rangle\!-\!BCl_2 \longrightarrow \begin{matrix} O \\ O \end{matrix}\!\!B\!-\!\langle\ \rangle\!-\!OMe$$

$$X\!\!\begin{matrix} NH_2 \\ NH_2 \end{matrix} + \begin{matrix} R\!-\!\langle\ \rangle\!-\!B\!\!\begin{matrix} OH \\ OH \end{matrix} \\ \text{or} \\ R\!-\!\langle\ \rangle\!-\!BCl_2 \end{matrix} \longrightarrow X\!\!\begin{matrix} NH \\ NH \end{matrix}\!\!B\!-\!\langle\ \rangle\!-\!R$$

The selective destruction of neoplastic cells in the brain by neutron capture therapy is naturally an urgent subject of investigation which has found ready support from the U.S. Atomic Energy Commission,

the American Cancer Society, the National Institute of Arthritis and
Metabolic Diseases, and other such institutes. Although at present
the procedure is not yet adaptable for malignant tumours other than
those of the brain, there should be no lack of incentive to extend
applicable knowledge. It is suggested by Nyilas *et al.* (1959) that the
preparation of boron-containing purine antimetabolite might extend
this treatment to other neoplasms. The antimetabolic properties may
competitively inhibit the growth of certain tumours. Furthermore,
the incorporation of certain molecules containing [10]B into the nucleic
acids of malignant cells would enable the cells to be destroyed by
thermal neutron irradiation. To this end, work has been directed to
the replacement of carbon 8 in the purines with a boron atom. Several
borimidazolines have recently been synthesized.

o-Phenylenediamine, for example, reacts with phenylboronic acid in
toluene or xylene which is being slowly distilled off with the reaction
water. 3,4-Diaminotoluene, 4-chloro-o-phenylenediamine, 4-nitro-o-
phenylenediamine, 2,3-diaminonaphthalene, 3,4-diaminobenzoic acid,
and 4-methoxy-o-phenyldiamine were also used.

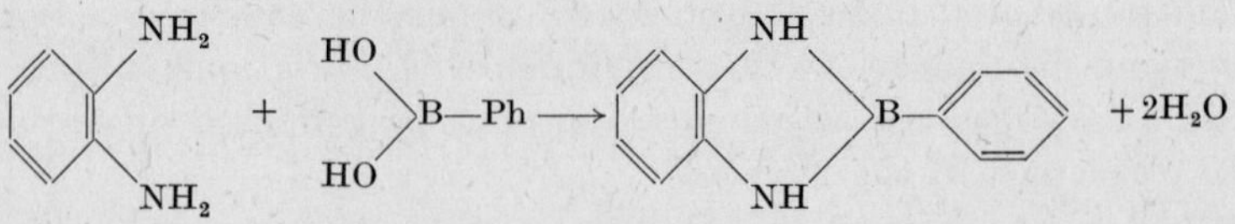

These boron heterocycle compounds are stable to hydrolysis; 2-phenyl-
borimidazoline was recovered after being heated in aqueous alcohol
under reflux; but in time hydrolysis becomes detectable. Compounds
such as

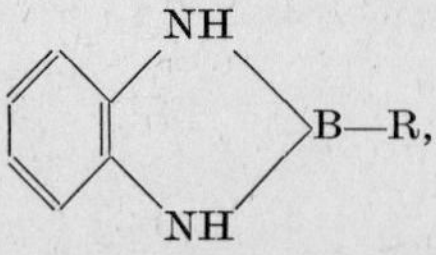

m-$NO_2C_6H_4$ or p-carboxyphenyl are mentioned, and some spectroscopic data, infra-red and ultra-violet, are given.

Chapter XV

OTHER BORON–HALIDE SYSTEMS

I. Introduction

THROUGHOUT this monograph fluorine compounds of boron have been referred to. For example, boron trifluoride has been a frequently used reagent for the attachment of boron to carbon, and for the preparation of diborane. As a Lewis acid, it has been extensively studied for its co-ordinating power with different bases such as amines, $RR'R''N{\rightarrow}BF_3$ (where R, etc., can be alkyl, aryl or hydrogen), ethers, $RR'O{\rightarrow}BF_3$, and so on. In this book there is little point in attempting a detailed scrutiny of this field of chemistry, because it has been well documented particularly with reference to acid-base systems. Evidence accrues, and this is dealt with elsewhere in this book, that the co-ordinating power is in the order $BF_3 < BCl_3 < BBr_3 < BI_3$, as shown recently by Brown et al. (1956) and in an entirely different system by Frazer et al. (1960). This order is contrary to that previously accepted, and is in the order of polarizability of the halogen atom.

A number of reactions mainly involving halogen exchange by means of antimony trihalides were studied by Long et al. (1953) in some detail, and emerging out of this work are a number of points which are rather difficult to summarize briefly. It is stated that the di-n-propylboron chloride ignites spontaneously in air, whereas the bromo-compound merely fumes, and the iodo-compound is rapidly oxidized, copiously liberating iodine. The relative ease of hydrolysis was in the order $Pr^n_2BCl > Pr^n_2BBr > Pr^n_2BI$, which raises an interesting point about the correlation of ease of hydrolysis and the electrophilic function of boron.

II. Reactions with Boron Trifluoride

Boron trifluoride has been extensively used as a catalyst in organic chemistry, and every week fresh examples come to light. Again, all this work has been well documented (see Topchiev et al. (1959), Booth et al. (1949)) and might be said to cover a vast area of organic chemistry. Obviously this aspect of the function of boron trifluoride must be passed over by mere mention in the present work. It is a different story when consideration is given to the boron trifluoride–alcohol systems, and it is plain that there are many distinguishing features between the boron trifluoride–alcohol systems and the boron trichloride–alcohol systems already described.

Gasselin (1894) claimed to have prepared methyl fluoro-esters by the interaction of boron trifluoride and sodium methoxide, but Meerwein *et al.* (1934) obtained the co-ordination compound, $BF_3.NaOMe$. Gasselin's second procedure, that of the interaction of boron trifluoride and trimethyl borate, did give a fluoro-ester, a result later confirmed presumably by Allen *et al.* (1932), and certainly by Goubeau *et al.* (1951). This system evidently entails a four-centre mutual replacement reaction.

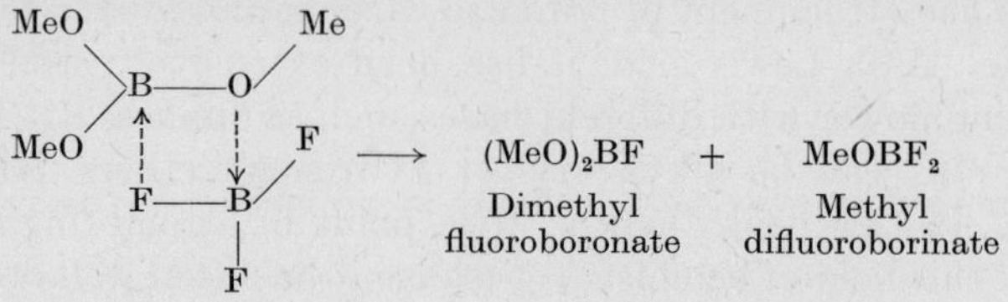

$$\longrightarrow \quad (MeO)_2BF \quad + \quad MeOBF_2$$

Dimethyl fluoroboronate   Methyl difluoroborinate

The parachor (Allen *et al.*, 1932) indicated the dimeric form of methyl difluoroborinate, and in this they were supported by Goubeau *et al.* (1951), who studied the Raman-spectra. Dimerization and trimerization is always a possibility with trivalent compounds of boron which contain an atom with lone pair electrons, and a boron atom which is sufficiently electrophilic. In the present example the structure is probably as shown.

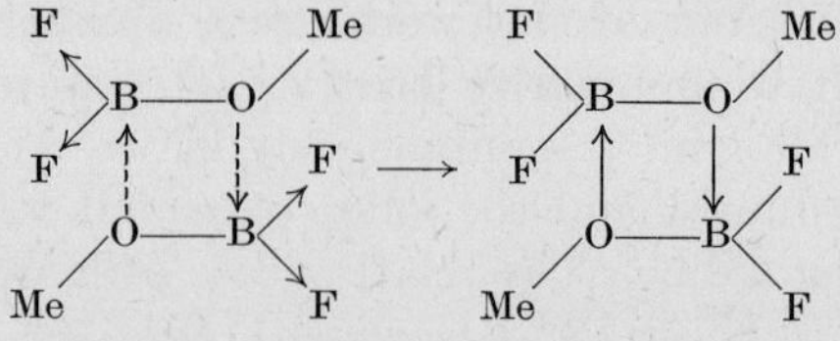

Goubeau *et al.* (1951) showed that dimethyl fluoroboronate tended to undergo mutual replacement of partners and afford trimethyl borate and methyl difluoroborinate.

$$\longrightarrow \quad MeOBF_2 \quad + \quad (MeO)_3B$$

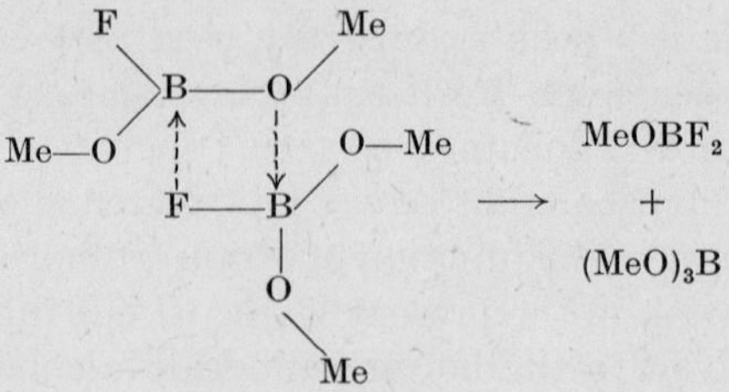

Vibration spectra of the methylboron fluorides have been discussed by Becher (1957), and the microwave spectrum of methylboron difluoride by Naylor *et al.* (1957).

Meerwein *et al.* (1939) believed that 1 : 1 and 2 : 1 boron trifluoride–trimethyl borate complexes were formed; but these are unstable and undergo mutual replacement reaction as shown. We have here an example of a point of general controversy; must one assume that in all examples of mutual replacement reactions the formation of a complex occurs before the actual exchange of partners? It is really a question of one or two transition states. If it be assumed that the complex forms as a chemical entity, potentially (at least) isolable, then one transition state has been involved. The compound may now react further in either or both of two ways. It may dissociate into the original compounds, and this process will involve a transition state; or it may undergo "irreversible decomposition", in the case in mind, mutual replacement involving a separate transition state.

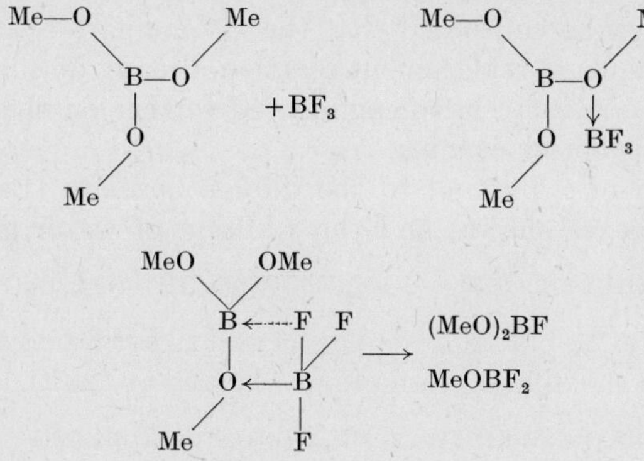

Cook *et al.* (1950) found the same trouble with mutual replacement in their attempts to prepare di-*n*-butyl fluoroboronate by the interaction of the boron trifluoride–diethyl ether complex, BF_3Et_2O, and tri-*n*-butyl borate.

In addition to the ether complexes, which are referred to in the review of Greenwood *et al.* (1954) (see also discussion by Gerrard *et al.*, 1959, 1960), boron trifluoride forms complexes with alcohols, $BF_3 . ROH$; $BF_3 . 2ROH$; which have been considered in the same review (1954). According to Greenwood *et al.* (1953), the 1:2 complexes, $BF_3 . 2ROH$, (R=Me or Pr^n) are probably of the ionic form $(ROH_2)^+$ $(ROBF_3)^-$, and this supports an earlier observation by Meerwein (1933) that the addition of boron trifluoride–di-*n*-butanol to nitrobenzene causes a marked increase in conductivity.

Meerwein *et al.* (1934) refer to the decomposition of boron trifluoride complexes of secondary alcohols, $BF_3 . 2ROH$, to give the corresponding olefin and hydroxonium fluorohydroxyborate, which shows the interplay amongst ions of a type not easily found in the boron trichloride systems.

During the progress of work by Gerrard *et al.* (1951, 1952) it became desirable to look further into the alcohol–boron trifluoride systems, and results for *n*-butanol are described by Lappert (1955).

At 0°C, 1 mole of tri-*n*-butyl borate readily absorbed 2 moles of boron trifluoride and gave *n*-butyl difluoroborinate. Di-*n*-butyl fluoroboronate, $(Bu^nO)_2BF$, could not be isolated, even when the correct

$$(BuO)_3B + 2BF_3 \longrightarrow 2BuOBF_2$$

stoicheiometric amounts of reagents were used, and the tri-*n*-butyl borate was mixed with *n*-butyl difluoroborinate. There appears to be no reasonable mechanism by which the difluoroborinate could be formed without the concurrent formation of the fluoroboronate, and therefore it must be concluded that the latter fluoro-ester very easily undergoes the mutual replacement reaction already discussed.

McCusker *et al.* (1960) have commented further on the constitution of alkoxydifluoroboron systems.

The addition of *n*-butanol to the difluoroborinate (Lappert, 1955) gave the complexes shown, and the addition of water gave the 1 : 2

$$3BuOH + 3BuOBF_2 \longrightarrow BF_3(BuO)_3B,BuOH + BF_3,2BuOH$$

boron trifluoride–2*n*-butanol complex as well as boric acid and hydrogen fluoride. When the difluoroborinate was heated at 120°C, redistribution

$$3H_2O + 2BuOBF_2 \longrightarrow BF_3,2Bu^nOH + (HO)_3B + HF$$

occurred giving boron trifluoride and tri-*n*-butyl borate.

$$3Bu^nOBF_2 \longrightarrow 2BF_3 + (Bu^nO)_3B$$

This reaction alone shows the essential difference between the boron trichloride and the boron trifluoride system, for no alkyl fluoride was formed. The addition of pyridine caused the formation of tri-*n*-butyl borate and the pyridine boron trifluoride complex.

$$3Bu^nOBF_2 + 2C_5H_5N \longrightarrow 2C_5H_5NBF_3 + (Bu^nO)_3B$$

Sufficient has been stated to show the complicated nature of the boron trifluoride–alcohol systems, and nothing short of the most detailed scrutiny, assisted by as many independent facts as can possibly be obtained from physico-chemical techniques, will be worthwhile in further work in this field of chemistry.

III. Reactions with Diboron Tetrachloride

Diboron tetrachloride was first prepared by Stock *et al.* (1925) by striking an arc across zinc electrodes in liquid boron trichloride; but the yields of pure material were less than 1%. Wartik *et al.* (1949) made very considerable improvements in the method of production, and although in the following years Schlesinger's school tried to effect still further desirable modifications, or find alternative methods, it had to be admitted (Ceron *et al.*, 1959) that apart from a certain degree of automation which was found possible, the method remained that described by Urry *et al.* in 1954. Gaseous boron trichloride at 1 to 2 mm is passed through a glow discharge between mercury electrodes. Diboron tetrachloride, B_2Cl_4, is trapped at $-78°C$, and unchanged boron trichloride is recyclized through the generator. The overall yield is 50%, and 5 to 10 grams a week were produced with little attention to the apparatus described.

It is unfortunate that production remains so slow; for there is no doubt that diboron tetrachloride is a compound having great and interesting potentialities; but such difficulties have been overcome in other examples, and one can hope for improvement if the demand is great enough.

Holliday *et al.* (1958, 1960) have effected some improvement by use of a direct current discharge, but an equally effective procedure is to use an alternating-current arc with a high current, a triple-jet diffusion pump to circulate the boron trichloride vapour from the trap at $-78°C$ to the water-cooled quartz H-shaped generating cell.

Diboron tetrachloride is a liquid having a vapour pressure of 44–46 mm at 0°C. It gradually decomposes, giving boron trichloride, a white solid, and a very slightly volatile red substance. With boron tribromide it gives diboron tetrabromide, but does not react with boron trifluoride.

$$3B_2Cl_4 + 4BBr_3 \longrightarrow 3B_2Br_4 + 4BCl_3$$

It reacts with hydrogen quickly, giving diborane, and it ignites in air. With lithium borohydride, tetraborane, B_4H_{10}, as well as diborane and lithium chloride are produced.

$$3B_2Cl_4 + 3H_2 \longrightarrow 4BCl_3 + B_2H_6$$

Although these points relate to the inorganic chemistry of boron, they may have future correlations with organic reactions.

Urry *et al.* (1954) showed that diboron tetrachloride reacts with ethylene at low temperature ($-80°C$) to give what they termed 1,2-bis-dichloroborylethane.

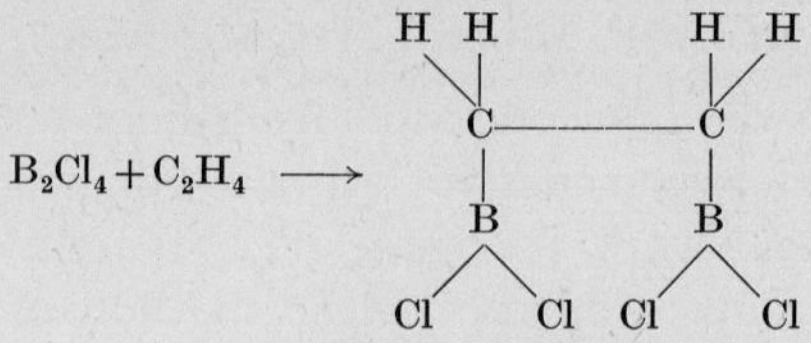

$$B_2Cl_4 + C_2H_4 \longrightarrow$$

m.p. $-28°C$
b.p. $142°C$

1,2-bis-dichloroborylethane
or diboron tetrachloride–ethylene
or dimethylene–bis-boron dichloride

The structure assigned by Urry *et al.* was later confirmed by Moore *et al.* (1956), and for vibrational spectrum and crystal structure, see Mann *et al.* (1957), Atoji *et al.* (1957) and Jacobson *et al.* (1958). In a more recent report (Ceron *et al.*, 1959) the extensions of this reaction to other olefins are described, and fascinating possibilities are revealed. Propene, and butene-2 react similarly, butadiene reacts with one or two molecules, but acetylene reacts with only one, because of the deactivating effect of boron attached to carbon.

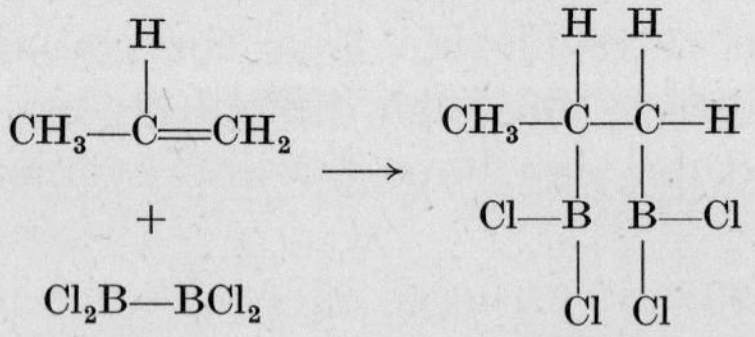

m.p. $-130°C$, b.p. $165°C$

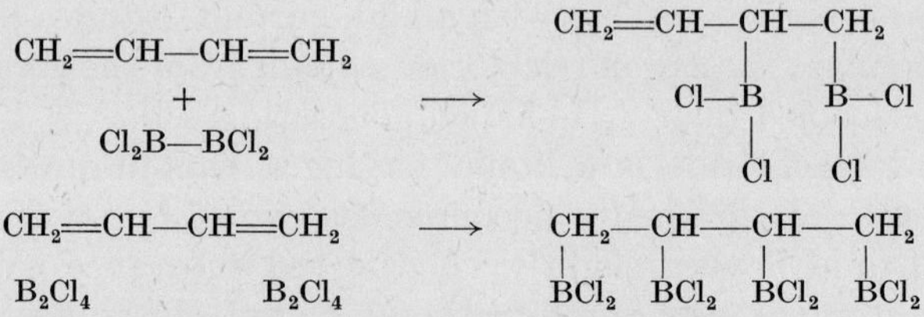

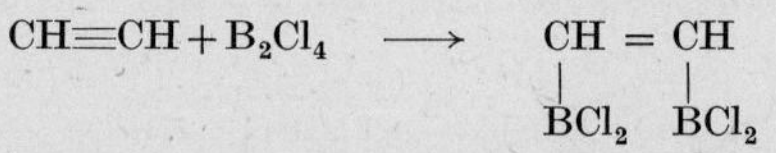

b.p. $144°C$

Allyl chloride, and 4-chlorobut-1-ene react, but vinyl halides and similar compounds do not, even at temperatures above $25°C$.

A very interesting extension of this procedure is the replacement of the chlorine atoms in the bis-compound by means of dimethylzinc,

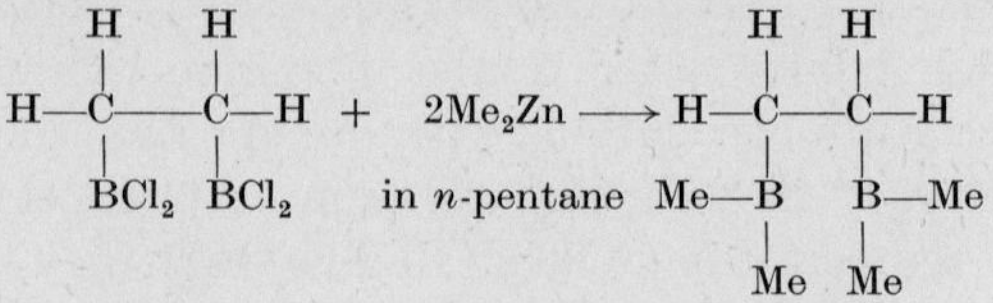

b.p. $98°C$

and by redistribution reactions the following structures appear to be amongst those formed.

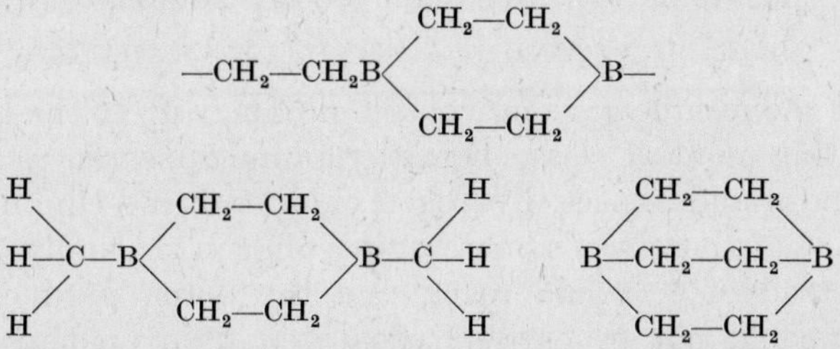

$$-CH_2-CH_2B{\overset{\displaystyle CH_2-CH_2}{\underset{\displaystyle CH_2-CH_2}{<\quad>}}}B-$$

$$\overset{H}{\underset{H}{\overset{\displaystyle |}{H-C}}}-B{\overset{\displaystyle CH_2-CH_2}{\underset{\displaystyle CH_2-CH_2}{<\quad>}}}B-\overset{H}{\underset{H}{\overset{\displaystyle |}{C-H}}} \qquad B{\overset{\displaystyle CH_2-CH_2}{\underset{\displaystyle CH_2-CH_2}{<\quad>}}}CH_2-CH_2-B$$

An amine-complex referred to as a tetramer is formed by the inter-action of diboron tetrachloride and trimethylamine, and has remark-able inactivity, yielding no hydrogen chloride when treated with ethanol. With dimethyl ether, a liquid monoetherate, $B_2Cl_4 . Me_2O$, and

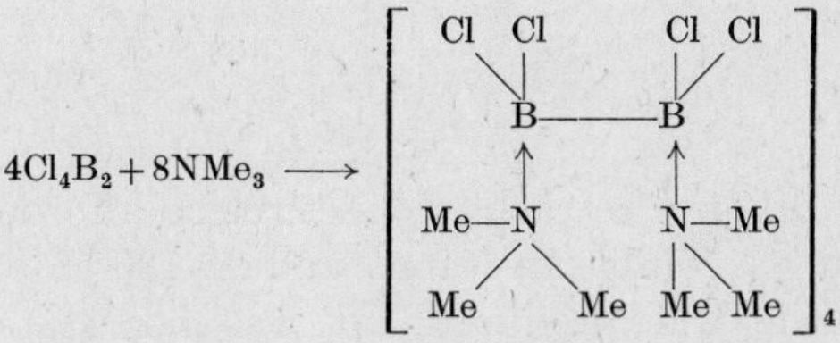

$$4Cl_4B_2 + 8NMe_3 \longrightarrow \left[\begin{array}{c} \overset{Cl}{\diagdown}\ \overset{Cl}{\diagup} \qquad \overset{Cl}{\diagdown}\ \overset{Cl}{\diagup} \\ B \text{------} B \\ \uparrow \qquad\quad \uparrow \\ Me-N \qquad N-Me \\ \diagup\ \diagdown \qquad |\ | \\ Me \quad Me \quad Me\ Me \end{array} \right]_4$$

Softens at 195°C, melts without
appreciable decomposition at 228°C

a solid dietherate are formed; and similar remarks may be made about interaction with diethyl ether. Methanol and ethanol gave the tetra-alkoxydiboron compound, described as being unstable (see also Wiberg *et al.*, 1937), and dimethylamine gave the tetra-amino com-pound, $(Me_2N)_4B_2$.

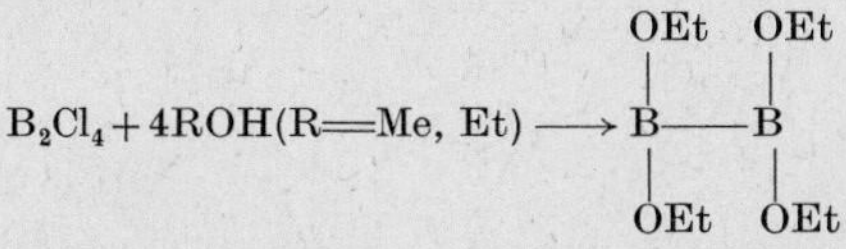

$$B_2Cl_4 + 4ROH(R\!=\!\!=\!\!Me, Et) \longrightarrow \overset{OEt \quad OEt}{\underset{OEt \quad OEt}{\overset{|\qquad |}{\underset{|\qquad |}{B\text{------}B}}}}$$

Holliday *et al.* (1960) have shown that diboron tetrachloride–ethylene reacts with trimethylamine to give a white solid which is thermally stable and not oxidized or hydrolysed at room temperature. Hydrogen chloride reacted with the complex, and gave a solid product which appeared to have the structure shown. A remarkably thermally stable

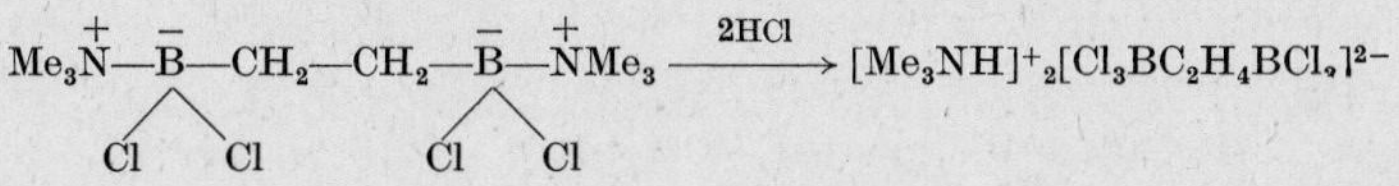

$$\overset{+}{Me_3N}-\overset{-}{\underset{\underset{Cl}{\diagup}\ \underset{Cl}{\diagdown}}{B}}-CH_2-CH_2-\overset{-}{\underset{\underset{Cl}{\diagup}\ \underset{Cl}{\diagdown}}{B}}-\overset{+}{NMe_3} \xrightarrow{2HCl} [Me_3NH]^+{}_2[Cl_3BC_2H_4BCl_3]^{2-}$$

8

compound formulated as shown was produced by the interaction of ammonia or methylamine (R=Me or H) and diboron tetrachloride–ethylene. The amine salt was removed by sublimation, and it was

$$xB_2Cl_4C_2H_4 + 6xRNH_2 \longrightarrow 4xRNH_3Cl + [B_2(NR)_2C_2H_4]_x$$

found that the residue gave no volatile products up to the temperature of the melting point of glass. Dimethylamine appeared to react in the main as one would expect, hydrogen chloride being eliminated as the amine salt; but there was something peculiar in that only 7.5 moles of amine, instead of 8, would react, and the boron product, although distillable, could not be entirely separated from unchanged diboron tetrachloride–ethylene.

$$B_2Cl_4C_2H_4 + 8Me_2NH \longrightarrow 4Me_2NH_2Cl + B_2(NMe_2)_4C_2H_4$$

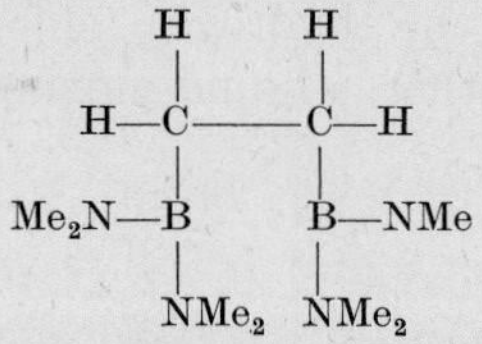

Hydrolysis of diboron tetrachloride–ethylene afforded a white solid which was sparingly soluble in water, and at 130°C, by loss of water, a glassy anhydride was produced. It was not stated whether the anhydride is considered as an internal one, or one resulting from condensation polymerization. It could be heated to 300°C without decomposition. Furthermore, the anhydride absorbed water to form the original acid. It is obvious that an interesting subject for study is hereby presented.

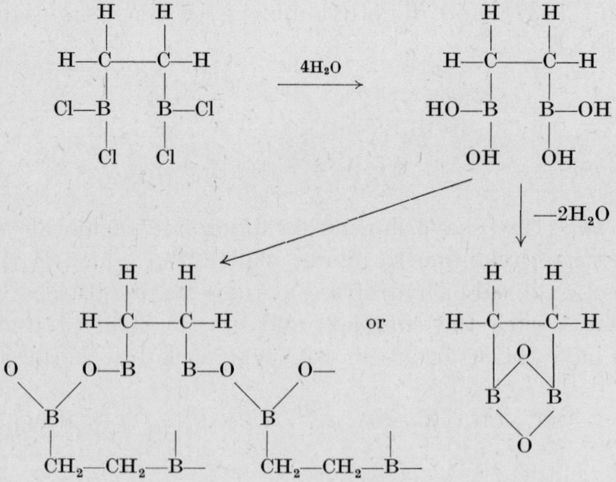

The oxidative fission of carbon and boron in the diboron tetrachloride–ethylene was investigated and discussed, and formation of intermediates such as that formulated here was postulated.

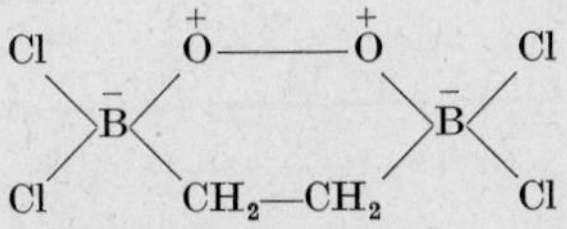

$$Cl \diagdown \overset{Cl}{\underset{Cl}{\diagup}}B \diagdown \overset{\overset{+}{O} - - - \overset{+}{O}}{\underset{CH_2 - CH_2}{}} \diagdown B \diagup \overset{Cl}{\underset{Cl}{}}$$

It would appear from this account that much more work is likely to be done on the diboron tetrachloride system.

IV. Reactions with Diboron Tetrafluoride

Diboron tetrafluoride was included in the programme of work by Urry *et al.* (1954), Ceron *et al.* (1959), and Finch *et al.* (1958). It appears that the electric discharge method is not suitable for the conversion of boron trifluoride into diboron tetrafluoride. A method used was the interaction of antimony trifluoride with diboron tetrachloride.

$$3B_2Cl_4 + 4SbF_3 \longrightarrow 3B_2F_4 + 4SbCl_3$$

It forms a stable tetramer with trimethyl amine, $[B_2F_4 . 2NMe_3]_4$, and reacts with olefins, but more slowly than diboron tetrachloride does, and the products are more stable. Furthermore, diboron tetrachloride–ethylene is converted into the tetrafluoride compound by interaction with antimony trifluoride.

$$Cl_2BCH_2CH_2BCl_2 \xrightarrow{SbF_3} F_2BCH_2CH_2BF_2 + SbCl_3$$

Allyl fluoride reacts with diboron tetrachloride to give allyl chloride and diboron tetrafluoride, and the tetrafluoride catalyses the polymerization of allyl compounds.

$$4C_3H_5F + B_2Cl_4 \longrightarrow 4C_3H_5Cl + B_2F_4$$

V. Reactions with Dichloroborane

Dichloroborane, $HBCl_2$ (Lynds *et al.*, 1959), is prepared by passing a mixture of hydrogen and boron trichloride over granules of magnesium at 400–450°C. Unreacted boron trichloride is condensed at -125°C, and dichloroborane at -135°C. At 25°C it undergoes redistribution to boron trichloride and diborane. It reacts energetically with olefins and alkynes at 10–30°C, without the aid of solvents, and the addition is by the Markownikoff rule.

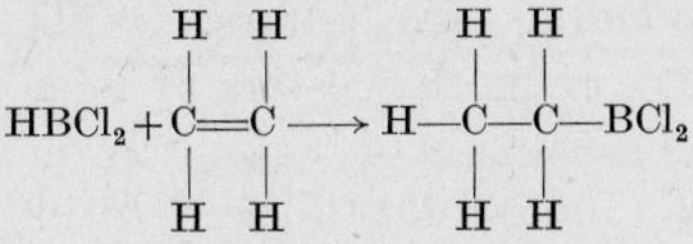

$$HBCl_2 + \overset{\overset{H}{|}}{C} = \overset{\overset{H}{|}}{\underset{\underset{H}{|}}{C}} \longrightarrow H - \overset{\overset{H}{|}}{\underset{\underset{H}{|}}{C}} - \overset{\overset{H}{|}}{\underset{\underset{H}{|}}{C}} - BCl_2$$

Propene gives i-propylboron dichloride, Pr^iBCl_2, isobutylene gives t-butylboron dichloride, Bu^tBCl_2, and $cyclo$hexene gives $C_6H_{11}BCl_2$. Butadiene reacted energetically, giving bis(dichloroborano)butane, b.p. 127–130°C/110 mm.

Acetylene reacted as shown in the equation.

$$2HBCl_2 + H - C \equiv C - H \longrightarrow Cl_2B\!-\!CH_2\!-\!CH_2\!-\!BCl_2$$
$$\text{b.p. } 39°C/35 \text{ mm}$$

Onak et $al.$ (1958) have reported on the formation and certain properties of the etherates of dichloroborane and deuteriodichloroborane.

VI. Reactions with Boron Tribromide

References have already been made to reactions involving boron tribromide and it remains merely to mention alcohol and ether systems, and to emphasize that boron tribromide has more functional resemblance to boron trichloride than the latter has to boron trifluoride. Boron tribromide has no special virtues as an economic reagent, except perhaps because of its lower volatility, which facilitates the maintenance of higher reaction temperatures at atmospheric pressure. It has greater reactivity due to the greater polarizability of the boron–bromine bond, and due also to the greater nucleophilic power of the bromide ion.

Boron tribromide (3 moles) and alcohols of ordinary reactivity immediately afford the alkyl borate and hydrogen bromide, whereas t-butanol, typical of an alcohol of much greater reactivity of the central carbon atom gives t-butyl bromide and boric oxide (Bujwid et $al.$, 1957).

$$3Bu^nOH + BBr_3 \longrightarrow (Bu^nO)_3B + 3HBr$$
$$6t\text{-}BuOH + 2BBr_3 \longrightarrow 6t\text{-}BuBr + B_2O_3$$

The greater nucleophilic power of the bromide ion is seen in the formation of some s-butyl bromide as well as s-butyl borate when s-butanol was used. By mixing the reagents at $-80°C$, and especially by the interaction of the trialkyl borate and boron tribromide, evidence of the formation of the n-butyl dibromoborinate, $ROBBr_2$, was obtained, and the pyridine complex of this compound was prepared as an unstable substance. Evidence of the formation of the bromoboronate, $(BuO)_2BBr$, was also discerned. The mechanisms of decomposition of these bromo-esters show strong analogy with those relating to the corresponding chloroesters. Goubeau et $al.$ (1955) have referred to the stability of methyl esters of mono- and dibromo-boric acid, and consider the interaction of boron tribromide with trimethoxyboroxine. Jander et $al.$ (1955) refer to the titration of boron tribromide with lithium ethoxide in the study of ionic reactions in ether. Schabacher *et al.* (1958) have reported on the preparation and vibration spectra of the methylborobromides.

Ether fission by boron tribromide has been investigated by Benton *et al.* (1942) and Bujwid *et al.* (1960); see also Cueilleron (1943). The tribromide is a more vigorous cleaving reagent than the trichloride, and even dialkyl bromoboronates, in contrast to the dialkyl chloroboronates, will still function in this way. It is suggested that there is evidence of an SN2 mechanism, as well as a concurrent SN1 reaction, the relative extent depending on the particular ether. It is quite likely that the essential factor is the difference in polarizability, which enables a complex to exist as an entity, albeit for a brief period, and therefore to have a chance to decompose by more than one mechanism.

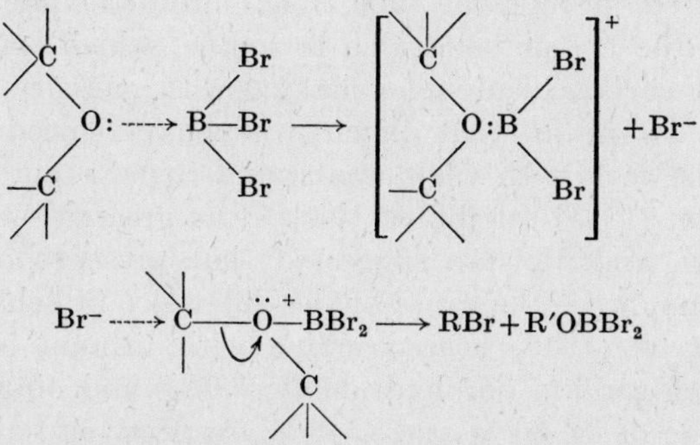

SN2, preferential selection of C with smaller electron release

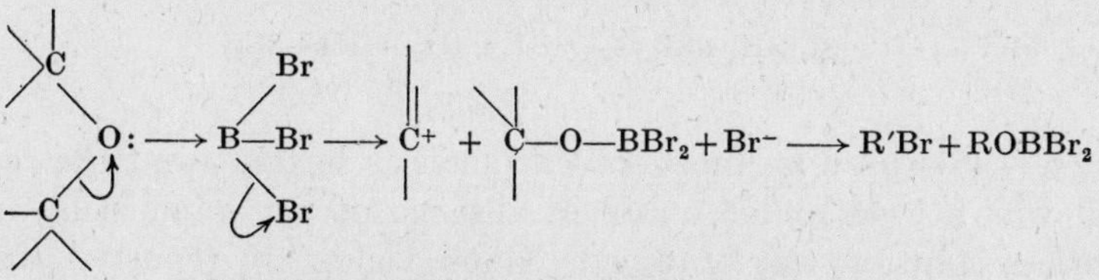

Selection according to greater electron release

No complex was formed, and neither was there fission with diphenyl ether, in accordance with the low electron density on the oxygen atom.

Henry *et al.* (1956) have studied conductometric titrations of boron tribromide with heterocyclic bases such as quinoline, *iso*quinoline and β-picoline in thionyl chloride. Most bases formed a 1 : 1 compound. Schuele *et al.* (1956) refer to the analysis of boron tribromide and its addition compounds, and Musgrave (1956) has reported on the preparation of some dimeric secondary amine (e.g. pyrrolidine, piperidine or morpholine)–boron dihalides with boron tribromide.

VII. Boron–Iodine Systems

Very little work has been reported on the organic reactions of boron triiodide. By passing iodine vapour over a heated mixture of carbon and boric oxide, Inglis (1835) obtained a yellow sublimate which he believed to be boron triiodide. Wohler *et al.* (1858) could not obtain the triiodide this way, nor from boron and silver iodide. The first reliable reports of preparation came from Moissan (1891, 1892, 1895), who obtained it in small yield by heating a mixture of boron trichloride and iodine vapours, by the action of iodine on amorphous boron at high temperatures, and by the action of hydrogen iodide on amorphous boron at red heat. In the last procedure boron was heated to 200°C in a stream of hydrogen, and then heated to a temperature just below the melting-point of glass. A stream of hydrogen iodide was then passed through the tube. The boron triiodide sublimed as a purple solid, the colour being due to iodine, which was removed by dissolution in carbon disulphide, shaking with mercury, filtration and evaporation. It appears that difficulty was experienced in attempting to repeat this work with what was stated to be a purer specimen of boron. Besson (1891) attributed this to the greater "activity" of the impure boron, and Moissan suggested that presence of sodium and iron borides might be the cause of the difference in behaviour.

Schumb *et al.* (1949) heated iodine with lithium borohydride at 125°C, or with sodium borohydride at 200°C, and obtained yields of boron triiodide of 64–66%, and 45–50% respectively, and the overall equation given is as follows:

$$3LiBH_4 + 8I_2 \longrightarrow 3BI_3 + 4H_2 + 4HI + 3LiI$$
(or Na)   (or Na)

Renner (1957) used an inert solvent such as hexane, and the reaction liquid was heated under reflux in absence of air or moisture until a colourless solution was obtained. Good yields are reported, and the following equation was given. Unreacted lithium borohydride (the

$$LiBH_4 + 4I_2 \longrightarrow LiI + BI_3 + 4HI$$

reagent being added in some excess) and lithium iodide were filtered off, and boron triiodide obtained from the solution by evaporation of the solvent.

Boron triiodide is a colourless solid, m.p. 49.9°C and b.p. 210°C. It is, of course, sensitive to moisture, and oxygen (air) liberates iodine, especially in sunlight. It burns in air or oxygen with the liberation of iodine.

The precursor to co-ordination chemistry, analogous to that outlined for the other boron trihalides, is the so-called ammonia–boron triiodide complex which Besson (1892) formulated as $BI_3.5NH_3$; but Joannis (1902) believed the system was one of a mixture of ammonium iodide and "boronamide". McDowell *et al.* (1956) represented the reaction thus:

$$2BI_3 + 9NH_3 \longrightarrow B_2(NH)_3 + 6NH_4I$$

Interactions with certain organic compounds are mentioned, and, as is to be expected, with alcohols (Moissan, 1891) boron triiodide gives alkyl iodide and boric acid, the reactivity of hydrogen iodide apparently dominating the system.

$$6C_2H_5OH + 2BI_3 \longrightarrow 6EtI + 2H_3BO_3$$

Diethyl ether reacts vigorously with boron triiodide, giving a liquid, which after treatment with water, affords ethyl iodide, boric acid and alcohol. The equation given is as follows:

$$3(C_2H_5)_2O + BI_3 \longrightarrow 3C_2H_5I + B(OC_2H_5)_3$$

A systematic investigation on organic compounds with iodine attached to boron has been reported by Long *et al.* (1953), who have studied the interaction of iodine and certain trialkylborons. Contrary to previous conclusion that trialkylborons do not react with iodine (Frankland, 1862; Johnson *et al.*, 1938), it was found that a slow reaction occurred above 140°C between tri-*n*-propylboron and iodine resulting in the formation of alkyl iodide and di-*n*-propylboron iodide. A four-centre broadside mechanism was suggested which may be depicted thus:

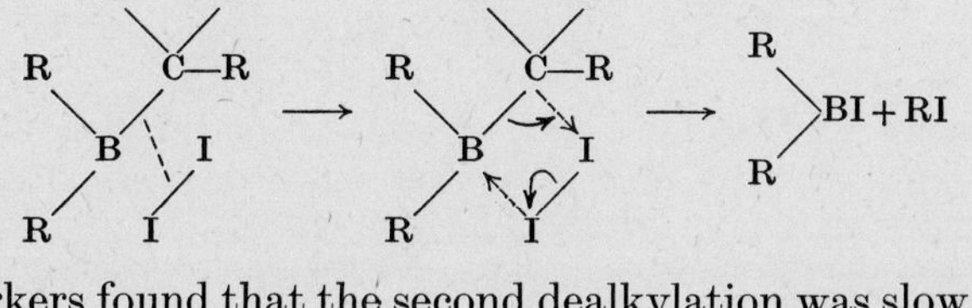

These workers found that the second dealkylation was slow, *n*-propylboron diiodide was not isolated, and this was attributed to the electronegative effect of iodine. Much more work will have to be done before conclusions such as this can be confidently drawn. The inductive effect of iodine in the dialkylboron iodide will reduce electron density on boron, and hence that electrophilic power which enables it to react with the δ- end of the iodine molecule as depicted in the diagram. This effect alone should increase the reactivity. On the other hand it will be more difficult for the B—C binding electrons to break towards the retiring carbon atom.

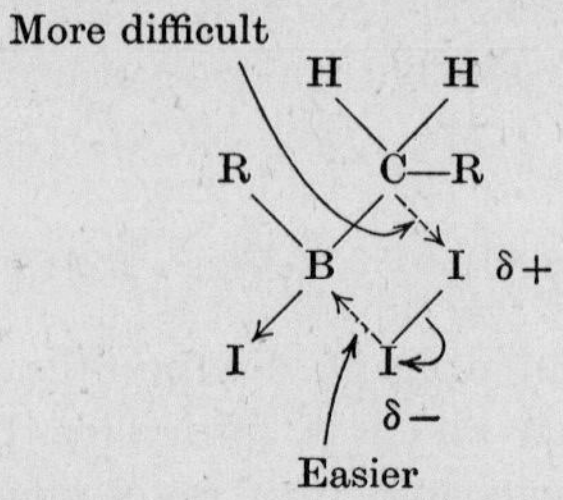

Di-n-propylboron iodide is a colourless liquid which can be kept for some time in the absence of air. It has a m.p. of $-87°C$ and a b.p. of $174°C$. It fumes in air and is rapidly oxidized, iodine being thereby liberated. It is, of course, quickly hydrolysed by water, giving di-n-propylboronic acid and hydrogen iodide.

$$\text{Pr}^n{}_2\text{BI} + \text{H}_2\text{O} \longrightarrow \text{Pr}^n{}_2\text{BOH} + \text{HI}$$

Hydrolysis in the presence of air gives n-propylboronic acid, $\text{Pr}^n\text{B(OH)}_2$, due to oxidative dealkylation, as already discussed.

Hartmann *et al.* (1959) have used dialkylboron iodides as reagents for the preparation of bis(dialkylboron)acetylene, $\text{R}_2\text{B}—\text{C}\equiv\text{C}—\text{BR}_2$. Tri-*iso*propylboron and iodine gave di-*iso*propylboron iodide which with $\text{BrMgC}\equiv\text{CMgBr}$ in chloroform gave the acetylene boron compound. Di-n-butylboron iodide and di-*iso*butylboron iodide are mentioned.

Dimethylboron iodide was mentioned by Goubeau (1948).

It is obvious that much remains to be discovered in this almost uncharted field of chemistry.

Chapter XVI

THE INFRA-RED SPECTRA OF BORON COMPOUNDS

I. Introduction

As a result of the rapidly extending studies of the infra-red spectra of boron-containing compounds, it is possible to derive a number of frequency assignments, e.g. for B—O, B—C, B—N bonds, and these data provide information on the structure of new boron compounds and the effect of substituent groups on the electron density on boron. The infra-red spectra of a number of organo-boron compounds and some boron hydrides have been reviewed by Bellamy *et al.* (1958), and a number of characteristic group frequencies have been assigned.

II. Characteristic Group Frequencies

A. B—O *Frequencies*

The spectra of fifty-six organic borates (I), boronates (II) and borinates (III) have been examined. They show a strong absorption band between 1350–1310 cm^{-1} with the majority absorbing in the narrower range of 1346–1316 cm^{-1}. This band has been assigned to the B—O stretching frequency in these compounds. This assignment agrees well with that previously suggested by Werner *et al.* (1955, 1956) for the

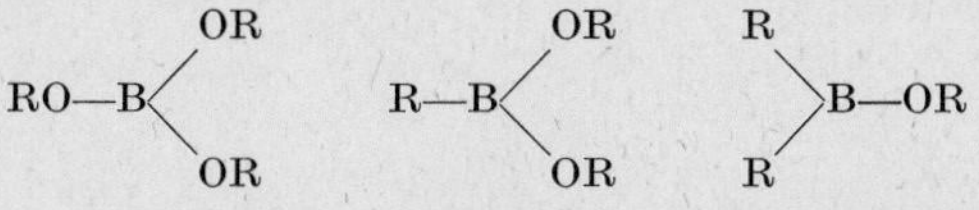

$$
\begin{array}{ccc}
\text{OR} & \text{OR} & \text{R} \\
\text{RO—B} & \text{R—B} & \text{B—OR} \\
\text{OR} & \text{OR} & \text{R} \\
\text{(I)} & \text{(II)} & \text{(III)}
\end{array}
$$

strong 1340 cm^{-1} band found in simple borates (I). The band is very strong and is consequently readily recognized; the strength of the band and its stability in position have led Werner *et al.* to suggest that the B—O bond has some double-bond character due to p_π—p_π bonding (i.e. back–co-ordination from oxygen to boron). With a simple compound such as methyl phenylchloroborinate, resonance is possible between a number of canonical structures, one of which (IV) can account for B—O double-bond character. This is supported by the comparison of the spectrum of the compound (IV) with that of a complex (V) made with a strong electron donor such as pyridine, where the boron octet is completed. The strong B—O band of (IV) occurring near 1340 cm^{-1} is stated to be absent in the complex (V).

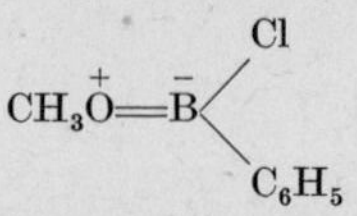

$$CH_3\overset{+}{O}=\overset{-}{B}\begin{cases} Cl \\ C_6H_5 \end{cases}$$

(IV)

$$CH_3O—B\begin{cases} Cl \\ C_6H_5 \end{cases} \longleftarrow :N$$

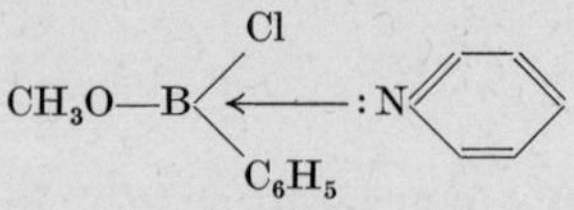

(V)

A similar situation is found by Letsinger *et al.* (1955) with the 2-amino-ethyldiphenylborinate which has a transannular complex structure (VI), thus completing the octet of the boron atom and reducing

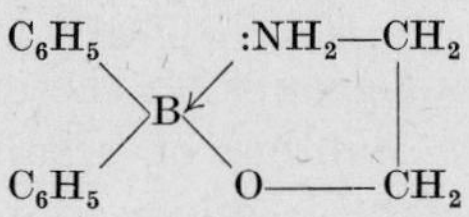

$$\begin{array}{cc} C_6H_5 & :NH_2—CH_2 \\ & B \\ C_6H_5 & O——CH_2 \end{array}$$

(VI)

pπ—pπ bonding in the B—O bond to a minimum.

In the alkyl metaborates (VII) higher frequency bands are found (Lappert, 1958) for the B—O stretching modes (e.g. (VII), R=Me, 1486 cm^{-1}), which suggests higher B—O bond order.

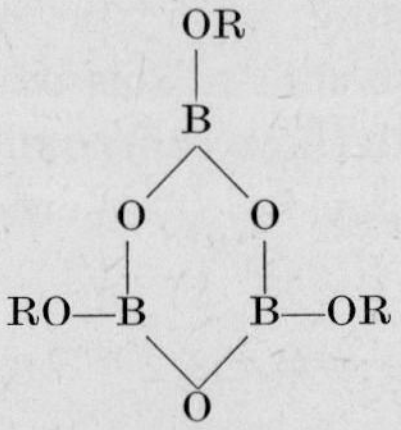

$$\begin{array}{c} OR \\ | \\ B \\ O \quad O \\ RO—B \quad B—OR \\ O \end{array}$$

(VII)

The two bands at 720 and 735 cm^{-1} which occur in these compounds are not present in the spectra of normal alkyl orthoborates. It is possible that these bands may arise from the out-of-plane vibrations of the B_3O_6 skeleton; whatever their origin, they are valuable for the identification of metaborates.

In a series of *B*-trialkoxy-*N*-trimethyl borazoles (VIII) a band between 1330–1318 cm^{-1} has been assigned by Bradley *et al.* (1959) to the B—O bond. It seems more likely, however, that this is the B—O stretching band, as in the boron esters. The bands occur at lower frequency than in the metaborates due to p$_\pi$—p$_\pi$ bonding between the nitrogen and boron atoms of the borazole nucleus, which reduces the B—O bond strength.

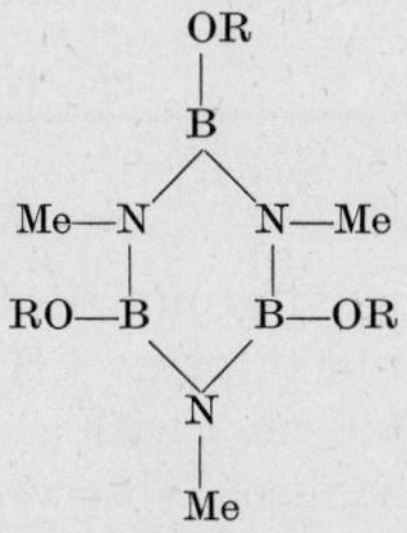

$$
\begin{array}{c}
\text{OR} \\
| \\
\text{B} \\
\diagup \quad \diagdown \\
\text{Me—N} \qquad \text{N—Me} \\
| \qquad\qquad | \\
\text{RO—B} \qquad \text{B—OR} \\
\diagdown \quad \diagup \\
\text{N} \\
| \\
\text{Me}
\end{array}
$$

(VIII)

Diphenylborinic anhydride $(Ph_2B)_2O$, which contains the B—O—B structure, has two bands (Abel *et al.*, 1956) at 1262 and 1378 cm^{-1}, attributed to the B—O stretching modes. These are most likely due to the asymmetric and symmetric stretching modes of the B—O—B group. A similar doublet has been observed by Gerrard *et al.* (1957, 1959) in the di-*n*-butylborinic anhydride $(Bu_2B)_2O$, in the tetra-acetoxydiborate (IX) and in glyceryl diborate (X), and it appears that this doubled B—O stretching frequency is a characteristic of the $>$B—O—B$<$ structure.

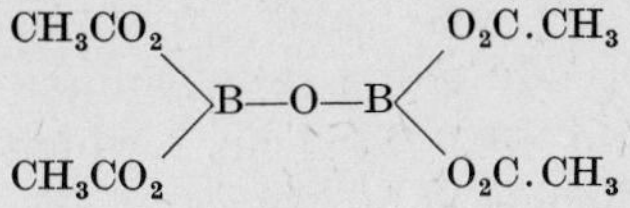

$$
\begin{array}{cc}
\text{CH}_3\text{CO}_2 \qquad\qquad \text{O}_2\text{C.CH}_3 \\
\diagdown \qquad\qquad \diagup \\
\text{B—O—B} \\
\diagup \qquad\qquad \diagdown \\
\text{CH}_3\text{CO}_2 \qquad\qquad \text{O}_2\text{C.CH}_3 \\
\\
\text{(IX)}
\end{array}
\qquad
\begin{array}{cc}
\text{CH}_2\text{—O} \qquad\qquad \text{O—CH}_2 \\
| \qquad\qquad\qquad | \\
\text{HO.CH} \quad >\text{B—O—B}< \quad \text{CHOH} \\
| \qquad\qquad\qquad | \\
\text{CH}_2\text{—O} \qquad\qquad \text{O—CH}_2 \\
\\
\text{(X)}
\end{array}
$$

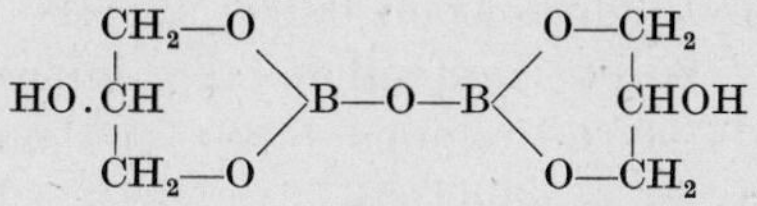

B. B–Aryl Frequencies

Investigation (Bellamy *et al.*, 1958) of forty-two B–Aryl compounds has shown a strong sharp absorption band between 1440 and 1430 cm^{-1}; a similar band has been noted in P–Aryl and Si–Aryl compounds at 1430 cm^{-1}. It seems likely, however, that this band is not characteristic of the non-metal–carbon bond, but is probably the 1470–1438 cm^{-1} band which occurs in the majority of mono substituted aromatic compounds (Josien *et al.*, 1956).

A series of sixteen compounds containing the Ph_2B group showed a strong absorption band between 1280 and 1250 cm^{-1} which can be assigned to the Ph—B group. This band is well defined in diphenyl-boron chloride and bromide (Ph_2BX). In the alkyl phenylchloro-borinates (XI) the band moves to lower frequency and is found between

1220 and 1198 cm⁻¹. In the dialkyl phenylboronates (XII) the band is found at still lower frequency, 1175–1125 cm⁻¹.

$$Ph\!-\!B\!\!\begin{array}{l} OR \\ \\ Cl \end{array} \qquad Ph\!-\!B\!\!\begin{array}{l} OR \\ \\ OR \end{array}$$

(XI)   (XII)

When more than one phenyl group is attached to the boron atom there is a splitting of the out-of-plane CH deformation bands of the aromatic nucleus at about 750 cm⁻¹ with a separation of about 20 cm⁻¹ (Abel *et al.*, 1957). This has been observed in many derivatives of diphenylborinic acid and also in triphenyl and tetraphenylboron derivatives.

C. B–Methyl Frequencies

Assignments for B—C (alkyl) frequencies have been mainly confined to the methyl group attached to boron. The positions of the asymmetric and symmetric deformations of the CH_3 group attached to element X are fairly constant and depend only on the electronegativity of X (Sheppard, 1955; Bellamy *et al.*, 1956). From consideration of the electronegativity of boron the symmetric deformation frequency would be expected to occur in the 1350–1300 cm⁻¹ region. Trimethylboron (Goubeau *et al.*, 1952) has asymmetric and symmetric methyl deformation bands at 1460 and 1305 cm⁻¹ and the spectra of other boron compounds containing the B–Me group show comparable bands in the regions 1460–1405 cm⁻¹ and 1320–1280 cm⁻¹ which are similarly assigned (Becher, 1953, 1957; Becher *et al.*, 1952).

D. B—Cl Frequencies

The asymmetric and symmetric B—Cl stretching frequencies of boron trichloride have been assigned by Lindeman *et al.* (1956) at 955 and 471 cm⁻¹ respectively. A band observed in the alkyl phenylchloroborinates (XI) at 900 cm⁻¹ was assigned by Bellamy *et al.* (1958) to the B—Cl stretching mode, and spectra of some thirteen compounds of this type showed in each case a strong band between 909 and 893 cm⁻¹; similarly, diphenylboron chloride (Ph₂BCl) absorbs strongly at 895 cm⁻¹. However, the exact frequency of the B—Cl stretching mode appears to vary with the nature of the other substituents on the boron atom. In di-*n*-butyl chloroboronate the B—Cl stretching falls to 858 cm⁻¹, probably due to the increased electron-donating power of the oxygen atoms. In the spectrum of bis(diethylamine)boron chloride (Et₂N)₂BCl, the B—Cl mode occurs at even lower frequency, and it is difficult to give a precise assignment owing to the large number of lower frequency bands. The 895 cm^{-1} B—Cl band of diphenylboron chloride is removed by complex formation with tertiary bases (Bellamy *et al.*, 1958). It is probable that the extent of p_π—p_π bonding between the boron and chlorine atoms is decreased on completing the boron octet by co-ordination (compare B—O stretching mode) with the result that the B—Cl stretching mode occurs at lower frequency.

A B—Cl mode of the arylamine boron dichlorides (ArNHBCl$_2$), which is presumably the asymmetrical stretching, has been observed by Gerrard *et al.* (1960) between 849 and 828 cm^{-1}. This is again in keeping with the idea that there would be p_π—p_π bonding between the nitrogen and boron atoms. The ionic complexes of boron trichloride with secondary amines also demonstrate the shift of the B—Cl modes. In an ionic complex of the type (XIII) the lone pair of the nitrogen atom is utilized in protonization and consequently p_π—p_π bonding between nitrogen and boron cannot occur. Hence there is an increasing

$$\left[\begin{array}{c} \text{Me} \\ \\ \text{Me} \end{array} \!\!\! \underset{\overset{|}{\text{H}}}{\overset{+}{>\!\text{N}}}\!\!-\!\text{B}\!\!<\!\! \begin{array}{c} \text{Cl} \\ \\ \text{Cl} \end{array} \right] \text{Cl}^-$$

(XIII)

tendency for p_π—p_π bonding between the boron and chlorine atoms, with the result that the B—Cl bond length is shortened and the B—Cl stretching mode rises to 905 cm^{-1}, as found by Gerrard *et al.* (1960). Similar observations have been made on more complex ionic structures (XIV) (Me, 893 cm^{-1}, and Pr^i, 943 cm^{-1}).

$$\left[\begin{array}{c} \text{R}_2 \quad \text{H} \\ \backslash \quad | \\ \text{N}_+ \\ \text{R}_2 \quad \backslash \\ \qquad >\!\text{B}\!\!<\!\! \begin{array}{c} \text{Cl} \\ \\ \text{Cl} \end{array} \\ \text{N}^+ \\ | \\ \text{H} \end{array} \right] \left[\text{BCl}_4 \right]^-$$

(XIV)

The *B*-trichloroborazoles have been investigated (Gerrard *et al.*, 1960) and tentative assignments have been made for the asymmetric B—Cl stretch in these compounds (Table XXVI). The assignments have been based on the disappearance of these bands on alkylation, and on thermal polymerization, which proceeds by elimination of hydrogen chloride; the intensity of the band attributed to ν_{B-Cl} slowly decreases. A band found between 1030 and 1160 cm^{-1} in these compounds is,

TABLE XXVI

B—Cl Asymmetric Stretching Frequencies for B-Trichloroborazoles

R in (RNBCl)$_3$	B-Cl ν asym cm^{-1}	Associated band cm^{-1}
H		1032
Me	987	1085
Ph	846	1152
p-BrC$_6$H$_4$†	833	1155
p-BrC$_6$H$_4$‡	833	1152
p-CH$_3$OC$_6$H$_4$§	834	1033
p-BrC$_6$H$_4$‖	797	1071
p-CH$_3$C$_6$H$_4$§	849	1155

† 1 mole benzene of crystallization.
‡ 1 mole toluene of crystallization.
§ 0·5 mole benzene of crystallization.
‖ 1 mole toluene crystallization and 3 moles tetrahydrofuran complexed to boron.

by the same criteria, also associated with the B—Cl vibrations, and is the stronger of the pair. A similar pair of bands occurs in compounds containing the N—BCl$_2$ group, but here the higher frequency band is the weaker (Gerrard *et al.*, 1960). It would appear that the relative intensities of these two bands should be of diagnostic value in differ-

entiating between —N—B—Cl and N—B$\bigg\langle{\overset{\displaystyle Cl}{\underset{\displaystyle Cl}{}}}$ groups.

It is observed that varying the group R in (RNBCl)$_3$ from H to Me to Ph— results in increasing B—Cl bond length, indicating less p$_\pi$—p$_\pi$ bonding between the chlorine and boron atom in the N-methyl and N-phenyl borazoles. In the former the + I effect of the methyl group supports the back–co-ordination from nitrogen to boron, and the + M effect of phenyl group supports the back–co-ordination to an even greater extent.

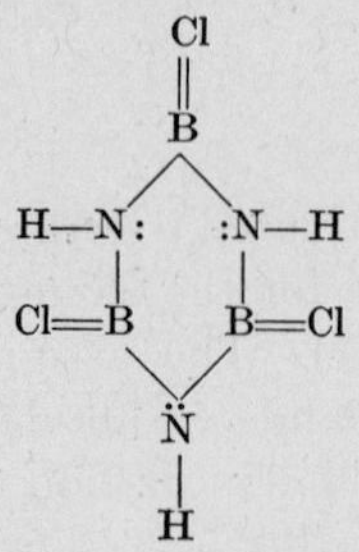

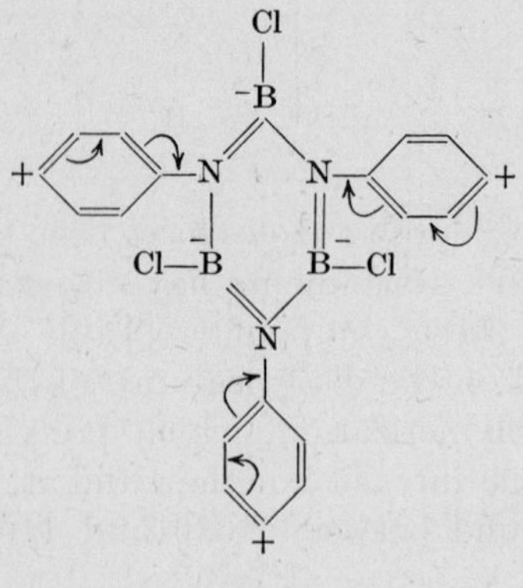

In the complex of tetrahydrofuran with B-trichloro-N-tri-p-bromo-phenylborazole the B—Cl asymmetric vibration occurs at even lower frequency. In this compound the octet of boron is completed by co-ordination of the lone-pair of the oxygen atom to the boron atom.

$$
\begin{array}{c}
\text{CH}_2 \\
\diagup \quad | \\
\text{CH}_2 \quad \text{CH}_2 \\
\quad \diagup \\
\text{Cl} \quad .\ddot{\text{O}}\text{—CH}_2 \\
| \diagdown .\diagup \\
\text{B} \\
p\text{-BrC}_6\text{H}_4\text{—N}: \qquad :\text{N C}_6\text{H}_4\text{Br-}p \\
| \qquad |
\end{array}
$$

E. B—N *Frequencies*

Thermodynamical considerations by Skinner *et al.* (1954) have shown that nitrogen is more effective than oxygen in back–co-ordination to boron, and consequently it would be anticipated that the boron–nitrogen bond would show considerable double-bond character. Co-ordination compounds of amines and boron compounds show bands in the region of 1100 cm^{-1}, as in the trimethylboron ammonia complex (Becher, 1953), but in compounds where p_π—p_π bonding is possible between boron and nitrogen atoms the B—N absorption occurs at higher frequency, as in phenylaminodimethylboron, $\text{Me}_2\text{B}.\text{NHC}_6\text{H}_5$ at 1332 cm^{-1} (Becher, 1957), and as in phenylaminodiphenylboron, $\text{Ph}_2\text{B}.\text{NHC}_6\text{H}_5$ at 1379 cm^{-1} (Gerrard *et al.*, 1960). In the former compound, compensation of electron deficiency of the boron atom by the inductive effect of the methyl groups is again demonstrated, as there is less B—N double-bond character in the dimethyl than in the diphenyl boron derivative.

The B—N ring frequency of borazole derivatives has received some attention, and has been a source of some controversy. Becher *et al.* (1958) examined a number of borazole derivatives in carbon tetra-chloride solution and found the B—N ring frequency to occur between 1373 cm^{-1} and 1472 cm^{-1}. Bradley *et al.* (1959) examined a series of B-tris-alkoxy-N-trimethyl borazoles, and located the B—N ring fre-quency between 1412–1458 cm^{-1}. However, these authors state that the B—N bands are of medium intensity, whereas the CH$_2$ deforma-tion and symmetrical CH$_3$ vibration modes give rise to rather stronger bands. This is contrary to the experience of Gerrard *et al.* (1960); for the B—N ring mode gave rise to the strongest band in the spectrum. Some of the B—N ring vibrations are given in Table XXVII.

TABLE XXVII

B—N Ring Vibrations in Substituted Borazoles

R in $(RNBX)_3$	X in $(RNBX)_3$	Frequency cm^{-1}	Reference
H	H	1465	Becher et al. (1958)
		1464	Price et al. (1950)
H	Cl	1445	Becher et al. (1958)
		1445	Gerrard et al. (1960)
Me	Cl	1458†	Bradley et al. (1959)
		1395	Gerrard et al. (1960)
		1392	Becher et al. (1958)
C_6H_5	Cl	1373	,, ,, ,,
		1376	Gerrard et al. (1960)
p-$CH_3C_6H_4$	Cl	1389	,, ,, ,, ,,
p-BrC_6H_4	Cl	1381	,, ,, ,, ,,
p-$CH_3OC_6H_4$	Cl	1404	,, ,, ,, ,,
H	C_6H_5	1472	Becher et al. (1958)
		1466‡	Gerrard et al. (1960)
		1473§	,, ,, ,, ,,
H	Bu^n	1475	,, ,, ,, ,,
		1475	Ruigh et al. (1955)
CH_3	C_6H_5	1405	Becher et al. (1958)
C_6H_5	H	1401	,, ,, ,, ,,
C_6H_5	CH_3	1380	,, ,, ,, ,,
		1380	Gerrard et al. (1960)
C_6H_5	C_2H_5	1393	,, ,, ,, ,,
p-$CH_3C_6H_4$	CH_3	1385	,, ,, ,, ,,
p-BrC_6H_4	CH_3	1385	,, ,, ,, ,,
C_6H_5	Et_2N	1285	,, ,, ,, ,,
H	Et_2N	1495	,, ,, ,, ,,
H	Pr^i_2N	1488	,, ,, ,, ,,
Bu^n	Et_2N	1414	,, ,, ,, ,,

† These authors quote a frequency of 1379 cm^{-1} for the sym. CH_3— bend, which is stated as being the strongest band in the spectrum.
‡ 1% solution in chloroform.
§ Nujol mull.

Some polymeric borazoles have been examined, and again the B—N ring vibration frequencies can be readily assigned (see Table XXVIII).

TABLE XXVIII

B—N Ring Vibrations in Polyborazoles

Polyborazole	Frequency cm^{-1}	Reference
$(HNBNHBNH)_x$	1410	Gerrard et al. (1960)
$(PhNBC_6H_4NBNC_6H_4)_x$	1429	,, ,, ,, ,,
$(CH_3C_6H_4NBC_6H_3(CH_3)NBC_6H_3(CH_3)—)_x$	1453	,, ,, ,, ,,

The nature of the B—N bond varies with the nature of the substituents R and X in $(RNBX)_3$. If we consider the borazole derivatives R=Ph, X=Et_2N, Cl, Me and H, which have electron-donating properties in the same decreasing order, we should expect the B=N character to increase in the same order. This is indeed observed, X=Et_2N—, 1285 cm^{-1}; Cl—, 1376 cm^{-1}; Me—, 1380 cm^{-1}; and H—, 1401 cm^{-1}.

F. B—H *Frequencies*

The B—H stretching and deformation frequencies have been readily recognized in the boranes, and the subject has been adequately reviewed by Bellamy *et al.* (1958). The correlations, however, refer only to the more complicated boranes, e.g. diborane, pentaborane, tetraborane, etc., and very few data are available on simple aminoboranes and the B—H frequencies of borazoles of the type $(RNBH)_3$ (R=H, alkyl or aryl).

III. Co-ordination of the Carbonyl Group to Boron

The observations on the free and bonded carbonyl frequencies have yielded useful information on the co-ordinating power of the carbonyl group to boron atoms. The carbonyl frequency is very useful for this purpose as the $\rangle C=O$ stretching band is readily recognized and has been widely studied in many types of organic compounds (Bellamy, 1958). Examples of the use of carbonyl bands include the study of acyloxy-derivatives of boron (Duncanson *et al.*, 1958). In tetra-acetyl diborate two $\rangle C=O$ stretching bands were observed, corresponding to the free and co-ordinated carbonyl groups (XV).

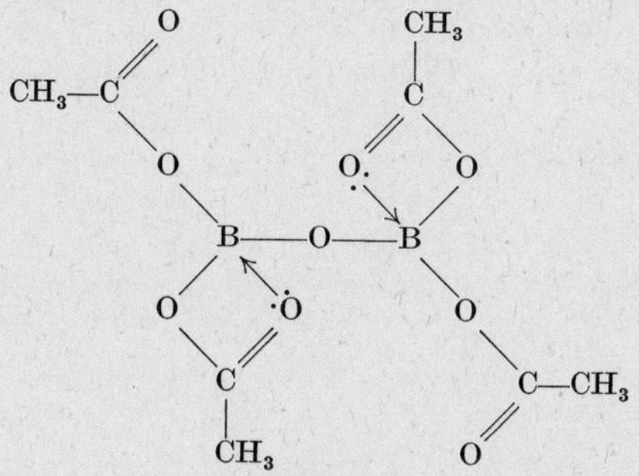

(XV)

Similarly, the effect of the acetoxy- and substituted acetoxy-groups on bonding to boron in acetoxy–ethylene borates (XVI) had been studied by Blau *et al.* (1960) and both free and bonded carbonyl bands were observed. This supports the idea of co-ordination from the carbonyl groups to complete the octet of boron. In the trifluoro-

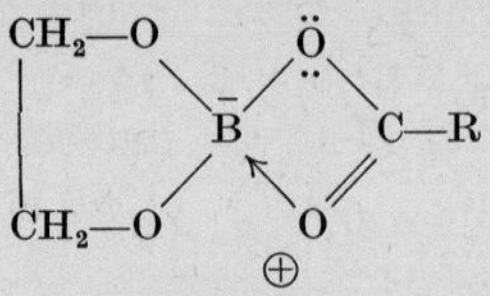

$$
\begin{array}{c}
\text{CH}_2\text{—O} \qquad \overset{\cdot\cdot}{\text{O}} \\
\diagdown \qquad \diagdown \\
\text{B} \qquad \text{C—R} \qquad\qquad \text{R}=\text{Me, CH}_2\text{Cl, CHCl}_2 \text{ or CCl}_3 \\
\diagup \qquad \diagdown \\
\text{CH}_2\text{—O} \qquad \text{O} \\
\oplus
\end{array}
$$

(XVI)

acetoxy-ethylene borate (XVI, $R=CF_3$) no bonded carbonyl stretching band was observed, and this observation is interpreted as being due to the relatively low basic strength of the carbonyl–oxygen atom because of the powerful electron withdrawing properties of the trifluoromethyl group.

APPENDIX

Contents

1. General Reviews of the Subject 234
2. Preparation of Lithium and Sodium Tetraphenylboron . . . 234
3. Transition Element Compounds Containing the Tetraphenylboron Ion 235
4. Physical Measurements on the Tetraphenylboron Ion . . . 236
5. Miscellaneous Non-analytical References to the Tetraphenylboron Ion 237
6. Compounds Related to the Tetraphenylborons and their Analytical Uses: Tetraphenyl Diboroxide (diphenylboronous anhydride) . 238
7. Triarylboron Complexes 238
8. Casignost (sodium triphenylboroncyanide) 239
9. Tetra-aryl Boron Compounds 239
10. Miscellaneous Analytical Uses of Sodium Tetraphenylboron . . 241
11. Determination of Thallium with Sodium Tetraphenylboron . . 241
12. Determination of Potassium and Nitrogen Bases (covered in the same paper) 242
13. More Elaborate Analytical Techniques as Applied to Potassium and Nitrogen Bases:
       A. Turbidimetric 242
       B. Polarographic 242
       C. Conductometric and Potentiometric, etc. 243
14. Analysis of Nitrogen Bases with Sodium Tetraphenylboron . . 244
15. Imides 246
16. Determination of Potassium as Potassium Tetraphenylboron . . 246
17. Determination of Traces of Potassium 249
18. Determination of Potassium in:
       A. Milk 249
       B. Wines 250
       C. Silicates Including Glass 250
       D. Gunpowders, etc. 250
       E. Blood 250
       F. Plants 251
       G. Sea water 251
       H. Fertilizers 251
19. Determination of Radioactive Potassium 251

233

1. General Reviews of the Subject

Barnard, A. J. (1955), *Chem. Anal.*; **44**, 104; (1956), **45**, 110; (1957), **46**, 16; (1958), **47**, 46; (1959), **48**, 44: Sodium tetraphenylboron.

Belcher, R., and Wilson, C. L. (1955), "New Methods in Analytical Chemistry", pp. 260–266. Chapman & Hall Ltd., London: Review of the use of sodium tetraphenylboron for potassium determination.

Coates, G. E. (1956), "Organometallic Compounds", pp. 66–70. Methuen, London: Tetraphenylboron.

Gerrard, W. (1957), *Chem. Prod.* December: Industrial actualities and potentialities of hydrido- and organo-compounds of boron.

Haegi, W. (1955), *Bull. soc. chim. Fr.*, 581: Recent progress in the chemistry of organic complexes of metalloids.

Heyl & Co. (1952), November brochure: Kalignost (sodium tetraphenylboron) reagent for the detection and estimation and isolation of potassium, other alkali metals, and nitrogen and oxygen bases.

Heyl & Co. (1954), November brochure: Kalignost (sodium tetraphenylboron), ideal specific precipitant, rationalizing the analysis and technique for monovalent cations with relevant methods of analysis and isolation.

Heyrovsky, A. (1957), *Chemie (Prague)* **9**, 100: Analytical uses of tetraphenylboron (review, 50 refs.).

Lappert, M. F. (1956), *Chem. Rev.* **56**, 959: Organic compounds of boron.

Mukoyama, T. (1956), *J. Jap. Chem.* **10**, 103: Application of sodium tetraphenylboron to analytical chemistry.

Nutten, A. J. (1954), *Industr. Chem. Mfr.* **29**, 57: Reviews of the analytical uses of sodium tetraphenylboron.

Nutten, A. J. (1956), *Chem. Prod.* **19**, 264: Reagents ripe for development.

Prodinger, W. (1954), "Organische Fällungsmittel in der qualitativen Analyse", pp. 15, 223. Ferdinand Enke, Stuttgart, Germany: Review of the literature up to the end of 1953.

Sykes, A., *Industr. Chem. Mfr.* (1955), 245, 305; (1956), 164–223: Reviews of the analytical uses of sodium tetraphenylboron.

2. Preparation of Lithium and Sodium Tetraphenylboron

Heyl & Co. (1953), French Patent, 1043: Process for the preparation of complex salts of boron tetraaryls, also the resulting application.

Heyl & Co. (1954), British Patent, 705, 719: Preparation of complex metal tetraphenylboron salts.

Myl, J., and Kvapil, J. (1959), *Chem. Prumysl.* **9**, 77: Preparation of sodium tetrapherylboron.

Nesmeyanov, A. N., and Sazanova, V. A. (1955), *Bull. Acad. Sci. U.R.S.S.*, *Classe sci. chim.*, 187: Sodium tetraphenylboron synthesis.

Nesmeyanov, A. N., Sazanova, V. A., Liberman, G. S., and Emelyanov, L. I. (1955), *Bull. Acad. Sci. U.R.S.S.*, *Classe sci. chim.* **1**, 48: Reactions of organomagnesium compounds with potassium fluoroborate and triethyl-oxonium fluoroborate (includes the syntheses and reactions of lithium, potassium, cuprous and triethyloxonium tetraphenylboron compounds).

Wittig, G. (1950), *Angew. Chem.* **62A**, 231: Metallo-organic complex compounds including lithium tetraphenylboron and derivatives.

Wittig, G., and Keicher, G. (1947), *Naturwissenschaften* **34**, 216: Reaction of lithium phenyl with triphenylboron.

Wittig, G., Keicher, G., Ruckert, A., and Raff, P. (1949), *Liebigs Ann.* **563**, 110: Boron alkali metallo-organic complex compounds (preparation and properties).

Wittig, G., Meyer, F. J., and Lange, G. (1951), *Liebigs Ann.* **571**, 167: Behaviour of diphenyl metals as complex formers.

Wittig, G., and Raff, P. (1951), *Liebigs Ann.* **573**, 195: Complex formation with triphenylboron.

Wittig, G., and Raff, P. (1958), U.S. Patent 2,853,525 assigned to Heyl & Co.: Production of sodium tetraphenylboron.

3. *Transition Element Compounds containing the Tetraphenylboron Ion*

Engelmann, F. (1953), *Z. Naturf.* **8b**, 775: Magnetic investigation of the structure of cyclopentadiene compounds of transitional metals, including the tetraphenylboron derivatives of the cobalt complex.

Fischer, E. O., and Jira, R. (1953), *Z. Naturf.* **8b**, 1: Dicyclopentadienyl cobalt and its tetraphenylboron derivatives.

Fischer, E. O., and Jira, R. (1953), *Z. Naturf.* **8b**, 217: Nickel dicyclopentadiene and its tetraphenylboron derivatives.

Fischer, E. O., and Jira, R. (1953), *Z. Naturf.* **8b**, 327: Dicyclopentadienyl cobalt (II) and its tetraphenylboron derivatives.

Fischer, E. O., Seus, D., and Jira, R. (1953), *Z. Naturf.* **8b**, 692: Metal complexes of cobalt including the tetraphenylboron derivative of di-indenyl cobalt.

Fischer, E. O., and Seus, D. (1954), *Z. Naturf.* **9b**, 386: Precipitation of the oxidized form of ditetrahydroindenyl iron as the tetraphenylboron salt.

Fischer, E. O., and Hafner, W. (1955), *Z. Naturf.* **106**, 665: Diphenyl chromium and its tetraphenylboron derivatives.

Fischer, E. O., and Seus, D. (1956), *Chem. Ber.* **89**, 1809: Structure of chromium-phenyl compounds and their tetraphenylboron derivatives.

Fischer, E. O., and Böttcher, R. (1956), *Chem. Ber.* **89**, 2397: Mesitylene complex of Fe(II) and the preparation of its tetraphenylboron derivative.

Fischer, E. O., and Piesbergen, U. (1956), *Z. Naturf.* **11b**, 758: Magnetic studies of aromatic complexes of metals including tetraphenylboron salts.

Fischer, E. O., and Böttcher, R. (1957), *Z. anorg. Chem.* **291**, 305. Complex of Ru(II) and mesitylene and $(Me_2C_6H_3)_2Ru.BPh_4$.

Fischer, E. O., and Wirzmüller, A. (1957), *Chem. Ber.* **90**, 1725: Re(I) complex of benzene and mesitylene and preparation of $Ph_2Re.BPh_4$.

Fischer, E. O., Joos, G., and Meer, W. (1958), *Z. Naturf.* **13b**, 456: Magnetic moments of aromatic-metal complexes including tetraphenylboron salts.

Hein, F., Kleinert, P., and Kurras, E. (1957), *Z. anorg. Chem.* **289**, 229: Benzenebiphenylchromium and the preparation of $(C_6H_6)(C_6H_5C_6H_5)$ $Cr.BPh_4$.

Hieber, W., and Freyer, W. (1958), *Chem. Ber.* **91**, 1230: Triphenylphosphine-containing cobalt carbonyls including $Co(CO)_3P(Ph)_3.BPh_4$.

Kauer, E. (1956), *Z. phys. Chem.* **6**, 105: Fine structure of K edge of X-radiation of covalent complex compounds of iron, cobalt and nickel, including $Ph_2Ni.BPh_4$.

Sacco, A. (1956), *Gazz. chim. ital.* **86**, 201; (1956), *Rec. Trav. chim., Pays-Bas* **75**, 646: Isonitrile complexes including $(p\text{-}MeOC_6H_4NC)_6Mn.BPh_4$.

APPENDIX

236

Sacco, A., and Freni, M. (1956), *Gazz. chim. ital.* **86**, 195: Isonitrile complexes including $(p\text{-}MeC_6H_4NC_4)Au.BPh_4$.

Sacco, A., and Naldini, L. (1956), *Gazz. chim. ital.* **86**, 207: Isonitrile complexes of Mn(I) including $Mn(CNR)_6.BPh_4$.

Sacco, A., and Naldini, L. (1957), *R. C. Ist. lombardo*, Pt. I, **91**, 286: Electric moment of isonitrile complexes of Mn(I) including tetraphenylboron salts.

Sazanova, V. A., and Nazanova, I. I. (1956), *J. gen. Chem.*, Moscow **26**, 3440: The reaction of copper tetraphenylboron with unsaturated keto compounds as a method of reduction of them.

Vohler, O. (1958), *Chem. Ber.* **91**, 1235: IR-absorption of cobalt carbonyl phosphine compounds including $Co(CO)_3P(Ph)_3.BPh_4$.

Zeiss, H. H., and Herwig, W. (1956), *J. Amer. chem. Soc.* **78**, 5959: Aryl chromium complexes including $Ph_2Cr.BPh_4$.

Zeiss, H. H., and Herwig, W. (1957), *Ann. Liebigs* **606**, 209: Substituted di-benzene chromium compounds including $Ph_2Cr.BPh_4$.

4. *Physical Measurements on the Tetraphenylboron Ion*

Amin, A. M. (1947), *Chem. Anal.* **46**, (1) 6: Sensitivity of the tetraphenylboron ion in the detection of potassium.

Arnott, S., and Abrahams, S. C. (1958), *Acta cryst.* **11**, 449: Lattice constants of alkali salts of tetraphenylboron.

Cooper, S. S. (1957), *Analyt. Chem.* **29** (3) 446: Stability of sodium tetraphenylboron solutions.

Cooper, S. S. (1957), *Chem. Anal.* **46** (3), 62: Stability of sodium tetraphenylboron solutions.

Davies, T., and Stavely, L. A. K. (1957), *Trans. Faraday Soc.* **53**, 10: Behaviour of the ammonium ion in ammonium tetraphenylboron by comparison of the heat capacities of the ammonium, rubidium and potassium salts.

Fuoss, R. M., Berkowitz, J. B., Hirsch, E., and Petrucci, S. (1958), *Proc. nat. Acad. Sci.* **44**, 27: Conductance of $Bu_4N.BPh_4$.

Geske, D. H. (1958), *Abst. 134th Meeting Amer. chem. Soc.*, Chicago, Sept., 598: Electro-oxidation of the tetraphenylboron ion.

Gloss, G. H., and Olson, B. (1954), *Chem. Anal.* **43**, 80: Stability of solutions of sodium tetraphenylboron.

Howick, L. C., and Pflaum, R. T. (1958), *Analyt. chim. acta* **19**, 342: Solubility and thermal decomposition of some amine tetraphenylboron compounds.

Ieviņš, A., Ozols, J., and Gudriniece, E. (1955), *Latv. PSR Zinat. Akad. Vestis* **7**, 136: Crystal structure of potassium tetraphenylboron.

Montequi, R., Doadrio, A., and Serrano, C. (1956), *An. Fis. Quim.* Ser. B. **52**, 597: Mechanism of mercuration of alkali tetraphenylboron salts.

Montequi, R., Doadrio, A., and Serrano, C. (1956), *Publ. Inst. Quim. Barba* **10**, 183: Mechanism of mercuration of alkali tetraphenylboron salts.

Orr, W. C. *et al.* (1955), Report CC.1024 of Univ. of Connecticut to U.S.A.E.C. Jan. 21st (in the *Nuclear Sci. Abstr.* 9, item 2613, 1955): Electrochemistry of boron compounds in fused salt-solvent systems including sodium tetraphenylboron in aluminum chloride–sodium chloride melt.

Pellerin, F. M. (1956), *Ann. pharm. franç.* **14**, 193: Behaviour of the salts of tetraphenylboronic acid in anhydrous acetic acid.

Rüdorff, W., and Zannier, H. (1953), *Z. Naturf.* **8b**, 611: Mobility of the tetraphenylboron ion and the solubility product of potassium tetraphenylboron.

Wendlandt, W. W. (1956), *Analyt. Chem.* **28**, 1001: Thermogravimetric pyrolysis of ammonium and alkali metal tetraphenylboron salts.

Wendlandt, W. W. (1957), *Chem. Anal.* **46**, 38: Thermogravimetric pyrolysis of some metal tetraphenylboron compounds (lithium, sodium, mercurous, mercuric, silver).

Wendlandt, W. W. (1958), *Chem. Anal.* **47**, 6: Thermal decomposition of some amine tetraphenylboron compounds.

Wendlandt, W. W., Dunham, R. (1958), *Analyt. chim. acta* **19**, 505: Effect of heating rate on the melting point of amine tetraphenylboron derivatives.

5. *Miscellaneous Non-analytical References to the Tetraphenylboron Ion*

Agfa, A. G. (1953), Photofabrikalion., German Patent, 932,533: Increasing sensitivity of silver halide emulsions by adding tetraphenylboron salts.

Belcher, R., MacDonald, A. M. G., and West, T. S. (1958), *Talanta* **1**, 408: Determination of metals in organic compounds by closed flask method including boron in sodium and potassium tetraphenylborons.

Fischer, E. O. (1955), *Angew Chem.* **67**, 475: Review of metallocompounds of cyclopentadiene and indene.

Flaschka, H., and Sadek, F. (1958), *Chem. Anal.* **47**, 30: Application of EDTA titration to the determination of the tetraphenylboron ion and its salts.

Hunziker, R. R. (1958), *Dissertation Abstr.* **19**, 11, *Microfilms, L.C. Card* No. Mic 58–2187, p. 122: Degradation of soils and micaceous minerals by removal of potassium with sodium tetraphenylboron.

Kuhn, R., Trischmann, H. (1958), *Liebigs Ann.* **611**, 117: Isolation of $Me_3SO.BPh_4$.

Montequi, A., Doadrio, A., and Serrano, C. (1958), *An. Fis. Quim. Ser. B.* **54**, 29: Bromatometry of alkali tetraphenylboron salts.

Nesmeyanov, A. N., Sazonova, V. A., Liberman, G. S., and Emelyanova, L. I. (1953), *Bull. Acad. Sci. U.R.S.S., Classe sci. chim.* (*English Translation*) **1**, 48: Reactions of organomagnesium compounds with potassium and triethyloxonium fluoborates (synthesis and reactions of tetraphenylboron compounds of lithium, potassium, copper and ethoxide).

Nesmeyanov, A. N., and Tolstaya, T. P. (1955), *C. R. Acad. Sci. U.R.S.S.* **150**, 94: Diphenylchloronium salts, including $Ph_2Cl.BPh_4$.

Nesmeyanov, A. N., Tolstaya, T. P., and Isaeva, L. S. (1955), *C. R. Acad. Sci. U.R.S.S.* **104**, 872: Diphenylbromonium salts, including $Ph_2Br.BPh_4$.

Nesmeyanov, A. N., Kruglova, N. V., Materikova, R. B., and Tolstaya, T. P. (1956), *J. gen. Chem., Moscow* **26**, 2211: Diarylbromonium and diarylchloronium salts, including the preparation of (p-MeC_6H_4)$PhBr.BPh_4$ and similar compounds.

Nesmeyanov, A. N., Makarova, L. G., and Tolstaya, T. P. (1957), *Tetrahedron* **1**, 145: Heterolytic decomposition of "onium" compounds. Salts precipitated by sodium tetraphenylboron.

Rabiant, J., and Wittig, G. (1957), *Bull. Soc. chim. Fr.*, 798: Modification of the Hoffmann degradation using tetraphenylboron salts.

Razuvaev, G. A., and Brilkina, T. G. (1952), *C. R. Acad. Sci. U.R.S.S.* **85**,815: Free radical reactions of metallic tetraphenylboron salts.

Razuvaev, G. A., and Brilkina, T. G. (1953), *C. R. Acad. Sci. U.R.S.S.* **91**, 861: Free radical reactions of complexes of the type MBPh$_4$.

Sazonova, V. A., and Konrod, N. Ya. (1956), *J. gen. Chem. U.S.S.R.* (*English Translation*) **26**, 2093: Reaction of potassium fluoroborate with styryl magnesium bromide and phenylethynyl magnesium bromide.

Schuele, W. J. (1956), *Dissertation Abstr.* **16**, 2302, *Microfilm Publ.* No. 17274, p. 183: Analytical studies of boron compounds.

Vol'pin, M. E., Zhdanov, S. I., and Kursanov, D. N. (1957), *C. R. Acad. Sci. U.R.S.S.* **112**, 264: New tropylium salts including tetraphenylborons. Polarography of the tropylium ion.

Wille, F., Dirr, K., Heitzer, E., and Schneidmeir, W. J. (1957), *Liebigs Ann.* **608**, 22: Fission of esters of $\alpha\alpha'$-dihalogenated adipic acid by means of secondary amine tetraphenylboron salts.

6. Compounds Related to the Tetraphenylborons and their Analytical Uses: Tetraphenyl Diboroxide (diphenylboronous anhydride)

Heyl & Co. (1958), R. Neu, inventor, German patent 1,018,864: Preparation of tetra-aryldiboroxides including tetraphenyldiboroxide from sodium tetraphenylboron.

Neu, R. (1954), *Chem. Ber.* **87**, 802: Preparation of tetraphenyldiboroxide from sodium tetraphenylboron.

Neu, R. (1954), *Z. anal. Chem.* **142**, 335: New organoboron compound as an analytical reagent. Use of tetraphenyldiboroxide in flavonol synthesis.

Neu, R. (1954), *Z. anal. Chem.* **143**, 30: Tetraphenyl diboroxide III. New tests for choline and acetylcholine and other quaternary ammonium compounds.

Neu, R. (1955), *Fette u. Seif.* **57**, 568: Differentiation of long-chain cation active compounds of ampholytes and amine salts (using above compound).

Neu, R. (1955), *Chem. Ber.* **88**, 1761 (5): Aromatic boron compounds, including a preparation of tetraphenyl diboroxide.

Neu, R. (1956), *Naturwissenschaften* **43**, 82: New reagent for detection and separation of flavones on paper chromatograms.

Neu, R. (1956), *Z. anal. Chem.* **151**, 328: Tetraphenyl diboroxide and aromatic boric acids as bathochromic reagents for flavones.

Neu, R. (1958), *Chem. Anal.* **47**, 106: Analytical use of diarylboric acids and especially of tetraphenyldiboroxide.

Neu, R. (1958), *Mikrochim. Acta* 715: Detection of flavones with tetra-phenyldiboroxide and long-chain quaternary ammonium compounds by spot tests.

Neu, R. (1958), *Naturwissenschaften* **45**, 311: Detection of hydroxyphenyl-benzo-γ-pyrones with diarylboric acids, including tetraphenyldiboroxide, in the presence of electron donors.

Pfeil, E., and Baier, A. (1958), *Angew. Chem.* **70**, 702: Paper chromatographic detection of glucose, fructose and mannose with each other and the use of 2-aminoethyl diphenylborate or tetraphenyldiboroxide.

7. Triarylboron Complexes

Bettman, B., Branch, G. E. K., and Yabroff, D. L. (1934), *J. Amer. chem. Soc.* **56**, 1865: 1:1 addition compounds of triphenylboron with alkali metals in ether.

Brown, H., and Dodson, C. (1957), *J. Amer. chem. Soc.* **79**, 2302: Preparation of trimesitylboron and evidence for the nonlocal nature of the odd electron in the triarylboron radical ion and related free radicals.

Chu, Ting Li (1953), *J. Amer. chem. Soc.* **75**, 1730: Disodium triphenylboron complex by reaction of triphenylboron with sodium amalgam in tetrahydrofuran.

Chu, Ting Li, and Weismann, T. J. (1956), *J. Amer. chem. Soc.* **78**, 3610: Triaryl boron anions (Tri-methyl β-naphthyl boron).

Chu, Ting Li, and Weismann, T. J. (1956), *J. Amer. chem. Soc.* **78**, 23: Triaryl boron anions. Magnetic and cryoscopic studies of univalent trimesityl boron anion solution.

Krause, E. (1924), *Ber. dtsch. chem. Ges.* **57**, 216: 1:1 addition compounds of triphenylboron with alkali metals.

Krause, E., and Nitsche, R. (1921), *Ber. dtsch. chem. Ges.* **54**, 2784: Preparation of triphenyl boron with boron trifluoride and a Grignard reagent.

Krause, E., and Polack, H. (1926), *Ber. dtsch. chem. Ges.* **59**, 777: 1:1 addition compounds of tritolylborons with alkali metals.

Krause, E., and Nobbe, P. (1930), *Ber. dtsch. chem. Ges.* **63**, 1261: Preparation of alpha naphthyl boron with a Grignard reagent.

Krause, E., and Nobbe, P. (1931), *Ber. dtsch. chem. Ges.* **64**, 2112: Preparation of $(p\text{-}CH_3O.C_6H_4)_3B$ using a Grignard reagent and boron trifluoride.

Mikhailov, B. M., and Vaver, V. (1957), *Bull. Acad. Sci. U.R.S.S. Classe sci. chim.* 812: New triaryl boron compounds and their complexes (unsymmetrical members).

Von Bohme, H., and Ball, E. (1957), *Z. anorg. Chem.* **291**, 160: Addition compounds of triphenylboron with sulphides and tertiary amines.

Wittig, G., Keicher, G., Ruckert, A., and Raff, P. (1949), *Liebigs Ann.* **563**, 110: Preparation of triphenylboron with lithium phenyl and boron trifluoride.

Wittig, G., and Raff, P. (1951), *Liebigs Ann.* **573**, 195. Complex formation with triphenylboron.

Wittig, G., and Herwig, W. (1955), *Chem. Ber.* **88**, 962: Preparation of all three tritolylborons using a Grignard reagent and boron trifluoride.

Wittig, G., Stilz, W., and Herwig, W. (1956), *Liebigs Ann.* **598**, 85: Triphenylboron to remove metalhydrides.

Wittig, G., and Stilz, W. (1956), *Liebigs Ann.* **598**, 93: New synthesis of cyclopolyenes using triphenylboron.

8. *Casignost* (*sodium triphenylborocyanide*)

Schultz, O. E., and Goerner, H. (1955), *Arch. Pharm. Berl.* **288**, 520: Qualitative and quantitative estimation of nitrogen—containing pharmaceuticals with "Casignost".

Wittig, G., and Raff, P. (1951), *Liebigs Ann.* **573**, 195: Preparation of "Casignost".

9. *Tetra-aryl Boron Compounds*

Brown, H. C., and Subba Rao, B. C. (1956), *J. Amer. chem. Soc.* **78**, 5694: New techniques in the conversion of olefins into organoboranes and related alcohols.

Hannig, E., and Haendler, H. (1957), *Arch. Pharm. Berl.* **290**, 131: Characterization of *N*-substituted monoethanolamines as tetraphenylboron and triphenylborocyanide salts.

Krüerke, U. (1956), *Z. Naturf.* **11b**, 364: New organoboron compounds. $NaB(C \equiv CPh)_4$.

Krüerke, U. (1956), *Z. Naturf.* **11b**, 606: Preparation of $NaB(C \equiv CPh)_4$.

Krüerke, U. (1957), *Z. Naturf.* **11b**, 676: New organic boron compounds (acetylene derivatives of boron including $NaB(C \equiv CPh)_4$).

Mikhailov, B. M., and Aronovich, P. M. (1954), *C. R. Acad. Sci. U.R.S.S.* **98**, 791: Complex organoboron compounds.

Mikhailov, B. M., and Aronovich, P. M. (1956), *Bull. Acad. Sci. U.R.S.S., Classe sci. chim.* 322: Preparation of *iso*-butyl esters of phenylboric and diphenylboric acids with aid of PhLi including the preparation of $Ph_2B(OBu\text{-}iso)_2Li$.

Mikhailov, B. M., Kozminskaya, T. K., Blokhina, A. N., and Schegoleva, T. A. (1956), *Bull. Acad. Sci. U.R.S.S., Classe sci. chim.* 692; *English Translation* 703: Complex nature of organic boron acids. Preparation of $PhB(OH)_3 \cdot Na$.

Mikhailov, B. M., and Vaver, V. A. (1956), *Bull. Acad. Sci. U.R.S.S., Classe sci. chim.* 451; *English Translation*, p. 441: Synthesis and properties of diarylboric acids. Preparation of $Ar_2B(OH)_2 \cdot Na$.

Mikhailov, B. M., and Vaver, V. (1956), *C. R. Acad. Sci. U.R.S.S.* **109**, 94: Synthesis of complex compounds of triarylborons.

Mikhailov, B. M., Kostroma, T. V., and Fedotov, N. S. (1957), *Bull. Acad. Sci. U.R.S.S., Classe sci. chim.* 589: Organoboron compounds with asymmetric boron atom. Preparation of $(o\text{-}MeC_6H_4)_2BPh_2 \cdot Li$.

Neu, R. (1956), *Mikrochim. Acta*, 1169: Microgravimetric determination of 3-hydroxyflavones with arylboric acids.

Neu, R. (1957), *Naturwissenschaften* **44**, 181: Chelates of diarylboric acids with aliphatic hydroxyalkylamines as reagents for detection of hydroxyphenylbenzo-γ-pyrones.

Nielsen, D. R., McEwen, W. E., and Van der Werf, C. A. (1957), *Chem. & Ind.* 1069: Formation of triphenylboron by the reduction of $PhBCl_2$.

Razuvaev, G. A., and Brilkina, T. G. (1953), *C. R. Acad. Sci. U.R.S.S.* **91**, 861: Preparation and properties of lithium triphenyl-α-naphthylboron (the phenyl tri-α-naphthylboron and triphenyl *p*-tolylboron salts also made).

Razuvaev, G. A., and Brilkina, T. G. (1954), *J. gen. Chem., Moscow* **24**, 1415; *English Translation*, **24**, 1397: Preparation of compounds of the type $MBPh_4$.

Rondestvedt, C. S., Scribner, R. H., and Wulfman, C. E. (1955), *J. org. Chem.* **20**, 9: Alcoholysis of triarylboron compounds.

Sazonova, V. A., and Sorokina, L. B. (1955), *C. R. Acad. Sci. U.R.S.S.* **105**, 993: Tetra (1-pyrryl.) boron salts.

Sazonova, V. A., and Kronrod, N. Ya. (1956), *J. gen. Chem., Moscow* **26**, 1876; *English Translation*, **26**, 2093: Reaction of potassium fluoroborate with styryl and phenylethynyl magnesium bromide.

Sazonova, V. A., Serebryakov, E. P., and Kovaleva, L. S. (1957), *C. R. Acad. Sci. U.R.S.S.* **113**, 1295: Analytical properties and production of tetra (α-thienyl) and tetra(anisyl) borates of alkali metals.

Sazonova, V. A., and Serebryakov, E. P. (1957), U.S.S.R. patent, 106,396: Potassium tetra(2-thienyl)boron.

U.S. Borax & Chem. Corp. (1958), Belg. Patent, 563,585: Reaction of sodium alkyls or aryls with boron compounds to form products with 1 to 4 B—C bonds.

Wendlandt, W. W. (1958), *Chem. Anal.* **47**, 6: Thermal decomposition of amine-BPh_4 and $CNBPh_3$ compounds.

Wendlandt, W. W. (1958), *Chem. Anal.* **47**, 38: Thermal decomposition of metal $CNBPh_3$ compounds.

Wittig, G., and Raff, P. (1951), *Liebigs Ann.* **573**, 195: Preparation and properties of Li $PhC \equiv CBPh_3$.

Wittig, G., and Torssell, K. (1953), *Acta chem. scand.* **7**, 1293: Preparation and properties of $(Me_4Sb)(MeBPh_3)$.

Wittig, G., and Herwig, W. (1955), *Chem. Ber.* **88**, 962: New sodium tetra-aryl boron salts and their properties.

Wittig, G., and Schloeder, H. (1955), *Liebigs Ann.* **592**, 38: Addition of $Ph_3C.Na$ to butadiene in the presence of triphenylboron.

Wittig, G., and Polster, R. (1956), *Liebigs Ann.* **599**, 1: Preparation of triphenyl dimethylphenylmethyl boron salts of potassium and tetramethylammonium.

Wittig, G. (1957), *XVIth Int. Congr. Pure Appl. Chem., Experientia Suppl. VII* 291; (1958), *Angew. Chem.* **70**, 65: Complex formation and reactivity in organo-metallic chemistry.

10. *Miscellaneous Analytical Uses of Sodium Tetraphenylboron*

Braeuiger, H., and Spangenberg, K. (1954), *Pharmazie* **9**, 623: Analysis of thiofalicaine (alkylthiophenyl keto bases) and preparation and titration of the tetraphenylboron derivatives.

Kahn, B., Smith, D. K., and Straub, C. P. (1957), *Analyt. Chem.* **29**, 1210: Determination of low concentrations of radioactive caesium in water. Mention of caesium tetraphenylboron preparation.

Montequi, R., Doadrio, A., and Serrano, C. (1957), *An. Fis. Quim.* **53B**, 447: Determination of alkali metal tetraphenylboron salts with mercury.

Montequi, R., Doadrio, A., and Serrano, C. (1957), *Inform. quim. anal. Suppl. Ion (Madrid)* **11**, 8: Direct alkalimetry of potassium tetraphenylboron in the presence of mercuric chloride.

Musil, A., and Reimer, H. (1956), *Z. anal. Chem.* **152**, 154: Determination of sodium *via* ion exchange with Lewatite S100 (potassium formed and precipitation of potassium tetraphenylboron in effluent).

Neu, R. (1957), *Fette u. Seif.* **59**, 823: Detection and determination of poly-alkylene oxides with sodium tetraphenylboron.

Neu, R. (1958), *Seifen-Öle* **84**, 167: Determination 1,ω-dihydroxypoly-ethyleneoxides with sodium tetraphenylboron.

Shinagawa, M., and Matsuo, H. (1956), *J. Jap. Chem.* **10**, 111: Organic "onium" compounds, application to analytical chemistry (review, in Japanese, of 100 references including mention of sodium tetraphenylboron).

11. *Determination of Thallium with Sodium Tetraphenylboron*

Alimarin, I. P., Gibalo, I. M., and Sirotina, I. A. (1957), *Intern. J. Appl. Radiation and Isotopes* **2**, 117: Radiometric titration of thallium (I) with sodium tetraphenylboron.

Geilmann, W. (1954), *Angew. Chem.* **66**, 454: Determination of thallium (I) as the tetraphenylboron compound.

Sant, B. R., and Mukherji, A. K. (1959), *Talanta* **2**, 154: Titrimetric determination of tetraphenylboron by thallium.

Sirotina, I. A., and Alimarin, I. P. (1957), *J. anal. Chem., Moscow* **12**, (3) 367: Determination of thallium (I) by radiometric titration with sodium tetraphenylboron.

Wendlandt, W. W. (1957), *Analyt. chim. acta* **16**, (3) 216: Gravimetric determination of thallium as tetraphenylboron salt.

Wendlandt, W. W. (1957), *Chem. Anal.* **46** (I), 8: Conductometric precipitation of thallium (I) with sodium tetraphenylboron.

Yamaguchi, R., Osawa, K., and Yamaguchi, M. (1958), *Hoshi Yakka Daigaku Kiyô*, **7**, 10, 14: Specific qualitative analysis of Ag, Tl(I), and Hg(II) ions and separatory determination of Ag and Tl(I) ions with sodium tetraphenylboron.

12. *Determination of Potassium and Nitrogen Bases (covered in the same paper)*

Gloss, G. H. (1953), *Chem. Anal.* **42**, 50: Sodium tetraphenylboron, a new analytical reagent for potassium, ammonium and some organo-nitrogen bases.

Hahn, F. L. (1954), *Ciencia, Méx* (published 1955), **14**, 249: "Politest" (NaBPh$_4$), reagent for determination of potassium, ammonium and organic bases.

Kohler, M. (1953), *Z. anal. Chem.* **138**, 9: Determination of potassium and ammonium with sodium tetraphenylboron in solutions containing mineral acids and determination of both ions in one operation.

Neu, R. (1954), *Z. anal. Chem.* **143**, 254: Simultaneous determination of potassium and triethanolamine with sodium tetraphenylboron.

Rüdorff, W., and Zannier, H. (1953), *Z. anal. Chem.* **140**, 241: Titrimetric determination of potassium and ammonium in the presence of one another.

Rüdorff, W., and Zannier, H. (1954), *Angew. Chem.* **66**, 638: Argentiometric method with sodium tetraphenylboron and potassium and nitrogen bases.

13. *More Elaborate Analytical Techniques as Applied to Potassium and Nitrogen Bases*

A. *Turbidimetric*

De La Rubia, P. J., and Blasco, L. R. F. (1955), *Inform. Quím. Anal. Suppl. Ion (Madrid)* **9**, 1, 21: Turbidimetric microdetermination of potassium with sodium tetraphenylboron.

De La Rubia, P. J., and Blasco, L. R. F. (1955), *Chem. Anal.* **44**, 58: Turbidimetric determination of potassium with sodium tetraphenylboron.

Power, M. H., and Ryan, C. (1956), paper *Intern. Congr. Clin. Chem.*, *N.Y.*, Sept. Abstract in *Clin. Chem.* (1956) **2**, 230: Turbidimetric estimation of potassium in biological fluids with sodium tetraphenylboron.

B. *Polarographic*

De Vries, T., and Findeis, A. F. (1955), *Abs. Meeting Amer. Chem. Soc.*, Minneapolis, Sept., 9b: Polarographic work on potassium tetraphenylboron in NN-dimethylformamide. Determination of potassium.

Findeis, A. F., and De Vries, T. (1956), *Analyt. Chem.* **28**, 209: Polarography of potassium tetraphenylboron in NN-dimethylformamide. (Normal solvents not satisfactory.)

Findeis, A. F., Jr. (1957), *Dissertation Abstr.* **17**, 1909, *Microfilms Publ.* No. 21,281, p. 188: Polarography in non-aqueous solvents. Polarography and amperometric titration of the tetraphenylboron ion.

Heyrovsky, A. (1956), *Vnitřni Lékařstri* **2**, 234: Polarographic determination of potassium in biological materials (*via* $KBPh_4$ pptn. and ignition to KBO_2).

Heyrovsky, A. (1956), *Chem. Listy* **50**, 69; (1956), *Coll. Trav. chim. Tchécosl.* **21**, 1150: Polarographic microdetermination of potassium (*via* $KBPh_4$ and ignition to KBO_2).

Heyrovsky, A. (1958), *Chem. Listy* **52**, 40; (1959), *Coll. Trav. chim. Tchécosl.* **24**, 170: Potentiometric and polarometric titrations with tetraphenylboron salts. Argentometric determinations of potassium, thalium (I) and organic bases.

Kemula, W., and Kornacki, J. (1954), *Roczn. Chem.* **28**, No. 4, 635: Indirect polarographic (amperometric) potassium estimation by sodium tetraphenylboron.

C. Conductometric and Potentiometric, etc.

Amin, A. M. (1956), *Chem. Anal.* **45** (3) 65: Determination of ammonium, potassium, rubidium and caesium *via* cationic resin exchange of their tetraphenylboron ions.

Clark, S. J. (1956), *Chem. Age, Lond.* **74**, 21: Spectrophotometric titrations including heterometric titrations of potassium as its tetraphenylboron salt.

Crane, F. E., Jr. (1957), *Analyt. chim. acta* **16**, (4) 370: Potentiometric determination of the tetraphenylboron ion with silver nitrate.

Findeis, A. F., and De Vries, T. (1956), *Analyt. Chem.*, **28**, (12), 1899: Amphometric titrations of the tetraphenylboron ion and a method for potassium.

Flaschka, H., and Sadek, F. (1956), *Chem. Anal.* **45**, 20: Titrimetric determination of potassium *via* precipitation as the tetraphenylboron salt and cationic exchange.

Franck, U. F. (1958), *Z. Elektrochem.* **62**, 245: Polarization titrations ($NaBPh_4$ *vs.* $AgNO_3$).

Havir, J. (1958), *Chem. Listy* **52**, 1274: Indirect potentiometric titration of potassium, rubidium and caesium with sodium tetraphenylboron.

Jander, G., and Anke, A. (1957), *Z. anal. Chem.* **154** (1) 8: Conductometric precipitation titration of potassium with sodium tetraphenylboron.

Kirsten, W. J., Berggren, A., and Nilsson, K. (1958), *Analyt. Chem.* **30**, 237: Potentiometric titrations of some organic and inorganic bases with sodium tetraphenylboron.

Lane, E. S. (1957), *Analyst* **82**, 406: 250-megacycle high-frequency titrimeter. Titration of potassium with sodium tetraphenylboron.

Neeb, K. H., and Gebauhr, W. (1958), *Z. anal. Chem.* **162**, 167: Flame photometric determination of potassium in sodium and its compounds after enrichment *via* precipitation with ammonium tetraphenylboron.

Pflaum, R. T., and Howick, L. C. (1956), *Analyt. Chem.* **28** (10), 1542: Spectrophotometric determination of potassium as potassium tetraphenylboron.

Schmidt, H. J. (1957), *Z. anal. Chem.* **157**, 321: Volumetric determination of potassium by "dead stop" end-point method.

Stephen, W. L. (1956), *Mikrochim. Acta* 1540: Spot colorimetry with the Weisz ring oven technique including potassium as its tetraphenylboron salt.

14. *Analysis of Nitrogen Bases with Sodium Tetraphenylboron*

Aklin, O. (1957), Thesis, Faculté de Pharmacie, Univ. de Strasbourg. Gravimetric and titrimetric determination of opium alkaloids and derivative using sodium tetraphenylboron.

Aklin, O., and Durst, J. (1956), *Pharm. Acta Helvet.* **31**, (10), 457: Gravimetrical assay of papaverine and strychnine.

Balenovic, K., Bregant, N., and Stefanac, Z. (1957), *Croat. Chem. Acta* **29**, 45: Characterization and isolation in muscarine series including the tetraphenylboron salts of muscarine and acetylcholine.

Belcher, R. (1956), *Chem. Age* **74**, 435: Qualitative organic analysis: Recent Advances.

Brauniger, H., and Hofmann, R. (1955), *Pharmazie* **10**, 644: Determination of antihistaminics including the precipitation of tetraphenylboron salts.

Brookes, P., Terry, R. J., and Walker, J. (1957), *J. chem. Soc.*, 3165: Tetraphenylboron salt of dipiperidyl.

Chatten, L. G., Pernarowski, M., and Levi, L. (1959), *J. Amer. pharm. Ass.* **48**, 276: The identification of some local anaesthetics as tetraphenylborons.

Corrodi, H., Hardegger, E., Kögl, F., and Zeller, P. (1957). *Experimentia* **13**, 138: Synthesis of stereoisomers of muscarine tetraphenylboron salt.

Cox, H. C., Hardegger, E., Kögl, F., Liechti, P., Lohse, F., and Salemink, C. A. (1958), *Helv. chim. acta* **41**, 229: Isolation of synthetic (—)-muscarine as tetraphenylboron salts.

Crane, F. E., Jr. (1956), *Analyt. Chem.* **28**, 1794: Identification of amines as tetraphenylborates.

Crane, F. E., Jr. (1958), *Analyt. Chem.* **30**, 1426: Tetraphenylboron spot test for detection of amines and their salts.

Espersen, T. (1958), *Dansk Tidskr. Farm.* **32**, 99: Identification of nitrogen-containing organic drugs *via* sodium tetraphenylboron.

Eugster, C. H., Häfliger, F., Denss, R., and Girod, E. (1958), *Helv. chim. acta* **41**, 583: *dl*-Allomuscarone and *dl*-allomuscarine and their tetraphenylboron salts.

Fischer, R., and Karawia, M. S. (1953), *Microchim. Acta*, 366: Detection of analgesics and alkaloids by means of sodium tetraphenylboron.

Flaschka, H., Holasek, A., and Amin, A. M. (1954), *Arzneimittel-Forsch* **4**, 38: Microtitrimetric determination of several drugs after precipitation as tetraphenylboron salts.

Friedrick, W., and Bernhauer, K. (1956), *Chem. Ber.* **89**, 2030: Sodium tetraphenylboron in column chromatography of vitamin B_{12} Factor III.

Gautier, J. A. (1956), "Hommage au Doyen Rene Fabre", p. 187. Sedes, Paris: Sodium tetraphenylboron as reagent for basic nitrogen in pharmaceutical and toxicological analysis.

Gautier, J. A., Renault, J., and Pellerin, F. (1955), *Ann. pharm. franç.* **13**, 725. Preliminary note in *Angew. Chem.* (1955). **67**, 755: Alkalimetric determination of salts of organic bases after precipitation with sodium tetraphenylboron (ion exchange). (I) Quaternary ammonium compounds with long chains.

Gautier, J. A., Renault, J., and Pellerin, F. (1956), *Ann. pharm. franç.* **14**, (5) 337: Alkalimetric determination of salts of organic bases after precipitation as tetraphenylborons. (II) Alkaloid salts and salts of organic bases of pharmaceutical interest.

Gayer, J. (1956), *Biochem. Z.* **328**, 39: Identification of biogenetic amines in plasma (*via* tetraphenylborons) electrophoresis.

Hädicke, K. (1958), *Pharm. Zentralh.* **97**, 365: Determination of piperazine in pharmaceuticals as tetraphenylboron derivatives.

Hardegger, E., and Lohse, F. (1957), *Helv. chim. acta* **40**, 2383: Isolation of synthetic muscarine as a tetraphenylboron salt.

Hardegger, E., Furter, H., and Kiss, J. (1958), *Helv. chim. acta* **41**, 2401: Isolation of (−)-muscarine as a tetraphenylboron salt.

Ievins, A. F., Gudriniece, E. Y. (1956), *J. anal. Chem., Moscow* **11**, 735: *English Translation, II*, 789. Volumetric determination of aliphatic and aromatic amines with sodium tetraphenylboron.

Keller, W., and Weiss, F. (1957), *Pharmazie* **12**, 19: Estimation of alkaloids especially ephedrine and other bases with sodium tetraphenylboron (also potassium ions in pharmaceuticals).

Kranjcevic, M., and Broz-Kajganovic, V. (1958), *Croat. Chem. Acta* **30**, 47: Determination of codeine and aminopyrine . . . (with sodium tetraphenylboron).

Kuhn, R., and Osswald, G. (1957), *Angew. Chem.* **69**, 60: dl-β-hydroxy pyrrolidine and N-methyl derivative isolated as the tetraphenylboron salt.

Marquardt, P., and Vogg, G. (1952), *Hoppe-Seyl. Z.* **291**, 143: Sensitive assay of choline and acetylcholine by means of sodium tetraphenylboron.

Meyer, H., Schmid, H., Waser, P., and Karrer, P. (1956), *Helv. chim. acta* **39**, 1214: Tetraphenylboron salt of a calabash alkaloid.

Montequi, M. R., and Santiso, F. (1957), *J. Méd. Bordeaux et Sud-Ouest* **133**, 650: Precision of two methods for the volumetric determination of strychnine including precipitation as tetraphenylboron salt and mercurization.

Neu, R. (1954), *Arzneimittel-Forsch.* **4**, 601: Compounds in Passiflora Incarnata (II) Basic Constituents (alkaloids with tetraphenylboron salt).

Neu, R. (1956), *Arzneimittel-Forsch.* **6**, 94: Compounds of Passiflora Incarnata (III), 3-methyl,4-carboline-alkaloid of passion flower, detection as tetraphenylboron salt.

Pahlow, M. (1953) *Dtsch. Apoth. Ztg.* **93**, 541: Kalignost for the analysis of pharmaceuticals.

Patel, D. M., and Anderson, R. A. (1958), *Drug Standards* **26**, 189: Determination of benzethonium chloride by tetraphenylboron method.

Rosenthaler, L., and Lüdy-Tenger, F. (1957), *Pharm. Acta Helvet.* **32**, 35: Potassium precipitants including sodium tetraphenylboron as alkaloid reagents.

Rutkowski, R. (1953), *Arzneimittel-Forsch.* **3**, 537: Quantitative assay of succinyl bis choline in injectable solutions.

Schultz, O. E., and Mayer, G. (1952), *Dtsch. Apoth. Ztg.* **92**, 358: Kalignost ($NaBPh_4$), new precipitant for the detection of nitrogen containing bases (pharmaceuticals).

Schultz, O. E., and Goerner, H. (1953), *Dtsch. Apoth. Ztg.* **93**, 585: Titrimetric determination of nitrogen containing pharmaceuticals with Kalignost.

Schultz, O. E., and Goerner, H. C. (1958), *Arch. Pharm., Berl.* **291**, 386: Isolation of alkaloids with Kalignost and regeneration of precipitant.

Schwaibold, J., and Kohler, M. (1953), *Landw. Jb. Bayern* 30, Heft. 1/2, 55–62: Application of new method of determination of K_2O and N content of fertilizers.

Scott, W. E., Doukas, H. M., and Schaffer, P. S. (1956), *J. Amer. pharm. Ass.*, (b) **45**, 568: Use of sodium tetraphenylboron as a means of identification and isolation of alkaloids.

Van Pinxteren, J. A. C., Verloop, M. E., and Westerink, D. (1956) *Pharm. Weekbl.* **91**, (24) 873: Determination of atropine and hyoscyamine with tetraphenylboron.

Wachsmuth, H., and Mertens, E. (1958), *J. Pharm. Belg.* **13**, 58: Tetraphenylboron color reaction applied to the determination of nitrogen bases.

Walter, W. (1958), *Angew. Chem.* **70**, 404: Thioacetamido-oxide, reaction with tetraphenylboron ion.

Wittig, G., and Ludwig, H. (1954), *Liebigs Ann.* **589**, 55: Preparation of 2 : 2-dimethyl-4 : 5-benzisoindolinium tetraphenylboron salt.

Wittig, G., and Polster, R. (1956), *Liebigs Ann.* **599**, 13: Sodium tetraphenylboron and quaternary ammonium bromide (used to characterize).

Worrell, L., and Ebert, W. R. (1956), *Drugs Standards* **24**, 153: Use of sodium tetraphenylboron for separation and determination of drugs with basic nitrogen groups.

Yanson, E., Ievins, A. F., and Gudriniece, E. (1957), *Uchen. Zap., Latv. Univ.* **14**, 9: Sodium tetraphenylboron in quantitative analysis. I. Volumetric determination of aliphatic amines.

Zeidler, L. (1952), Hoppe-Seyl. *Z.* **291**, 177: Tetraphenylboron compounds for the identification of hystamine and similar bases.

Zief, M., Woodside, R., and Huber, E. (1957), *Antibiotics & Chemotherapy* **7**, 604: Identification of antibiotics as tetraphenylboron derivatives.

15. *Imides*

Becke-Goehring, M., and Schwarz, R. (1958), *Z. anorg. Chem.* **296**, 3: Reaction of $S_2N_2 . NH_3$ with $NaMeBPh_3$.

Berg, W., Goehring, M. Z., and Malz, H. (1950), *Z. anorg. Chem.* **283**, 13: Di-imides of tetravalent sulphur and their potassium analysis as potassium tetraphenylboron.

Berg, W., and Goehring, M. (1954), *Z. anorg. Chem.* **257**, 273: Potassium imides of divalent and tetravalent sulphur (potassium analysis *via* sodium tetraphenylboron).

16. *Determination of Potassium as Potassium Tetraphenylboron*

Amin, A. M. (1953), *Resalet-el-Elm* **19**, 177: Kalignost new reagent for microdetermination of potassium.

Amin, A. M. (1954), *Risalatul-Kimia* **2**, No. 6, 407: Kalignost new reagent for microdetermination of potassium. Review.

Belcher, R., Nutten, A. J., and Thomas, H. (1954), *Anal. chim. acta* **11**, 120: Determination of sodium and potassium in coal ash.

Benzi, L. J. (1957), *Rev. bras. cir.* **34**, 189: New method of determination of potassium using lithium or sodium tetraphenylboron.

Berkhout, H. W. (1952), *Chem. Weekbl.* **48**, 909: Potassium determination as potassium tetraphenylboron.

Berkhout, H. W., and Jongen, G. H. (1955), *Chem. Weekbl.* **51**, 607: Potassium determination as potassium tetraphenylboron.

Berkhout, H. W., and Jongen, G. H. (1956), *Chem. Anal.* **45**, (1) 6: Determination of potassium by precipitation of tetraphenylboron salt using a simple filtration assembly.

Doadrio, A., and Serrano, C. (1957), *Inf. quím. anal. Suppl. Ion (Madrid)* **11**, (1), 8: Direct alkalimetry of potassium tetraphenylboron in the presence of mercuric chloride (presence of acetone as used in some other procedures was found disadvantageous).

Elster, K., and Otto, H. (1956), *Klin. Wschr.* **34**, 1139: Potassium content (by tetraphenylboron method) of human cadaver hearts in energetic dynamic cardiac insufficiency.

Engelbrecht, R. M., and McCoy, F. A. (1956), *Analyt. Chem.* **28**, (11), 1772: Determination of potassium by tetraphenylboron method.

Erdey, L., Buzas, I., and Vigh, K. (1958), *Talanta* **1**, 377: Argentometric titrations with redox indication including potassium tetraphenylboron.

Flaschka, H. (1952), *Z. anal. Chem.* **136**, 99: Kalignost (sodium tetraphenylboron) for microgravimetric determination of potassium as potassium tetraphenylboron.

Flaschka, H. (1955), *Chem. Anal.* **44**, (3), 60: Determination of potassium by precipitation and non-aqueous titration of its tetraphenylboron salts.

Flaschka, H., and Amin, A. M. (1953), *Chem. Anal.* **42**, 78: Rapid volumetric method for determination of micro amounts of sodium and potassium.

Flaschka, H., Amin, A. M., and Holasek, A. (1953), *Z. anal. Chem.* **138**, 241: New methods for titrimetric determination of potassium after precipitation as potassium tetraphenylboron.

Flaschka, H., Holasek, A., and Amin, A. M. (1953), *Angew. Chem.* **65**, 258: Microtitrimetric determination of potassium as potassium tetraphenylboron. (Abstract of paper Chemikertreffen, Innsbruck, 1953.)

Flaschka, H., Holasek, A., and Amin, A. M. (1953), *Z. anal. Chem.* **138**, 161: Acidimetric determination (micro) of potassium after precipitation with the tetraphenylboron ion.

Flaschka, H., and Abdine, H. (1955), *Z. anal. Chem.* **144**, 415: Rapid alkalimetric determination of potassium after precipitation with sodium tetraphenylboron.

Gagliardi, E., and Reimers, H. (1958), *Z. anal. Chem.* **160**, 1: Complexometric determination of sodium, potassium and their sum *via* ion-exchange (with magnesium; determination of potassium by tetraphenylboron precipitation).

Gilman, W., and Geibauhr, W. (1953), *Z. anal. Chem.* **139**, 161. Brief note of this paper in *Angew. Chem.* **65**, 568: Precipitation of alkali metals as tetraphenylboron compounds.

Gilman, W., and Geibauhr, W. (1954), *Z. anal. Chem.* **142**, 241: Detection of potassium, rubidium and caesium in the wet way (a critical study).

Gübeli, O. (1954), Potassium symposium, *Ann. Meeting Board Tech. Advisers Internatl. Potash Inst.*, Zurich, 269: Chemical methods for the determination of potassium.

Gübeli, O. (1956), *Mitt. Lebensm. Hyg.*, Bern. **47**, 305: Modern methods of mineral water study (including potassium *via* potassium tetraphenylboron).

Hahn, F. L. (1955), *Z. anal. Chem.* **145**, 97: Titrimetric potassium determination *via* potassium tetraphenylboron.

Havir, J. (1956), *Voda* **12**, 402: The determination of potassium in water using sodium tetraphenylboron.

Havir, J., and Trkan, M. (1956), *Kvasný průmsyl* **2**, 274: The determination of potassium in barley and malt.

Hegemann, F., and Pfab, B. (1955), *Glastech. Ber.* **28**, 242: Procedure for rapid and exact determination of sodium and potassium involving the precipitation of potassium tetraphenylboron.

Howick, L. C. (1957), *Dissertation Abstr.* **17**, 2841, *Microfilms Publ.*, No. 23751: Analytical aspects of tetraphenylboron salts.

Ieviņš, A. F., and Gudriniece, E. Yu. (1954), *J. anal. Chem., Moscow* **9**, 270: Determination of potassium with sodium tetraphenylboron.

Ieviņš, A. F., and Gudriniece, E. Yu. (1954), *Nachr. Akad. Wiss. Lett.* **8**, 131: Determination of potassium with, and the preparation of, sodium tetraphenylboron.

Jongen, G. H., and Berkhout, H. W. (1955), *Chem. Weekbl.* **51**, 607: Determination of potassium with sodium tetraphenylboron.

Kingsley, W. K., and Wolfram, W. (1957), *Analyt. Chem.* **29**, (6), 939: Determination of trace of potassium in reagent chemicals with sodium tetraphenylboron and sodium cobaltinitrite.

Klevstrand, R. (1955), *Medd. norsk. farm. Selsk.* **17**, (4–5), 190: Determination of sodium and potassium in compound injections. Titrimetric and gravimetric methods.

Kornacki, J. (1954), *Wiadomości Chem.* **8**, 538: Application of lithium and sodium tetraphenylboron to the determination of potassium. Review.

Mevel, N., and Lacruche, G. (1958), *Mikrochim. Acta* 241: Titrimetric determination of potassium after precipitation as the tetraphenylboron salt.

Muraca, R. F. (1954), *Chem. Anal.* **43**, 69: Student performance in the determination of potassium with sodium tetraphenylboron.

Muraca, R. F., Collier, H. E., Bonsock, J. P., and Jacobs, E. S. (1954), *Chem. Anal.* **43**, 102: Sodium tetraphenylboron as used for the detection of potassium in systematic qualitative procedures.

Muraire, M. (1957), *Chim. anal.* **39**, (5), 184: Qualitative examination and detection of potassium with sodium tetraphenylboron.

Neumann, F. (1953), *Papier, Darmstadt* **7**, 388: Rapid hydrogen peroxide oxidation of sulphite of spent liquors in determination of inorganic components including potassium determined as potassium tetraphenylboron.

Oak Ridge National Lab. (1958), "Master Analytical Manual", U.S.A.E.C. Rept. TID-7015, Proc. 1-216451: Potassium by the gravimetric tetraphenylboron method.

Pilleri, R. (1957), *Z. anal. Chem.* **157**, 1: Complexometric determination of potassium.

Raff, P., and Brotz, W. (1951), *Z. anal. Chem.* **133**, 241: Potassium determination as potassium tetraphenylboron.

Reimers, H. (1957), *Chemikerztg.* **81**, 357: Reconversion of potassium tetraphenylboron.

Renault, J. (1958), *Mises au point chim. anal. pure et appl. et anal. bromatol.*
6, 109: Sodium tetraphenylboron as an analytical reagent.

Riva, B. (1958), *Ann. chim. Roma* **48,** 50: The determination of potassium in
common salt using sodium tetraphenylboron.

Rüdorff, W., and Zannier, H. (1952), *Angew. Chem.* **64,** 613: Brief abstract
of paper "Sudwestdeutsche Dozeutentagung Freiburg", Oct. 6th (1951).
Titrimetric determination of potassium *via* sodium tetraphenylboron.

Rüdorff, W., and Zannier, H. (1952), *Z. anal. Chem.* **137,** 1: Titrimetric
determination of potassium *via* sodium tetraphenylboron.

Rüdorff, W., and Zannier, H. (1953), *Z. anal. Chem.* **140,** 1: Rapid titrimetric
determination of potassium with sodium tetraphenylboron.

Schall, E. D. (1957), *Analyt. Chem.* **29,** 1044: Volumetric determination of
potassium.

Schmeid, W., and Stegmüller, L. (1957), *Ber. dtsch. keram. Ges.* **34,** 135:
Use of Komplexon III (EDTA) for analysis of ceramic materials (and
sodium tetraphenylboron for potassium).

Schneyder, J. (1957), *Fachliche Mitt. Österr. Tabakregie* 25..: Rapid analysis of
tobacco ash including potassium precipitated as its tetraphenyl boron salt.

Spier, H. W. (1952), *Biochem. Z.* **322,** 467: Microdetermination of potassium
(oxidimetric determination of potassium *via* potassium tetraphenylboron.

Spier, H. W., and Wagner, G. (1952), *Klin. Wschr.* **30,** 757: Tetraphenylboron
ion as a potassium blocking anion.

Sporek, K., and Williams, A. F. (1955), *Analyst* **80,** 347: Quantitative
determination of potassium as its tetraphenylboron salt.

Sunderman, F. W., Jr., and Sunderman, F. A. (1958), *Amer. J. clin. Path.*
29, 95: Rapid, reliable tetraphenylboron method for serum potassium.

Teijgeler, C. A. (1957), *Chem. pharm. Tech.* **13,** 48, 57, 75, 90: Review of
sodium tetraphenylboron in analysis.

Zymny, E. (1955), *Prakt. Chem.* **6,** (12), 327: Potassium in water and effluents
as potassium tetraphenylboron.

17. *Determination of Traces of Potassium*

Kingsley, W. K., Wolfram, W., and Wolf, G. (1956), *Abs. Meeting Amer.
Chem. Soc.* Atlantic City 23B: A comparative study of trace potassium
estimation in reagent chemicals by tetraphenylboron and cobaltinitrite
methods.

Kingsley, W. K., Wolf, G. E., and Wolfram, W. E. (1957), *Analyt. Chem.* **29,**
939: Detection of traces of potassium in reagent chemicals using sodium
tetraphenylboron.

Tendrille, C. (1955), *Ann. Inst. nat. Rech. agron., Paris Sér. A.* **6,** 1055.
(Through *Chim. anal.* (1956), **38,** 371): Potassium determination with
sodium tetraphenylboron.

Vittori, O. (1956), *Tech. Notes Cloud Physics Lab.* (Univ. of Chicago) No. 5,
18019: Chemical composition of atmospheric particles including the
detection of potassium as potassium tetraphenylboron.

18. *Determination of Potassium in:*

A. *Milk*

Schober, I. R., and Fricker, A. (1952), *Z. Lebensmitt.-Untersuch* **95,** 107: New
methods for the determination of potassium in milk.

Schober, I. R., and Fricker, A. (1953), *Z. Lebensmitt.-Untersuch* **97**, 177: New methods for the determination of potassium in milk (II). Potassium content of allgau milk from normal and mastitis infected secretion.

B. *Wines*

Bonastre, J. (1955), *Ann. Falsif., Paris* **48**, 347: Comparison of some methods for the determination of potassium in wines.

Garino-Canina, E. (1944–5), *Ann. Accad. Agric. Torino*, **90**, 1: Rapid determination of potassium with Kalignost, applied to wine-growing.

Reichard, O. (1953), *Z. anal. Chem.* **140**, 188: Determination of potassium in wines—modification of official procedure.

Reichard, O. (1954), *Wein u. Rebe* **32**, 668: Chemical determination of potassium in wines and its diagnostic significance.

C. *Silicates Including Glass*

Cluley, H. J. (1955), *Analyst* **80**, 354: Determination of potassium by the precipitation of potassium tetraphenylboron and its application to silicates (a good summary of early gravimetric methods).

Flaschka, H. (1955), *Sprechsaal* **88**, (9), 188: Rapid analysis in glass technology laboratory.

Levina, N. D., and Panteleeva, L. I. (1957), *Fact. Lab., Moscow* **23**, (3), 285: Use of sodium tetraphenylboron for determination of potassium in glass.

Lieber, W. (1957), *Zement-Kalk-Gips* **10**, 62: Quantitative determination of potassium with sodium tetraphenylboron.

Mashika, Y., and Kaurajii, Y. (1956), *J. pharm. Soc. Japan* **76**, 689: Mineral Analysis Part (III) Determination of potassium.

Yankov, H. F. (1959), *Chem. Anal.* **48**, 38: Gravimetric tetraphenylboron method for potassium in glass and silicates.

Zymny, E. (1955), *Glas-Email-Keramo-Tech.* **6**, 236: Determination of alkalis in silicates (titration of potassium tetraphenylboron with silver nitrate).

D. *Gunpowders, etc.*

Emeury, J. M. (1956), *Mémor. Poud.* **38**, 357: Gravimetric determination of potassium salts in propellants.

Havir, J., and Vrestal, J. (1957), *Chem. zvesti* **11**, 35: Estimation of potassium in gunpowder by sodium tetraphenylboron.

Parpaillon, M. (1957), *Mémor. Poud.* **39**, 417: The determination, in explosives, of potassium as its tetraphenylboron salt *via* attack with sodium peroxide.

Tranchant, J., and Marvillet, L. (1956), *Mémor. Poud.* **38**, 337: Determination of potassium sulphate in dinitrotoluene. A comparison of several methods.

E. *Blood*

Flaschka, H., and Holasek, A. (1956), *Hoppe-Seyl. Z.* **303**, 9: Complexometric determination of potassium in blood serum.

Jancic, M. S. (1955), *Bull. soc. chimistes rép. populaire Bosnie et Herzégovine* **3**, 37: Method of micridetermination of potassium in blood serum.

Teeri, A. E., and Sesin, P. G. (1957), *Tech. Bull. Registry Med. Technologists* **27**, 280: Determination of potassium, turbidimetrically as potassium tetraphenylboron, in blood serum.

Teeri, A. E., and Sesin, P. G. (1958), *Amer. J. clin. Path.* **29**, 86: Determination of potassium in blood serum as potassium tetraphenylboron.

F. Plants

Amin, A. M. (1954), *Chem. Anal.* **43**, 4: Rapid microvolumetric method of determination of potassium in plants using sodium tetraphenylboron as precipitant.

G. Sea Water

Sporek, K. (1956), *Analyst* **81**, 540: Gravimetric determination of potassium in sea water as potassium tetraphenylboron.

H. Fertilizers

Anon. (1958), *J. Ass. off. agric. Chem., Wash.* **41**, 32: Potassium in fertilizer by tetraphenylboron method.

Epps, E. M., and Burden, J. C. (1958), *Analyt. Chem.* **30**, 1882: Volumetric determination of potassium in fertilizers.

Fontana, P., and Zanetti, B. (1956), *Pubbl. univ. cattolica S. Cuore, Ann. fac. agrari* **60**, 201: Potassium determination in fertilizers *via* tetraphenylboron method.

Ford, O. W. (1956), *J. Ass. off. agric. Chem. Wash.* **39**, 598: Comparison of the tetraphenylboron method for potassium with other methods in fertilizer analysis.

Ford, O. W. (1958), *J. Ass. off. agric. Chem. Wash.* **41**, 533: Report on potassium in fertilizers.

Solari, J. (1956), *Industr. agric.* **73**, (1), 25: Rapid determination of nitrogen, phosphorus and potassium in fertilizers.

Solari, J. (1956), *Chim. anal.* **38**, 91: Rapid determination of nitrogen, phosphorus and potassium in fertilizers.

19. Determination of Radioactive Potassium

Munroe, D. S., Reuschler, H., and Wilson, G. M. (1955), *J. Physiol* **128**, 68P: Uses of physical methods and sodium tetraphenylboron in separation of radioactive potassium (42) and sodium (24) in biological fluids.

Munroe, D. S., Renschler, H., and Wilson, G. M. (1958), *Phys. in Med. Biol.* **2**, 239: Assay of mixt. of ^{24}Na and ^{42}K in clinical tracer studies; measurement of exchangeable sodium and potassium (Potassium enrichment *via* tetraphenylboron precipitation).

REFERENCES

Abdel-Akher, M., Hamilton, J. K., and Smith, F. (1951), *J. Amer. chem. Soc.* **73**, 4691.

Abel, E. W., Dandegaonker, S. H., Gerrard, W., and Lappert, M. F. (1956), *J. chem. Soc.* 4697.

Abel, E. W., Edwards, J. D., Gerrard, W., and Lappert, M. F. (1957), *J. chem. Soc.* 501.

Abel, E. W., Gerrard, W., and Lappert, M. F. (1957), *J. chem. Soc.* 112, 3833, 5051.

Abel, E. W., Gerrard, W., Lappert, M. F. (1958), *J. chem. Soc.* 1451.

Abel, E. W., Gerrard, W., and Lappert, M. F. (1958), *Chem. & Ind.* 158.

Abraham, M. H., and Davies, A. G. (1957), *Chem. & Ind.* 1622.

Abraham, M. H., and Davies, A. G. (1959), *J. chem. Soc.* 429.

Accascina, F., Petrucci, S., and Fuoss, R. M. (1959), *J. Amer. chem. Soc.* **81**, 1301.

Ahmad, T., Haider, S. Z., and Khundkar, M. H. (1954), *J. appl. Chem.* **4**, 543.

Ahmad, T., and Khundkar, M. H. (1954), *Chem. & Ind.* 248.

Ainley, A. D., and Challenger, F. (1930), *J. chem. Soc.* 2171.

Aklin, O., and Dürst, J. (1956), *Pharm. Acta Helvet.* **31**, 457.

Allen, E. C., and Sugden, S. (1932), *J. chem. Soc.* 760.

Allen, S., Bonner, T. G., Bourne, E. J., and Saville, N. M. (1958) *Chem. & Ind.* 630.

Alton, E. R., Brown, R. D., Carter, J. C., and Taylor, R. C. (1959), *J. Amer. chem. Soc.* **81**, 3550.

Altshuller, A. P. (1955), *J. Amer. chem. Soc.* **77**, 5455.

Amin, A. M. (1954), *Chem. Anal.* **43**, 4.

Ananthakrishnan, R. (1937), *Proc. Indian Acad. Sci.* **5A**, 200.

Anderson, H. C., and Belz, L. H. (1953), *J. Amer. chem. Soc.* **75**, 4828.

Anderson, J. R., O'Brien, K. G., and Reuter, F. H. (1952), *J. appl. Chem.* **2**, 241.

Anderson, T. F., and Burg, A. B. (1938), *J. chem. Phys.* **6**, 586.

Anderson, W. E., and Barker, E. F. (1950), *J. chem. Phys.* **18**, 698.

Angyal, S. J., and Macdonald, C. G. (1952), *J. chem. Soc.* 686.

Angyal, S. J., and McHugh, D. J. (1956), *Chem. & Ind.* 1147.

Angyal, S. J., and McHugh, D. J. (1957), *J. chem. Soc.* 1423.

Antikainen, P. J. (1955), *Acta chem. scand.* **9**, 1008.

Antikainen, P. J. (1956), *Suomen Kemistilehti* **29B**, 179.

Antikainen, P. J. (1957), *Suomen Kemistilehti* **30B**, 185.

Antikainen, P. J. (1958), *Suomen Kemistilehti* **31B**, 255.

Appel, F. J. (1940), *U.S.P.* 2,217,354.

Arbuzov, B. A., and Vinogradova, V. S. (1947), *C. R. Acad. Sci. U.R.S.S.* **55**, 411.

Arimoto, F. S. (1955), *U.S.P.* 2,720,448; 2,720,449.

Arnold, H. R. (1946), *U.S.P.* 2,402,589; 2,402,590.

Arnold, J. R. (1954) *J. chem. Phys.* **22**, 757.

Asahara, T., and Kanabu, K. (1952), *J. chem. Soc. Japan, Ind. Chem. Sect.* **55**, 589.

Ashby, E. C. (1959), *J. Amer. chem. Soc.* 81, 4791.

Ashikari, N. (1958), *J. Polym. Sci.* **28**, 250.

Atoji, M., Wheatley, P. J., and Lipscomb, W. N. (1957), *J. chem. Phys.* **27**, 196.

Atteberry, R. W. (1958), *J. phys. Chem.* **62**, 1458.

Auten, R. W., and Kraus, C. A. (1952), *J. Amer. chem. Soc.* **74**, 3398.

Balacco, F. (1950), *Ann. Chim. Roma* **40**, 707.

Ballard, S. A. (1948), *U.S.P.* 2,431,224.

Baltimore Paint and Varnish Production Club (1955), *Off. Dig. Fed. Paint Varn. Prod. Cl.* **27**, 779.

Bamford, C. H., Levi, D. L., and Newitt, D. M. (1946), *J. chem. Soc.* 468.

Bamford, C. H., and Newitt, D. M. (1946), *J. chem. Soc.* 695.

Bannister, W. J. (1928), *U.S.P.* 1,668,797.

Banus, M. D., and Gibbs, T. R. P. (1951), *U.S.P.* 2,542,746.

Banus, M. D., Bragdon, R. W., and Gibbs, T. R. P. (1952), *J. Amer. chem. Soc.* **74**, 2346.

Banus, M. D., Gibbs, T. R. P., and Bragdon, R. W. (1954), *U.S.P.* 2,678,949.

Banus, M. D., Bragdon, R. W., and Hinckley, A. A. (1954), *J. Amer. chem. Soc.* **76**, 3848.

Barker, S. A., and Bourne, E. J. (1952), *J. chem. Soc.* 905.

Barnes, R. F., Diamond, H., and Fields, P. R. (1956), *U.S.P.* 2,739,979.

Barnes, R. P., Graham, J. H., and Taylor, M. D. (1958), *J. org. Chem.* **23**, 1561.

Bashkirov, A. N. (1956), *Khim. Nauka i. Prom.* **1**, 273.

Bauer, S. H. (1937), *J. Amer. chem. Soc.* **59**, 1096, 1804.

Bauer, S. H. (1938), *J. Amer. chem. Soc.* **60**, 524, 805.

Bauer, S. H. (1942), *Chem. Rev.* **31**, 43.

Bauer, S. H. (1950), *J. Amer. chem. Soc.* **72**, 622.

Bauer, S. H. (1956), *J. Amer. chem. Soc.* **78**, 5775.

Bauer, S. H., and Pauling, L. (1936), *J. Amer. chem. Soc.* **58**, 2403.

Bauer, S. H., and Beach, J. Y. (1941), *J. Amer. chem. Soc.* **63**, 1394.

Bauer, S. H., and Hastings, J. M. (1942), *J. Amer. chem. Soc.* **64**, 2686.

Bauer, S. H., Shepp, A., and McCoy, R. E. (1953), *J. Amer. chem. Soc.* **75**, 1003.

Bauer, S. H., and McCoy, R. E. (1956), *J. phys. Chem.* **60**, 1529.

Bawn, C. E. H., and Ledwith, A. (1958), *Chem. & Ind.* 1329.

Bawn, C. E. H., Margerison, D., and Richardson, N. M. (1959), *Proc. chem. Soc., Lond.* 397.

Bax, C. M., Katritzky, A. R., and Sutton, L. E. (1958), *J. chem. Soc.* 1254, 1258.

Beach, J. Y., and Bauer, S. H. (1940), *J. Amer. chem. Soc.* **62**, 3440.

Beachell, H. C., and Lange, K. R. (1956), *J. phys. Chem.* **60**, 307.

Beachell, H. C., and Meeker, T. R. (1956), *J. Amer. chem. Soc.* **78**, 1796.

Beachell, H. C., and Veloric, H. S. (1956), *J. phys. Chem.* **60**, 102.

Beachell, H. C., and Schar, W. C. (1958), *J. Amer. chem. Soc.* **80**, 2943.

Bean, F. R., and Johnson, J. R. (1932), *J. Amer. chem. Soc.* **54**, 4415.

Becher, H. J. (1952), *Z. anorg. Chem.* **270**, 273.

Becher, H. J. (1953), *Z. anorg. Chem.* **271**, 243.

Becher, H. J. (1954), *Z. physik. Chem. (Frankfurt)* **2**, 276.

Becher, H. J. (1956), *Z. anorg. Chem.* **287**, 235, 285.

Becher, H. J. (1957), *Z. anorg. Chem.* **289**, 262.

Becher, H. J. (1957), *Z. anorg. Chem.* **291**, 151.

Becher, H. J., and Frick, S. (1957), *Z. physik. Chem. (Frankfurt)* **12**, 241.

Becher, H. J., and Goubeau, J. (1952), *Z. anorg. Chem.* **268**, 133.

Becher, H. J., and Frick, S. (1958), *Z. anorg. Chem.* **295**, 83.

Bell, R. P., and Eméleus, H. J. (1948), *Quart. Rev. chem. Soc., Lond.* **2**, 132.

Bellamy, L. J. (1958), "Infra-red Spectra of Complex Molecules", pp. 125, 132, 161, 178. Methuen Co. Ltd., London.

Bellamy, L. J., and Williams, R. L. (1956), *J. chem. Soc.* 2753.

Bellamy, L. J., Gerrard, W., Lappert, M. F., and Williams, R. L. (1958), *J. chem. Soc.* 2412.

Bennett, H. (1934), *U.S.P.* 1,953,741.

Bent, H. E., and Dorfman, M. (1932), *J. Amer. chem. Soc.* **54**, 2132.

Bent, H. E., and Dorfman, M. (1935), *J. Amer. chem. Soc.* **57**, 1259, 1924.

Benton, F. L., and Dillon, T. E. (1942), *J. Amer. chem. Soc.* **64**, 1128.

Bera, B. C., Forster, A. B., and Stacey, M. (1956), *J. chem. Soc.* 4531.

Beran, F., Prey, V., and Böhm, H. (1952), *Mitt. chem. Forschinst. Ind. Wien* **6**, 54.

Berkhout, H. W. (1952), *Chem. Weekbl.* **48**, 909.

Berl, W. G., Gayhart, E. L., Olsen, H. L., Broida, H. P., and Schuler, K. E. (1956), *J. chem. Phys.* **25**, 797.

Berzelius, J. J. (1824), *Ann. Phys., Lpz.* **78**, 113.

Besson, A. (1891), *C. R. Acad. Sci. Paris* **112**, 1001.

Besson, A. (1892), *C. R. Acad. Sci. Paris* **114**, 542.

Bethke, G. W., and Kent Wilson, M. (1957), *J. chem. Phys.* **26**, 1118.

Bettman, B., and Branch, G. E. K. (1934), *J. Amer. chem. Soc.* **56**, 1616.

Bettman, B., Branch, G. E. K., and Yabroff, D. L. (1934), *J. Amer. chem. Soc.* **56**, 1865.

Bissot, T. C., and Parry, R. W. (1955), *J. Amer. chem. Soc.* **77**, 3481.

Bissot, T. C., Campbell, D. H., and Parry, R. W. (1958), *J. Amer. chem. Soc.* **80**, 1868.

Blau, E. J., and Mulligan, B. W. (1957), *J. chem. Phys.* **26**, 1085.

Blau, J. A., Gerrard, W., and Lappert, M. F. (1957), *J. chem. Soc.* 4116.

Blau, J. A., Gerrard, W., Lappert, M. F., Mountfield, B. A., and Pyszora, H. (1960), *J. chem. Soc.* 380.

Blau, J. A., Gerrard, W., and Lappert, M. F. (1960), *J. chem. Soc.* 667.

Blum, E., and Herzberg, G. (1937), *J. phys. Chem.* **41**, 91.

Böhme, H., and Boll, E. (1957), *Z. anorg. Chem.* **291**, 160.

Böeseken, J. (1913), *Ber. dtsch. chem. Ges.* **46**, 2612.

Böeseken, J. (1921), *Rec. Trav. chim., Pays-Bas* **40**, 553.

Böeseken, J. (1928), *Rec. Trav. chim., Pays-Bas* **47**, 683.

Böeseken, J. (1933), *Bull. Soc. chim. Fr.* **53**, 1333.

Böeseken, J. (1949), *Advanc. Carbohyd. Chem. IV*, 189.

Böeseken, J., and Meulenhoff, J. (1924), *Proc. Acad. Sci. Amst.* **27**, 174.

Böeseken, J., and Mijs, J. A. (1925), *Rec. Trav. chim., Pays-Bas* **44**, 758.

Böeseken, J., and Julius, A. (1926), *Rec. Trav. chim., Pays-Bas* **45**, 489.

Böeseken, J., Vermaas, N., and Küchlin, A. T. (1930), *Rec. Trav. chim., Pays-Bas* **49**, 711.

Böeseken, J., and Vermaas, N. (1931), *J. phys. Chem.* **35**, 1477.

Böeseken, J., de Bruin, J. A., and Jong, W. E. van R. de (1939), *Rec. Trav. chim., Pays-Bas* **58**, 3.

Bogaty, H., and Brown, A. E. (1956), *U.S.P.* 2,766,760.

Booth, H. S., and Martin, D. R. (1949), "Boron Trifluoride and its Derivatives", Wiley/Chapman and Hall.

Booth, R. B., and Kraus, C. A. (1952), *J. Amer. chem. Soc.* **74**, 1415.

Borax Consolidated Ltd. (1957), *B.P.* 775,418; 765,515.

Borisov, A. E. (1951), *Bull. Acad. Sci. U.R.S.S., Classe sci. chem.* 402.

Bowlus, H., and Nieuwland, J. A. (1931), *J. Amer. chem. Soc.* **53**, 3835.

Bowman, C. M. (1958), *Dissertation Abstr.* **18**, 773, *Microfilm.* Publ. No. 24369.

Boyer, J. H., and Ellzey, S. E., Jr. (1958), *J. org. Chem.* **23**, 127.

Bradley, M. J. (1959), *Dissertation Abstr.* **19**, 1553, *Microfilm* L.C. Card. no. Mic. 587–131.

Bradley, M. J., Ryschkewitsch, G. E., and Sisler, H. H. (1959), *J. Amer. chem. Soc.* **81**, 2635.

Bragdon, R. W. (1956), *U.S.P.* 2,756,259.

Bragdon, R. W. (1957), *U.S.P.* 2,813,115.

Bragg, J. K., McCarty, L. V., and Norton, F. J. (1951), *J. Amer. chem. Soc.* **73**. 2134.

Branch. G. E. K., Yabroff, D. L., and Bettman, B. (1934), *J. Amer. chem. Soc.* **56**, 937.

Brandenberg, W., and Galat, A. (1950), *J. Amer. chem. Soc.* **72**, 3275.

Bremer, C. (1941), *U.S.P.* 2,223,949.

Brinckman, F. E., and Stone, F. G. A. (1959), *Chem. & Ind.* 254.

Brindley, P. B., Gerrard, W., and Lappert, M. F. (1955), *J. chem. Soc.* 2956.

Brindley, P. B., Gerrard, W., and Lappert, M. F. (1956), *J. chem. Soc.* 824, 1540.

Brockway, L. O. (1936), *Rev. mod. Phys.* **8**, 231.

Brokow, R. S., Badin, E. J., and Pease, R. N. (1950), *J. Amer. chem. Soc.* **72**, 1793.

Brokow, R. S., and Pease, R. N. (1950), *J. Amer. chem. Soc.* **72**, 3237, 5263.

Brown, C. A., and Osthoff, R. C. (1952), *J. Amer. chem. Soc.* **74**, 2340.

Brown, C. A., Muetterties, E. L., and Rochow, E. G. (1954), *J. Amer. chem. Soc.* **76**, 2537.

Brown, C. A., and Laubengayer, A. W. (1955), *J. Amer. chem. Soc.* **77**, 3699.

Brown, H. C. (1945), *J. Amer. chem. Soc.* **67**, 374, 378, 1452.

Brown, H. C. (1955), *U.S.P.* 2,709,704.

Brown, H. C. (1956), *J. chem. Soc.*, 1248.

Brown, H. C. (1958), *U.S.P.* 2,860,167.

Brown, H. C., Schlesinger, H. I., and Burg, A. B. (1939), *J. Amer. chem. Soc.* **61**, 673.

Brown, H. C., Schlesinger, H. I., and Cardon, S. Z. (1942), *J. Amer. chem. Soc.* **64**, 325.

Brown, H. C., and Adams, R. M. (1942), *J. Amer. chem. Soc.* **64**, 2557.

Brown, H. C., and Adams, R. M. (1943), *J. Amer. chem. Soc.* **65**, 2253.

Brown, H. C., Taylor, M. D., and Gerstein, M. (1944), *J. Amer. chem. Soc.* **66**, 431.

Brown, H. C., Bartholomay, L. H., Jr., and Taylor, M. D. (1944), *J. Amer. chem. Soc.* **66**, 435.

Brown, H. C., and Pearsall, H. (1945), *J. Amer. chem. Soc.* **67**, 1765.

Brown, H. C., and Barbaras, G. K. (1947), *J. Amer. chem. Soc.* **69**, 1137.

Brown, H. C., and Taylor, M. D. (1947), *J. Amer. chem. Soc.* **69**, 1332.

Brown, H. C., and Sujishi, S. (1948), *J. Amer. chem. Soc.* **70**, 2793, 2878.

Brown, H. C., and Gerstein, M. (1950), *J. Amer. chem. Soc.* **72**, 2923.

Brown, H. C., and Johannesen, R. B. (1950), *J. Amer. chem. Soc.* **72**, 2934.

Brown, H. C., Taylor, M. D., and Sujishi, S. (1951), *J. Amer. chem. Soc.* **73**, 2464.

Brown, H. C., and Fletcher, E. A. (1951), *J. Amer. chem. Soc.* **73**, 2808.

Brown, H. C., Barbaras, G. K., Berneis, H. L., Bonner, W. H., Johannesen, R. B., Grayson, M., and Le Roi Nelson, K. (1953), *J. Amer. chem. Soc.* **75**, 1.

Brown, H. C., and Barbaras, G. K. (1953), *J. Amer. chem. Soc.* **75**, 6.

Brown, H. C., and Johannesen, R. B. (1953), *J. Amer. chem. Soc.* **75**, 16, 19.

Brown, H. C., Schlesinger, H. I., Sheft, I., and Ritter, D. M. (1953), *J. Amer. chem. Soc.* **75**, 192.

Brown, H. C., Schlesinger, H. I., Gilbreath, J. R., and Katz, J. J. (1953), *J. Amer. chem. Soc.* **75**, 195.

Brown, H. C., and Mead, E. J. (1953), *J. Amer. chem. Soc.* **75**, 6263.

Brown, H. C., and Johnson, S. (1954), *J. Amer. chem. Soc.* **76**, 1978.

Brown, H. C., and Boyd, A. C. (1955), *Analyt. Chem.* **27**, 156.

Brown, H. C., and Horowitz, R. H. (1955), *J. Amer. chem. Soc.* **77**, 1730, 1733.

Brown, H. C., and Subba Rao, B. C. (1955), *J. Amer. chem. Soc.* **77**, 3164.

Brown, H. C., Mead, E. J., and Subba Rao, B. C. (1955), *J. Amer. chem. Soc.* **77**, 6209.

Brown, H. C., and Holmes, R. R. (1956), *J. Amer. chem. Soc.* **78**, 2173.

Brown, H. C., and Subba Rao, B. C. (1956), *J. Amer. chem. Soc.* **78**, 2582.

Brown, H. C., Mead, E. J., and Shoaf, C. J. (1956), *J. Amer. chem. Soc.* **78**, 3613, 3616.

Brown, H. C., and Mead, E. J. (1956), *J. Amer. chem. Soc.* **78**, 3614.

Brown, H. C., Gintis, D., and Podall, H. (1956), *J. Amer. chem. Soc.* **78**, 5375.

Brown, H. C., and Gintis, D. (1956), *J. Amer. chem. Soc.* **78**, 5378.

Brown, H. C., and Domash, L. (1956), *J. Amer. chem. Soc.* **78**, 5384.

Brown, H. C., Gintis, D., and Domash, L. (1956), *J. Amer. chem. Soc.* **78**, 5387.

Brown, H. C., Stehle, P. F., and Tierney, P. A. (1957), *J. Amer. chem. Soc.* **79**, 2020.

Brown, H. C., and Dodson, V. H. (1957), *J. Amer. chem. Soc.* **79**, 2302.

Brown, H. C., Mead, E. J., and Tierney, P. A. (1957), *J. Amer. chem. Soc.* **79**, 5400.

Brown, H. C., and Subba Rao, B. C. (1957), *J. org. Chem.* **22**, 1136.

Brown, H. C., and Tierney, P. A. (1958), *J. Amer. chem. Soc.* **80**, 1552.

Brown, H. C., and Murray, K. (1959), *J. Amer. chem. Soc.* **81**, 4108.

Brown, H. C., and Tierney, P. A. (1959), *J. Inorg. & Nuclear Chem.* **9**, 51.

Brown, H. C., and Zweifel, G. (1959), *J. Amer. chem. Soc.* **81**, 1512.

Brown, H. C., and Subba Rao, B. C. (1959), *J. Amer. chem. Soc.* **81**, 6423–6434.

Brown, J. F. (1952), *J. Amer. chem. Soc.* **74**, 1219.

Brumberger, H., and Marcus, R. A. (1956), *J. chem. Phys.*, **24**, 741.

Bryusova, L. Ya., and Ioffe, M. L. (1941), *J. gen. Chem., Moscow (Eng. Transl.)*, **11**, 722.

Bujwid, Z. J., Gerrard, W., and Lappert, M. F. (1957), *Chem. & Ind.* 1386.

Bujwid, Z. J., Gerrard, W., and Lappert, M. F. (1959), *Chem. & Ind.* 1091.

Bujwid, Z. J., Gerrard, W., and Lappert, M. F. (1960), unpublished work.

Buls, V. W., Davis, O. L., and Thomas, R. I. (1957), *J. Amer. chem. Soc.* **79**, 337.

Burg, A. B. (1934), *J. Amer. chem. Soc.* **56**, 499.

Burg, A. B. (1940), *J. Amer. chem. Soc.*, **62**, 2228.

Burg, A. B. (1943), *J. Amer. chem. Soc.* **65**, 1635.

Burg, A. B. (1947), *J. Amer. chem. Soc.* **69**, 747.

Burg, A. B. (1952), *J. Amer. chem. Soc.* **74**, 1340, 3482.

Burg, A. B. (1954), *Rec. Chem. Progr.* **15**, 159.

Burg, A. B. (1957), *J. Amer. chem. Soc.* **79**, 2129.

Burg, A. B., and Schlesinger, H. I. (1933), *J. Amer. chem. Soc.*, **55**, 4009, 4020.

Burg, A. B., and Schlesinger, H. I. (1937), *J. Amer. chem. Soc.* **59**, 780.

Burg, A. B., and Schlesinger, H. I. (1940), *J. Amer. chem. Soc.* **62**, 3425.

Burg, A. B., and Martin, L. V. L. (1943), *J. Amer. chem. Soc.* **65**, 1635.

Burg, A. B., and Roso, M. K. (1943), *J. Amer. chem. Soc.* **65**, 1637.

Burg, A. B., and Green, A. A. (1943), *J. Amer. chem. Soc.* **65**, 1838.

Burg, A. B., and Bickerton, J. H. (1945), *J. Amer. chem. Soc.* **67**, 2261.

Burg, A. B., and Randolph, C. L. (1949), *J. Amer. chem. Soc.* **71**, 3451.

Burg, A. B., and Kuljian, E. S. (1950), *J. Amer. chem. Soc.* **72**, 3103.

Burg, A. B., and Randolph, C. L. (1951), *J. Amer. chem. Soc.* **73**, 953.

Burg, A. B., and McKee, W. E. (1951), *J. Amer. chem. Soc.* **73**, 4590.

Burg, A. B., and Campbell, G. W. (1952), *J. Amer. chem. Soc.* **74**, 3744.

Burg, A. B., and Stone, F. G. A. (1953), *J. Amer. chem. Soc.* **75**, 228.

Burg, A. B., and Wagner, R. I. (1953), *J. Amer. chem. Soc.* **75**, 3872.

Burg, A. B., and Woodrow, H. W. (1954), *J. Amer. chem. Soc.* **76**, 219.

Burg, A. B., and Wagner, R. I. (1954), *J. Amer. chem. Soc.* **76**, 3307.

Burg, A. B., and Banus, J. (1954), *J. Amer. chem. Soc.* **76**, 3903.

Burg, A. B., and Boone, J. L. (1956), *J. Amer. chem. Soc.* **78**, 1521.

Burg, A. B., and Graber, F. M. (1956), *J. Amer. chem. Soc.* **78**, 1523.

Burg, A. B., and Good, C. D. (1956), *J. Inorg. & Nuclear Chem.* **2**, 237.

Burg, A. B., and Brendel, G. (1958), *J. Amer. chem. Soc.* **80**, 3198.

Burg, A. B., and Grant, L. R. (1959), *J. Amer. chem. Soc.* **81**, 1.

Burkhardt, L., and Fetter, N. R. (1959), *Chem. & Ind.* 1191.

Burns, J. J., and Schaeffer, G. W. (1958), *J. phys. Chem.* **62**, 380.

Cahours, A. (1873), *C. R. Acad. Sci. Paris* **76**, 1383.

Callaghan, J., and Siegel, B. (1959), *J. Amer. chem. Soc.* **81**, 504.

Callery Chem. Co. (1955) (Pittsburgh), *U.S.P.* 2,708,152.

Calvert, R. P., and Thomas, D. L. (1919), *U.S.P.* 1,308,567.

Cambi, L. (1914), *R. C. Accad. Lincei*, [v], **23**, i, 244.

Campbell, D. H., Bissot, T. C., and Parry, R. W. (1958), *J. Amer. chem. Soc* **80**, 1549.

Campbell, G. W., Jr. (1957), *J. Amer. chem. Soc.* **79**, 4023.

Carpenter, R. A. (1957), *Ind. Eng. Chem.* **49** (No. 4), 42A.

Carpmael, A. (1930), *B.P.* 358,491.

Caujolle, F., Gayel, P., Roux, G., and Moscarella, C. (1951), *Bull. Acad. nat. Méd.* **135**, 314.

Cazes, J. (1958), *C. R. Acad. Sci. Paris* **247**, 2019.

Ceron, P., Finch, A., Frey, J., Kerrigan, J., Parsons, T. D., Urry, G., and Schlessinger, H. I. (1959), *J. Amer. chem. Soc.* **81**, 6368.

Chaikin, S. W. (1953), *Analyt. Chem.* **25**, 831.

Chaikin, S. W., and Brown, W. G. (1949), *J. Amer. chem. Soc.* **71**, 122.

Chainani, G. R., and Gerrard, W. (1960), *J. chem. Soc.* 3168.

Challenger, F., and Richards, O. V. (1934), *J. chem. Soc.* 405.

Charnley, T., Skinner, H. A., and Smith, N. B. (1952), *J. chem. Soc.* 2288.

Chatt, J. (1949), *J. chem. Soc.* 3340.

Chaudhuri, T. C. (1920), *J. chem. Soc.* **117**, 1081.

Cherbuliez, E., Leber, J. P., and Ulrich, A. M. (1953), *Helv. chim. acta* **36**, 910.

Chu, Ting Li (1953), *J. Amer. chem. Soc.* **75**, 1730.

Chu, Ting Li, and Weismann, T. J. (1956), *J. Amer. chem. Soc.* **78**, 23.

Clark, M. M. (1956), *U.S.P.* 2,769,746.

Clark, S. L. (1958), *Chem. Engng. News*, April 28th, 56.

Clark, S. L., and Jones, J. R. (1958), *Abstracts, 133rd Meeting, Amer. chem. Soc.*, April, p. 34.

Clarke, R. P., and Pease, R. N. (1951), *J. Amer. chem. Soc.* **73**, 2132.

Clear, C. G., and Branch, G. E. K. (1938), *J. org. Chem.* **2**, 522.

Clifford, J. (1957), *Dissertation Abstr.* **17**, 2805, *Microfilm* Publ. No. 22625.

Cluley, H. J. (1955), *Analyst* **80**, 354.

Coates, G. E. (1950), *J. chem. Soc.* 3481.

Coates, G. E. (1956), "Organo-metallic Compounds", p. 68, Methuen, London.

Coerver, H. J., and Curran, C. (1958), *J. Amer. chem. Soc.* **80**, 3522.

Coffin, K. P., and Bauer, S. H. (1955), *J. phys. Chem.* **59**, 193.

Cohn, G. (1911), *Pharm. Zentralh.* **52**, 479.

Colclough, T., Gerrard, W., and Lappert, M. F. (1955), *J. chem. Soc.* 907.

Colclough, T., Gerrard, W., and Lappert, M. F. (1956), *J. chem. Soc.* 3006.

Commerford, J. D. (1957), *131st Meeting Amer. chem. Soc., Division of Industr. and Engng. Chem.*, Miami.

Consden, R., and Stainier, W. M. (1952), *Nature* **169**, 783.

Cook, H. G., Ilett, J. D., Saunders, B. C., and Stacey, G. J. (1950), *J. chem. Soc.* 3125.

Coolidge, A. S., and Bent, H. E. (1936), *J. Amer. chem. Soc.* **58**, 505.

Cooper, S., Frazer, M. J., and Gerrard, W., unpublished work.

Cooper, S. S. (1957), *Analyt. Chem.* **29**, (3), 446.

Copaux, H. (1898), *C. R. Acad. Sci. Paris* **127**, 719.

Core, A. F. (1927), *Chem. & Ind.* **5**, 642.

Cornwell, C. D. (1950), *J. chem. Phys.* **18**, 1118.

Councler, C. (1871), *J. prakt. Chem.* **18**, 371.

Councler, C. (1876), *Ber. dtsch. chem. Ges.* **9**, 485.

Councler, C. (1877), *Ber. dtsch. chem. Ges.* **10**, 1655.

Councler, C. (1878), *Ber. dtsch. chem. Ges.* **11**, 1106.

Cowan, R. D. (1949), *J. chem. Phys.* **17**, 218.

Cowie, W. P., Jackson, A. H., and Musgrave, O. C. (1959), *Chem. & Ind.* 1248.

Cowley, E. G., and Partington, J. R. (1935), *Nature* **136**, 643.

Coyle, T. D., Kaesz, H. D., and Stone, F. G. A. (1959), *J. Amer. chem. Soc.* **81**, 2989.

Crane, F. E., Jr. (1956), *Analyt. Chem.* **28**, 1794.

Crawford, B. L., Jr. and Edsall, J. T. (1939), *J. chem. Phys.* **7**, 223.

Crawhall, J. C., and Elliott, D. F. (1955), *Nature* **175**, 299.

Csapo, F., Bihari, M., Gilde, M., and Sztanko, E. (1956), *Z. anal. Chem.* **151**, 273.

Cueilleron, J. (1943), *C. R. Acad. Sci. Paris* **217**, 112.

Curran, C., McCusker, P. A., and Makowski, H. S. (1957), *J. Amer. chem. Soc.* **79**, 5188.

Currell, B. R., Frazer, M. J., and Gerrard, W. (1960), *J. chem. Soc.* 2776.

Dandegaonker, S. H., Gerrard, W., and Lappert, M. F. (1957), *J. chem. Soc.* 2872, 2893.

Dandegaonker, S. H., Gerrard, W., Lappert, M. F., and Mountfield, B. A. (1959), *J. chem. Soc.*, 1529.

Dandegaonker, S. H., Gerrard, W., and Lappert, M. F. (1959), *J. chem. Soc.* 2076.

Darling, S. M. (1955), *U.S.P.* 2,710,251; 2,710,252.

Darling, S. M., Fay, P. S., and Szabo, L. S. (1956), *U.S.P.* 2,741,548.

Das, T. P. (1957), *J. chem. Phys.* **27**, 1.

Davidson, J. M., and French, C. M. (1958), *J. chem. Soc.* 114.

Davidson, J. M., and French, C. M. (1959), *Chem. & Ind.* 750.

Davies, A. G., and Moodie, R. B. (1957), *Chem. & Ind.* 1622.

Davies, A. G., and Moodie, R. B. (1958), *J. chem. Soc.* 2372.

Davies, A. G., and Hare, D. G. (1959), *J. chem. Soc.* 438.

Davies, A. G., Hare, D. G., and White, R. F. M. (1959), *Chem. & Ind.* 1315.

Davies, T., and Staveley, L. A. K. (1957), *Trans. Faraday Soc.* **53**, 19.

Davis, M. (1956), *J. chem. Soc.* 3981.

Davis, W. D., Mason, L. S., and Stegeman, G. (1949), *J. Amer. chem. Soc.* **71**, 2775.

Dehmelt, H. G. (1952), *Z. Phys.* **133**, 528.

Dehmelt, H. G. (1953), *Z. Phys.* **134**, 642.

Denson, C. L., and Crowell, T. I. (1957), *J. Amer. chem. Soc.* **79**, 5656.

Dewar, M. J. S., and Dietz, R. (1959), *J. chem. Soc.* 2728.

Dewar, M. J. S., Kubba, V. P., and Pettit, R. (1958), *J. chem. Soc.* 3073, 3076.

Dickens, P. G., and Linnett, J. W. (1957), *Quart. Rev. chem. Soc.*, *Lond.* **11**, 291.

Dickerson, R. E., and Lipscomb, W. N. (1957), *J. chem. Phys.* **27**, 212.

Dilthey, W. (1906), *Liebigs Ann.* **344**, 300.

Dimroth, O. (1925), *Liebigs Ann.* **446**, 97.

Dimroth, O., and Faust, T. (1921), *Ber. dtsch. chem. Ges.* **54**, 3020.

Dodson, R. M., and Sollman, P. B. (1956), *U.S.P.* 2,763,669.

Dollimore, D., and Long, L. H. (1953), *J. chem. Soc.* 3906.

Dornow, A., and Gehrt, H. H. (1956), *Angew. Chem.* **68**, 619.

Duncanson, L. A., Gerrard, W., Lappert, M. F., Pyszora, H., and Shafferman, R. (1958), *J. chem. Soc.* 3652.

Dupire, A. (1936), *C. R. Acad. Sci. Paris* **202**, 2086.

Dworkin, A. S., and Van Artsdalen, E. R. (1954), *J. Amer. chem. Soc.* **76**, 4316.

Ebelman, and Bouquet (1846), *Ann. Chim.* (*Phys.*) (3), **17**, 54.

Ebelman, and Bouquet (1846), *Liebigs Ann.* **60**, 251.

Eberhardt, W. H., Crawford, B., Jr., and Lipscomb, W. N. (1954), *J. chem. Phys.* **22**, 989.

Eddy, L. B., Smith, S. H., and Miller, R. R. (1955), *J. Amer. chem. Soc.* **77**, 2105.

Edmonds, J. T., Jr. (1956), *U.S.P.* 2,731,454.

Edwards, J. D., Gerrard, W., and Lappert, M. F. (1955), *J. chem. Soc.* 1470.

Edwards, J. O., Morrison, G. C., Ross, V. F., and Schultz, J. W. (1955), *J. Amer. chem. Soc.* **77**, 266.

Edwards, J. D., Gerrard, W., and Lappert, M. F. (1957), *J. chem. Soc.* 348, 377.

Ehrlich, P., and Keil, T. (1959), *Z. anal. Chem.* **165**, 188.

Ellis, C. (1935), "The Chemistry of Synthetic Resins", Vol. II, p. 893, Reinhold, New York.

Elson, R. E., Hornig, H. C., Jolly, W. L., Kury, J. W., Ramsey, W. J., and Zalkin, A. (1956), *U.S. Atomic Energy Comm.* **UCRL**-4519.

Eméleus, H. J., and Stone, F. G. A. (1950), *J. chem. Soc.* 2755.

Eméleus, H. J., and Videla, G. J. (1957), *Proc. chem. Soc.*, *Lond.* 288.

Eméleus, H. J., and Onyszchuk, M. (1958), *J. chem. Soc.* 604.

Eméleus, H. J., and Videla, G. H. (1959), *J. chem. Soc.* 1306.

Engelmann, F. (1953), *Z. Naturf.* **8b**, 775.

Etherington, T. L., and McCarty, L. V. (1952), *Arch. Industr. Hyg.* **5**, 447.

Etridge, J. J., and Sugden, S. (1928), *J. chem. Soc.* 989.

Farkas, L., and Sachsee, H. (1934), *Trans. Faraday Soc.* **30**, 331.

Ferguson, G. A., Jr., and Jablonski, F. E. (1957), *Rev. sci. Instrum.* **28**, 893.

Finch, A., and Schlesinger, H. I. (1958), *J. Amer. chem. Soc.* **80**, 3573.

Finholt, A. E. (1946), Ph.D. Dissertation. The University of Chicago.

Finholt, A. E., Bond, A. C., Jr., and Schlesinger, H. I. (1947), *J. Amer. chem. Soc.* **69**, 1199.

Finholt, A. E., Bond, A. C., Jr., Wilzbach, K. E., and Schlesinger, H. I. (1947), *J. Amer. chem. Soc.* **69**, 2692.

Finholt, A. E., Schlesinger, H. I., Barbaras, G. D., Dillard, C., Wartik, T., and Wilzbach, K. E. (1951), *J. Amer. chem. Soc.* **73**, 4585.

Fischer, E. O., Seus, D., and Jira, R. (1953), *Z. Naturf.* **8b**, 692.

Fischer, R., and Karawia, M. S. (1953), *Mikrochem. Acta* 366.

Fitch, S. J., and Laubengayer, A. W. (1958), *J. Amer. chem. Soc.* **80**, 5911.

Flaschka, H. (1952), *Z. anal. Chem.* **136**, 99.

Flaschka, H., and Abdine, H. (1955), *Z. anal. Chem.* **144**, 415.

Flaschka, H., Amin, A. M., and Holasek, A. (1953), *Z. anal. Chem.* **138**, 161.

Florin, R. E., Wall, L. A., Mohler, F. L., and Quinn, E. (1954), *J. Amer. chem. Soc.* **76**, 3344.

Fornaseri, M. (1949), *Period. miner.* **18**, 103.

Fornaseri, M. (1950), *Period. miner.* **19**, 157.

Foster, A. B. (1953), *J. chem. Soc.* 982.

Foster, A. B. (1957), *J. chem. Soc.* 1395, 4214.

Foster, A. B., and Stacey, M. (1955), *J. chem. Soc.* 1778.

Fowler, D. L., and Kraus, C. A. (1940), *J. Amer. chem. Soc.* **62**, 1143, 2237.

Frankland, E. (1876), *Proc. roy. Soc.* **25**, 165.

Frankland, E. (1862), *J. chem. Soc.* **15**, 363.

Frankland, E. (1862), *Proc. roy. Soc.* **12**, 123.

Frankland, E. (1862), *Liebigs Ann.* **124**, 129.

Frankland, E., and Duppa, B. F. (1860), *Proc. roy. Soc.* **10**, 568.

Frankland, E., and Duppa, B. F. (1860), *Ann. Chim. (Phys.)* **60**, 374.

Frankland, E., and Duppa, B. F. (1860), *Liebigs Ann.* **115**, 319.

Frazer, M. J., and Gerrard, W. (1955), *J. chem. Soc.* 2959.

Frazer, M. J., Gerrard, W., and Lappert, M. F. (1957), *J. chem. Soc.* 739.

Frazer, M. J., Gerrard, W., and Mistry, S. N. (1958), *Chem. & Ind.* 1263.

Frazer, M. J., Gerrard, W., and Patel, J. K. (1959), *Chem. & Ind.* **90**, 728.

Frazer, M. J., Gerrard, W., and Strickson J. A. (1960), *J. chem. Soc.*, in press.

Frazer, M. J., Gerrard, W., and Patel, J. K. (1960), *J. chem. Soc.*, 726.

French, H. E., and Fine, S. D. (1938), *J. Amer. chem. Soc.* **60**, 352.

French, F. A., and Rasmussen, R. S. (1946), *J. chem. Phys.* **14**, 389.

Frohnsdorff, R. S. M., and Lewin, J. U. (1956), *U.S.P.* 2,745,788.

Frush, H. L., and Isbell, H. S. (1956), *J. Amer. chem. Soc.* **78**, 2844.

Fuchs, R. (1956), *J. Amer. chem. Soc.* **78**, 5612.

Fuchs, R., and VanderWerf, C. A. (1954), *J. Amer. chem. Soc.* **76**, 1631.

Furukawa, G. T., and Park, R. P. (1955), *J. Res. nat. Bur. Stand.* **55**, 255.

Furukawa, G. T., McCoskey, R. E., Reilly, M. L., and Harman, A. (1955), *J. Res. nat. Bur. Stand.* **55**, 205.

Furukawa, J., Tsuruta, T., and Inoue, S. (1957), *J. Polym. Sci.* **26**, 234.

Furukawa, J., and Tsuruta, T. (1958), *J. Polym. Sci.* **28**, 227.

Gakle, P. S., and Tannenbaum, S. (1955), *J. Amer. chem. Soc.* **77**, 5289.

Gamble, E. L., and Gilmont, P. (1940), *J. Amer. chem. Soc.* **62**, 717.

Gamble. E. L., Gilmont, P., and Stiff, J. F. (1940), *J. Amer. chem. Soc.* **62**, 1257.

Gardner, D. M. (1955), *Dissertation Abstr.* **15**, 693.

Garner, P. J. (1955), *B.P.* 722,537; 722,538.

Garrett, E. R., and Lyttle, D. A. (1953), *J. Amer. chem. Soc.* **75**, 6051.

Gasselin, V. (1894), *Ann. Chim. (Phys.)* [7], **3**, 5.

Gates, N. V. (1956), *U.S.P.* 2,762,115.

Gattermann, L. (1889), *Ber. dtsch. chem. Ges.* **22**, 186.

Gautier, M. (1866), *C. R. Acad. Sci. Paris* **63**, 923.

Gautier, J. A., Renault, J., and Pellerin, F. (1956), *Ann. pharm. franç.* **14**, 337.

Geilmann, W., and Gebauhr, W. (1953), *Z. anal. Chem.* **139**, 161.

Geller, S., and Salmon, O. N. (1951), *Acta cryst.* **4**, 379.

Geller, S. (1955), *Acta cryst.* **8**, 120.

Gel'Perin, N. I., and Solopenkov, K. N. (1956), *Khim. Nauka i Prom.* **1**, 324.

George, P. D., and Ladd, J. R. (1955), *J. Amer. chem. Soc.* **77**, 1900.

Gerrard, W. (1939), *J. chem. Soc.* 99.

Gerrard, W. (1940), *J. chem. Soc.* 218, 1464.

Gerrard, W. (1944), *J. chem. Soc.* 85.

Gerrard, W. (1945), *J. chem. Soc.* 106, 848.

Gerrard, W. (1946), *J. chem. Soc.* 741.

Gerrard, W. (1951), *Chem. & Ind.* 463.

Gerrard, W. (1956), *Chem. & Ind.* 25.

Gerrard, W. (1957), *Chem. Prod.* **20**, 489.

Gerrard, W. (1959), *J. Oil. Col. Chem. Ass.* **42**, 625.

Gerrard, W., and French, R. H. V. (1947), *Nature* **159**, 263.

Gerrard, W., and Wyvill, P. L. (1949), *Research, Lond.* **2**, 536.

Gerrard, W., Nechvatal, A., and Wilson, B. M. (1950), *J. chem. Soc.*, 2088.

Gerrard, W., and Woodhead, A. H. (1951), *J. chem. Soc.* 519.

Gerrard, W., and Lappert, M. F. (1951), *J. chem. Soc.* 1020, 2545.

Gerrard, W., and Lappert, M. F. (1952), *Chem. & Ind.* 53.

Gerrard, W., and Lappert, M. F. (1952), *J. chem. Soc.* 1486.

Gerrard, W., and Jones, J. V. (1952), *J. chem. Soc.* 1690.

Gerrard, W., and Phillips, J. R. (1952), *Chem. & Ind.* 540.

Gerrard, W., and Wheelans, M. A. (1954), *Chem. & Ind.* 758.

Gerrard, W., and Howe, B. K. (1955), *J. chem. Soc.* 505.

Gerrard, W., and Lappert, M. F. (1955), *J. chem. Soc.* 3084.

Gerrard, W., Lappert, M. F., and Silver, H. B. (1956), *J. chem. Soc.* 3285, 4987.

Gerrard, W., and Macklen, E. D. (1956), *J. appl. Chem. (London)* **6**, 241.

Gerrard, W., and Wheelans, M. A. (1956), *J. chem. Soc.* 4296.

Gerrard, W., Lappert, M. F., and Silver, H. B. (1957), *Proc. chem. Soc., Lond.* 19.

Gerrard, W., Lappert, M. F., and Pearce, C. A. (1957), *J. chem. Soc.* 381.

Gerrard, W., Lappert, M. F., and Silver, H. B. (1957), *J. chem. Soc.* 1647.

Gerrard, W., Lappert, M. F., and Shafferman, R. (1957), *J. chem. Soc.* 3828.

Gerrard, W., Lappert, M. F., and Shafferman, R. (1958), *Chem. & Ind.* 722.

Gerrard, W., Lappert, M. F., and Shafferman, R. (1958), *J. chem. Soc.* 3648.

Gerrard, W., and Strickson, J. A. (1958), *Chem. & Ind.* 860.

Gerrard, W., and Mooney, E. F. (1958), *Chem. & Ind.* 227, 1259.

Gerrard, W., and Macklen, E. D. (1959), *Chem. & Ind.* 1549.

Gerrard, W., and Griffey, P. F. (1959), *Chem. & Ind.* 55.

Gerrard, W., Hudson, H. R., and Mooney, E. F. (1959), *Chem. & Ind.* 432.

Gerrard, W., Lappert, M. F., and Mountfield, B. A. (1959), *J. chem. Soc.* 1529.

Gerrard, W., and Macklen, E. D. (1959), *J. appl. Chem. (London)* **9**, 85.

Gerrard, W., Mincer, A. M. A., and Wyvill, P. L. (1959), *J. appl. Chem. (London)* **9**, 89.

Gerrard, W., Lappert, M. F., and Wallis, J. W. (1960), *J. chem. Soc.* 2141, 2178.

Gerrard, W., Lappert, M. F., Pyszora, H., and Wallis, J. W. (1960), *J. chem. Soc.* 2144, 2182.

Gerrard, W., and Lindsay, M. (1960), *Chem. & Ind.* 152.

Gerrard, W., and Macklen, E. D. (1960), *J. appl. Chem.* (*London*) **10**, 57.

Gerrard, W., Mincer, A. M. A., and Wyvill, P. L. (1960), *J. appl. Chem.* (*London*) **10**, 115.

Gerrard, W., and Griffey, P. F. (1960), *J. chem. Soc.* 3170.

Gerrard, W., and Gillett, V., unpublished work.

Gerrard, W., and Bedell, R., unpublished work.

Gerrard, W., and co-workers, unpublished work.

Gerrard, W., Hudson, H. R., and Mooney, E. F., unpublished work.

Gerrard, W., Lappert, M. F., and Shafferman, R., unpublished work.

Gerrard, W., and Mooney, E. F., unpublished work.

Gerrard, W., Mooney, E. F., and Pratt, D., unpublished work.

Gerrard, W., Mooney, E. F., and Rothenbury, R. A., unpublished work.

Gillette Industries Ltd. (1957), *B.P.* 773,794.

Gilman, H., and Vernon, C. C. (1926), *J. Amer. chem. Soc.* **48**, 1063.

Gilman, H., and Marple, K. E. (1936), *Rec. Trav. chim.*, *Pays-Bas* **55**, 76.

Gilman, H., and Nelson, J. F. (1936), *Rec. Trav. chim.*, *Pays-Bas* **55**, 518.

Gilman, H., and Honeycutt, J. B. (1957), *J. org. Chem.* **22**, 562.

Gilman, H., Santucci, L., Swayampati, D. R., and Ranck, R. O. (1957), *J. Amer. chem. Soc.* **79**, 2898, 3077.

Gilman, H., and Moore, L. O. (1958), *J. Amer. chem. Soc.* **80**, 3609.

Gilman, H., Swayampati, D. R., and Ranck, R. O. (1958), *J. Amer. chem. Soc.* **80**, 1355.

Gloss, G. H., and Olson, B. (1954), *Chem. Anal.* **43**, 70.

Goodspeed, N. C., and Sanderson, R. T. (1956), *J. Inorg. & Nuclear Chem.* **2**, 266.

Gorin, P. A. J., and Perlin, A. S. (1956), *Can. J. Chem.* **34**, 693.

Goubeau, J. (1948), F.I.A.T. Review of German Science Part I, 215.

Goubeau, J., unpublished work cited by K. Torssell.

Goubeau, J., Jacobshagen, U., and Rahtz, M. (1950), *Z. anorg. Chem.* **263**, 63.

Goubeau, J., and Bergmann, R. (1950), *Z. anorg. Chem.* **263**, 69.

Goubeau, J., and Keller, H. (1951), *Z. anorg. Chem.* **265**, 73.

Goubeau, J., and Böhm, U. (1951), *Z. anorg. Chem.* **266**, 161.

Goubeau, J., and Keller, H. (1951), *Z. anorg. Chem.* **267**, 1.

Goubeau, J., and Link, R. (1951), *Z. anorg. Chem.* **267**, 27.

Goubeau, J., and Becher, H. J. (1952), *Z. anorg. Chem.* **268**, 1.

Goubeau, J., and Ekhoff, E. (1952), *Z. anorg. Chem.* **268**, 145.

Goubeau, J., and Bues, W. (1952), *Z. anorg. Chem.* **268**, 221.

Goubeau, J., and Wittmeier, H. W. (1952), *Z. anorg. Chem.* **270**, 16.

Goubeau, J., and Lücke, K. E. (1952), *Liebigs Ann.* **575**, 37.

Goubeau, J., and Keller, H. (1953), *Z. anorg. Chem.* **272**, 303.

Goubeau, J., Rahtz, M., and Becher, H. J. (1954), *Z. anorg. Chem.* **275**, 161.

Goubeau, J., Richter, D. E., and Becher, H. J. (1955), *Z. anorg. Chem.* **278**, 12.

Goubeau, J., and Zappel, A. (1955), *Z. anorg. Chem.* **279**, 38.

Goubeau, J., Becher, H. J., and Griffel, F. (1955), *Z. anorg. Chem.* **282**, 86.

Goubeau, J., Bues, W., and Kampmann, F. W. (1956), *Z. anorg. Chem.* **283**, 123.

Goubeau, J., and Epple, R. (1957), *Ber. Chem.* **90**, 171.

Goubeau, J., and Rohwedder, K. H. (1957), *Liebigs Ann.* **604**, 168.

Goubeau, J., and Kallfass, H. (1959), *Z. anorg. Chem.* **299**, 160.

Gould, J. R. (1956), *U.S.P.* 2,754,177.

Graham, W. A. G., and Stone, F. G. A. (1956), *J. Inorg. & Nuclear Chem.* **3**, 164.

Graham, W. A. G., and Stone, F. G. A. (1956), *Chem. & Ind.* 319.

Graham, W. A. G., and Stone, F. G. A. (1957), *Chem. & Ind.* 1096.

Graves, G. D., and Werntz, J. H. (1936), *U.S.P.* 2,053,474.

Greenwood, N. N. (1958), *J. Inorg. & Nuclear Chem.* **5**, 224, 229.

Greenwood, N. N., Martin, R. L., and Eméleus, H. J. (1950), *J. chem. Soc.* 3030.

Greenwood, N. N., Martin, R. L., and Eméleus, H. J. (1951), *J. chem. Soc.* 1328.

Greenwood, N. N., and Martin, R. L. (1951), *J. chem. Soc.* 1795.

Greenwood, N. N., and Martin, R. L. (1953), *J. chem. Soc.* 751, 757, 4132.

Greenwood, N. N., and Martin, R. L. (1954), *Quart. Rev. chem. Soc., Lond.* **8**, 1.

Greenwood, N. N., and Morris, J. H. (1960), *J. chem. Soc.* 2922.

Grimley, J., and Holliday, A. K. (1954), *J. chem. Soc.* 1212.

Groszos, S. J., and Stafiej, S. F. (1958), *J. Amer. chem. Soc.* **80**, 1357.

Grummitt, O. (1942), *J. Amer. chem. Soc.* **64**, 1811.

Gunn, S. R., and Green, L. G. (1955), *J. Amer. chem. Soc.* **77**, 6197.

Gustavson, G. (1870), *Ber. dtsch. chem. Ges.* **3**, 426.

Guter, G. A., and Schaeffer, G. W. (1956), *J. Amer. chem. Soc.* **78**, 3546.

Haber, R. G., Nikuni, Z., Schmid, H., and Yogi, K. (1956), *Helv. chim. acta* **39**, 1654.

Haider, S. Z., Khundkar, M. H., and Siddiqullah, Md. (1954), *J. appl. Chem.* (*London*) **4**, 93.

Hamilton, J. K., and Smith, F. (1956), *J. Amer. chem. Soc.* **78**, 5907, 5910.

Hamilton, W. C. (1956), *Proc. roy. Soc.* **A.235**, 395.

Hamlen, R. P., and Koski, W. S. (1956), *Anal. Chem.* **28**, 1631.

Harris, J. J. (1959), *Dissertation Abstr.* **19**, 1554, *Microfilm* L.C. Card No. Mic. 58–7136.

Harris, P. M., and Meibohm, E. P. (1947), *J. Amer. chem. Soc.* **69**, 1231.

Hartmann, H., and Birr, K. H. (1959), *Z. anorg. Chem.* **299**, 174.

Hauser, C. R., and Hoffenberg, D. S. (1955), *J. org. Chem.* **20**, 1448.

Havir, J., and Vrestal, J. (1957), *Chem. zvesti*, 1957, **11**, 35.

Haworth, D. T., and Hohnstedt, L. F. (1959), *J. Amer. chem. Soc.* **81**, 842.

Hawthorne, M. F. (1957), *Chem. & Ind.* 1242.

Hawthorne, M. F. (1957), *J. org. Chem.* **22**, 1001.

Hawthorne, M. F. (1958), *J. org. Chem.* **23**, 1579.

Hawthorne, M. F. (1958), *J. org. Chem.* **23**, 1788.

Hawthorne, M. F. (1958), *J. Amer. chem. Soc.* **80**, 3480, 4291, 4293.

Hawthorne, M. F. (1959), *J. Amer. chem. Soc.* **81**, 5836.

Hawthorne, M. F., and Miller, J. J. (1958), *J. Amer. chem. Soc.* **80**, 754.

Hawthorne, M. F., and Lewis, E. S. (1958), *J. Amer. chem. Soc.* **80**, 4296.

Hawthorne, M. F., and Dupont, J. A. (1958), *J. Amer. chem. Soc.* **80**, 5830.

Hawthorne, M. F., and Pitochelli, A. R. (1958), *J. Amer. chem. Soc.* **80**, 6685.

Hawthorne, M. F., and Pitochelli, A. R. (1959), *J. Amer. chem. Soc.* **81**, 5519.

Hawthorne, M. F., and Miller, J. J. (1960), *J. Amer. chem. Soc.* **82**, 500.

Hayter, R. G., Laubengayer, A. W., and Thompson, P. G. (1957), *J. Amer. chem. Soc.* **79**, 4243.

Hedberg, K., and Stosick, A. J. (1952), *J. Amer. chem. Soc.* **74**, 954.

Hein, F., and Burkhardt, R. (1952), *Z. anorg. Chem.* **268**, 159.

Henkel and Cie, G.m.b.H. (1933), *B.P.* 398,064.

Henkel and Cie, G.m.b.H. (1933), *Fr.P.* 746,954.

Hennion, G. F., McCusker, P. A., Ashby, E. C., and Rutkowski, A. J. (1957), *J. Amer. chem. Soc.* **79**, 5190, 5194.

Hennion, G. F., McCusker, P. A., and Rutkowski, A. J. (1958), *J. Amer. chem. Soc.* **80**, 617.

Hennion, G. F., McCusker, P. A., and Marra, J. V. (1958), *J. Amer. chem. Soc.* **80**, 3481.

Hennion, G. F., McCusker, P. A., and Marra, J. V. (1959), *J. Amer. chem. Soc.* **81**, 1768.

Henry, M. C., Hazel, J. F., and McNabb, W. M. (1956), *Analyt. chim. acta* **15**, 187.

Hermans, P. H. (1925), *Z. anorg. Chem.* **142**, 83.

Hermans, P. H. (1938), *Rec. Trav. chim., Pays-Bas* **57**, 333.

Herzog, H. L., Jevnik, M. A., Perlman, P. L., Nobile, A., and Hershberg, E. B. (1953), *J. Amer. chem. Soc.* **75**, 266.

Hewitt, F., and Holliday, A. K. (1953), *J. chem. Soc.* 530.

Heyl, W. (1954), *B.P.* 705,719.

Hill, W. H., Levinskas, G. J., and Merrill, J. M. (1958), *A.M.A. Arch. industr. Hyg.* **17**, 124.

Hipple, J. A., Jr. (1940), *Phys. Rev.* **57**, 350.

Hirao, N., and Yagi, S. (1953), *J. chem. Soc. Japan, Ind. Chem. Sect.* **56**, 371.

Hoard, J. L., Owen, T. B., Buzzell, A., and Salmon, O. N. (1950), *Acta cryst.* **3**, 130.

Hochstein, F. A., and Brown, W. G. (1948), *J. Amer. chem. Soc.* **70**, 3484.

Hoekstra, H. R., and Katz, J. J., Declassified Document AECD 1894, Oak Ridge, Tennessee.

Hoekstra, H. R., and Katz, J. J. (1949), *J. Amer. chem. Soc.* **71**, 2488.

Hörhammer, L., and Hänsel, R. (1955), *Arch. Pharm. Berl.* **288**, 315.

Hoffmann, A. K., and Thomas, W. M. (1959), *J. Amer. chem. Soc.* **81**, 580.

Holliday, A. K., and Sowler, J. (1952), *J. chem. Soc.* 11.

Holliday, A. K., and Massey, A. G. (1958), *J. Amer. chem. Soc.* **80**, 4744.

Holliday, A. K., and Massey, A. G. (1960), *J. chem. Soc.* 43.

Holzbecher, Z. (1952), *Chem. Listy* **46**, 17.

Holzbecher, Z. (1952), *Chem. Listy*, **46**, 20.

Horii, Z., Sakai, T., and Inoi, T. (1955), *J. pharm. Soc. Japan* **75**, 1161.

Horner, L., and Scherf, K. (1951), *Liebigs Ann.* **573**, 35.

Hough, L., Jones, J. K. N., and Richards, E. L. (1953), *Chem. & Ind.* 1064.

Hough, W. V., Schaeffer, G. W., Dzurus, M., and Stewart, A. C. (1955), *J. Amer. chem. Soc.* **77**, 864.

Hough, W. V., Edwards, L. J., McElroy, A. D. (1956), *J. Amer. chem. Soc.* **78**, 689.

Hu, J. H., and MacWood, G. E. (1956), *J. Phys. chem.* **60**, 1483.

Hughes, E. C. (1955), *128th Meeting Amer. chem. Soc. Pet. Div.*, Minneapolis.

Hughes, E. C., Darling, S. M., Bartleson, J. D., and Klingel, A. R., Jr. (1951), *Industr. Engng. Chem.* **43**, 2841.

Hughes, E. C., Fay, P. S., Szabo, L. S., and Tupa, R. C. (1956), *Industr. Engng. Chem.* **48**, 1858.

Hughes, E. W. (1956), *J. Amer. chem. Soc.* **78**, 502.

Hunter, D. L., and Fajans, E. W. (1957), *B.P.* 787,146.

Hurd, D. T. (1948), *J. org. Chem.* **13**, 711.

Hurd, D. T. (1948), *J. Amer. chem. Soc.* **70**, 2053.

Hurd, D. T. (1948), *U.S.P.* 2,446,008.

Hurd, D. T. (1949), *J. Amer. chem. Soc.* **71**, 20.

Hurd, D. T. (1952), "The Chemistry of Hydrides", John Wiley, New York.

Hyman, M., and West, C. D. (1948), *U.S.P.* 2,445,579.

Ieviņš, A., and Gudriniece, E. (1954), *J. anal. Chem., Moscow* **9**, 270.

Ieviņš, A., Ozols, J., and Gudriniece, E. (1955), *Latv. PSR Zinat. Akad. Vestis No. 7*, 135.

Iffland, D. C., and Criner, G. X. (1953), *J. Amer. chem. Soc.* **75**, 4047.

Iffland, D. C., and Teh-Fu, Yen (1954), *J. Amer. chem. Soc.* **76**, 4083.

I. G. Farbenind, A-G. (1933), *Fr. P.* 743,942.

Iliceto, A., and Patron, G. (1956), *Ann. Chim. Roma* **46**, 267.

Inglis, J. (1835), *Phil. Mag.* (3), **7**, 441.

Inglis, J. (1836), *Phil. Mag.* (3), **8**, 191.

Inglis, J. (1836), *J. prakt. Chem.* **7**, 405.

Ingold, C. K. (1953), "Structure and Mechanism in Organic Chemistry", p. 738. Bell & Sons, London.

Irany, E. P. (1946), *Colloid Chem.* **6**, 1089.

Isbell, H. S., Brewster, J. F., Holt, N. B., and Frush, H. L. (1948), *J. Res. nat. Bur. Stand.* **40**, 129.

Ishibashi, M., Emi, K., Katroka, H., and Kitadani, S. (1955), *Rec. Oceanogr. Wks. Jap.* **2**, No. 2, 127.

Jablonski, E. M. (1939), *B.P.* 511,473.

Jacobson, R. A., and Lipscomb, W. N. (1958), *J. Amer. chem. Soc.* **80**, 5571.

Jander, G., and Kraffczyk, K. (1955), *Z. anorg. Chem.* **282**, 121.

Jansons, E., and Ieviņš, A. (1958), *Latv. Valsts. Univ. Kim. Fak. Zināt. Raksti* **22**, No. 6, 85.

Jaulmes, P., and Galhac, E. (1937), *Bull. Soc. chim. Fr.* (5) **4**, 139, 149.

Jensen, E. H. (1954), "A Study on Sodium Borohydride", Nyt Nordisk Forlag, Arnold Busck, Copenhagen.

Joannis, A. (1902), *C. R. Acad. Sci. Paris* **135**, 1106.

Joglekar, M. S., and Thatte, V. N. (1936), *Z. Phys.* **98**, 692.

Johnson, A. R. (1912), *J. phys. Chem.* **16**, 1.

Johnson, J. R., and Tompkins, S. W. (1933), *Org. Synth.* **13**, 16.

Johnson, F. W. (1936), *U.S.P.* 2,040,997.

Johnson, J. R., Van Campen, M. G., and Grummit, O. (1938), *J. Amer. chem. Soc.* **60**, 111.

Johnson, J. R., Snyder, H. R., and Van Campen, M. G. (1938), *J. Amer. chem. Soc.* **60**, 115.

Johnson, J. R., and Van Campen, M. G. (1938), *J. Amer. chem. Soc.* **60**, 121.

Johnson, H. L., and Hullet, N. C. (1953), *J. Amer. chem. Soc.* **75**, 1467.

Jones, F., and Taylor, R. L. (1881), *J. chem. Soc. Transactions* **39**, 213.

Jones, L. C. (1898), *Amer. J. Sci.* **7**, 147.

Jones, R. C., and Kinney, C. R. (1939), *J. Amer. chem. Soc.* **61**, 1378.

Jones, W. J., Thomas, L. H., Pritchard, E. H., and Bowden, S. T. (1946), *J. chem. Soc.* 824.

Josien, M. L., and Lebas, J. M. (1956), *Bull. Soc. chim. Fr.* **53**, 53.

Kahovec, L. (1938), *Z. phys. Chem.* **B40**, 135.

Kahovec, L. (1939), *Z. phys. Chem.* **B43**, 109.

Karrer, P. (1926), *Helv. chim. acta* **9**, 116.

Kaspar, J. S., McCarty, L. V., and Newkirk, A. E. (1949), *J. Amer. chem. Soc.* **71**, 2583.

Katz, J. R., and Selman, J. (1928), *Z. Phys.* **46**, 392.

Kaufmann, A. (1930), *Ger. P.*, 555,403.

Kaufman, J. J., and Koski, W. S. (1956), *J. chem. Phys.* **24**, 403.

Keenan, C. W., and McDowell, W. J. (1954), *J. Amer. chem. Soc.* **76**, 2839.

Kehiaian, H. V. (1957), *Acad. rep. Populare Romíne, Studii cercertâri chim.* **5**, No. 1, 51.

Keller, W., and Weiss, F. (1957), *Pharmazie* **12**, 19.

Kemula, W., and Kornacki, J. (1954), *Roczn. Chem.* **28**, 635.

Khan, I. A., and Sen, D. (1959), *Proc. Indian Acad. Sci* **49A**, 226.

Khotinsky, E., and Melamed, M. (1909), *Ber. dtsch. chem. Ges.* **42**, 3090.

Khotinskii, E. S., and Pupko, S. L. (1929), *J. chim. Ukr.* **4**, 13.

King, G. W. (1938), *J. chem. Phys.* **6**, 378.

Kinney, C. R., and Kolbezen, M. J. (1942), *J. Amer. chem. Soc.* **64**, 1584.

Kinney, C. R., and Mahoney, C. L. (1943), *J. org. Chem.* **8**, 526.

Kinney, C. R., Thompson, H. T., and Cheney, L. C. (1935), *J. Amer. chem. Soc.* **57**, 2396.

Kinney, C. R., and Pontz, D. F. (1935), *J. Amer. chem. Soc.* **57**, 1128.

Kinney, C. R., and Pontz, D. F. (1936), *J. Amer. chem. Soc.* **58**, 196.

Kinney, C. R., and Pontz, D. F. (1936), *J. Amer. chem. Soc.* **58**, 197.

Kinoshita, Y., Koike, H., Kuriyama, S., Hsami, Y., Okamoto, M. (1957), *Nagoya Shiritsu Daigaku Yakugakubu Kiyô* **5**, 32.

Kirkpatrick, W. H. (1956), *U.S.P.*, 2,755,296.

Klemm, L., and Klemm, W. (1935), *Z. anorg. Chem.* **225**, 258.

König, W., and Scharrnbeck, W. (1930), *J. prakt. Chem.* **128**, 153.

Köster, R. (1957), *Angew. Chem.* **69**, 94.

Köster, R., and Ziegler, K. (1957), *Angew. Chem.* **69**, 94.

Köster, R. (1958), *Angew. Chem.* **70**, 371.

Köster, R. (1958), *Liebigs Ann.* **618**, 31.

Köster, R. (1959), *Angew. Chem.* **71**, 520.

Köster, R., and Reinert, K. (1959), *Angew. Chem.* **71**, 521.

Kohler, M. (1953), *Z. anal. Chem.* **138**, 9.

Kolesnikov, G. S., and Fedorova, L. S. (1957), *Bull. Acad. Sci. U.R.S.S., Classe sci. chim.* 236.

Kolesnikov, G. S., and Fedorova, L. S. (1958), *Bull. Acad. Sci. U.R.S.S., Classe sci. chim.* 906.

Kolesnikov, G. S., and Klimentova, N. V. (1957), *Bull. Acad. Sci. U.R.S.S. Classe sci. chim.* 652.

Kolesnikov, G. S., and Soboleva, T. A. (1957), *Khim. Nauka i. Prom.* **2**, 663.

Kollonitsch, J., Fuchs, O., and Gabor, V. (1954), *Nature* **173**, 125.

Kollonitsch, J., and Fuchs, O. (1955), *Nature* **176**, 1081.

Kolski, T. L. (1958), *Dissertation Abstr.* **18**, 1607, *Microfilm*, Publ. No. 25128, p. 90.

Kolski, T. L., Moore, H. B., Roth, L. E., Martin, K. J., and Schaeffer, G. W. (1958), *J. Amer. chem. Soc.* **80**, 549.

Kornacki, J. (1954), *Wiadomości Chemi.* **8**, 538.

Koski, W. S., Kaufman, J. J., and Lauterbur, P. C. (1957), *J. Amer. chem. Soc.* **79**, 2382.

Krantz, J. C., Beck, F. F., Carr, C. J., and Evans, H. E. (1938), *J. phys. Chem.* **42**, 507.

Kraus, C. A. (1936), *Nucleus* **13**, 213.

Kraus, C. A., and Hawes, W. W. (1933), *J. Amer. chem. Soc.* **55**, 2776.

Krause, E. (1923), *Chem. Zbl.* **11**, 1089.

Krause, E. (1923), *Ger. P.* 371,467.

Krause, E. (1924), *Ber. dtsch. chem. Ges.* **57**, 216.

Krause, E. (1924), *Ber. dtsch. chem. Ges.* **57**, 813.

Krause, E., and Dittmar, P. (1921), *Ber. dtsch. chem. Ges.* **54**, 2784.

Krause, E., and Nitsche, R. (1921), *Ber. dtsch. chem. Ges.* **54**, 2784.

Krause, E., and Nitsche, R. (1922), *Ber. dtsch. chem. Ges.* **55**, 1261.

Krause, E., and Polack, H. (1926), *Ber. dtsch. chem. Ges.* **59**, 777.

Krause, E., and Polack, H. (1928), *Ber. dtsch. chem. Ges.* **61**, 271.

Krause, E., and Dittmar, P. (1930), *Ber. dtsch. chem. Ges.* **63B**, 2347.

Krause, E., and Nobbe, P. (1930), *Ber. dtsch. chem. Ges.* **63**, 934.

Krause, E., and Nobbe, P. (1931), *Ber. dtsch. chem. Ges.* **64**, 2112.

Krause, E., and Von Grosse, A. (1937), "Die Chemie der Metallorganischen Verbindungen", p. 207, Borntraeger, Berlin.

Kreshkov, A. P. (1950), *Zh. prikl. Khim., Mosk.* **23**, 545.

Krüerke, U. (1956), *Z. Naturf.* **116**, 364.

Kuck, J. A., and Grim, E. C. (1959), *Microchem. J.* **3**, 35.

Kuemmel, D. F., and Mellon, M. G. (1956), *J. Amer. chem. Soc.* **78**, 4572.

Kuivila, H. G. (1954), *J. Amer. chem. Soc.* **76**, 870.

Kuivila, H. G. (1955), *J. Amer. chem. Soc.* **77**, 4014.

Kuivila, H. G., Slack, S. C., and Siiteri, P. K. (1951), *J. Amer. chem. Soc.* **73**, 123.

Kuivila, H. G., and Easterbrook, E. K. (1951), *J. Amer. chem. Soc.* **73**, 4629.

Kuivila, H. G., and Hendrickson, A. R. (1952), *J. Amer. chem. Soc.* **74**, 5068.

Kuivila, H. G., Keough, A. H., and Soboczenski, E. J. (1954), *J. org. Chem.* **19** 780.

Kuivila, H. G., and Soboczenski, E. J. (1954), *J. Amer. chem. Soc.* **76**, 2675.

Kuivila, H. G., and Williams, R. M. (1954), *J. Amer. chem. Soc.* **76**, 2679.

Kuivila, H. G., and Wiles, R. A. (1955), *J. Amer. chem. Soc.* **77**, 4830.

Kuivila, H. G., and Benjamin, L. E. (1955), *J. Amer. chem. Soc.* **77**, 4834.

Kuivila, H. G., and Armour, A. G. (1957), *J. Amer. chem. Soc.* **79**, 5659.

Kurz, P. F. (1956), *Fuel* **35**, 318.

Kuskov, V. K. (1956), *C. R. Acad. Sci. U.R.S.S.* **110**, 223.

Kuskov, V. K., and Zhukova, V. A. (1956), *Bull. Acad. Sci. U.R.S.S., Classe sci. chim.* **733**; *English Translation*, 743.

Kuskov, V. K., and Sheiman, B. M. (1956), *Proc. Acad. Sci. U.S.S.R., Sect: Chem.* **106**, 83.

Kuskov, V. K., and Sheiman, B. M. (1956), *C. R. Acad. Sci. U.R.S.S.* **106**, 479.

Kuskov, V. K., and Yur'eva, L. P. (1956), *C. R. Acad. Sci. U.R.S.S.* **109**, 319.

Kuskov, V. K., Sheiman, B. M., and Maksimova, Z. G. (1957), *J. gen. Chem., Moscow (English Translation)* **27**, 1528.

Kynaston, W., and Turner, H. S. (1958), *Proc. chem. Soc.* 304.

Lane, T. J., McCusker, P. A., and Curran, B. C. (1942), *J. Amer. chem. Soc.* **64**, 2076.

Langer, S. H., and Ebling, I. N. (1957), *Industr. Engng Chem.* **49**, 1113.

Lappert, M. F. (1953), *J. chem. Soc.* **667**, 2784.

Lappert, M. F. (1955), *J. chem. Soc.* 784.

Lappert, M. F. (1956), *J. chem. Soc.* 1768.

Lappert, M. F. (1956), *Chem. Rev.* **56**, 959.

Lappert, M. F. (1957), *Proc. Chem. Soc.* 121.

Lappert, M. F. (1958), *J. chem. Soc.* 2790, 3256.

Laubengayer, A. W., and Sears, D. S. (1945), *J. Amer. chem. Soc.* **67**, 164.

Laubengayer, A. W., Ferguson, R. P., and Newkirk, A. E. (1941), *J. Amer. chem. Soc.* **63**, 559.

Lawesson, S. O. (1956), *Ark. Kemi.* **10**, 171.

Lawesson, S. O. (1957), *Ark. Kemi.* **11**, 387.

Lawrence, F. I. L., Smith, R. K., and Pohorilla, M. J. (1955), *U.S.P.* 2,721,180; 2,721,181; 2,721,121.

Lazier, W. A., and Salzberg, P. L. (1946), *U.S.P.* 2,402,591.

Leets, K. V., Shumeiko, A. K., Rozenoer, A. A., Kudryasheva, N. V., and Pilyavskaya, A. I. (1957), *J. gen. Chem., Moscow* **27**, 1510.

Letsinger, R. L., and Skoog, I. H. (1953), *J. org. Chem.* **18**, 895.

Letsinger, R. L., Skoog, I. H., and Remes, N. (1954), *J. Amer. chem. Soc.* **76**, 4047.

Letsinger, R. L., and Skoog, I. H. (1954), *J. Amer. chem. Soc.* **76**, 4174.

Letsinger, R. L., and Remes, N. (1955), *J. Amer. chem. Soc.* **77**, 2489.

Letsinger, R. L., and Skoog, I. H. (1955), *J. Amer. chem. Soc.* **77**, 2491, 5176.

Letsinger, R. L., and Nazy, J. R. (1958), *J. org. Chem.* **23**, 914.

Letsinger, R. L., and Hamilton, S. B. (1958), *J. Amer. chem. Soc.* **80**, 5411.

Letsinger, R. L., and Dandegaonker, S. H. (1959), *J. Amer. chem. Soc.* **81**, 498.

Letsinger, R. L., and Hamilton, S. B. (1959), *J. Amer. chem. Soc.* **81**, 3009.

Letsinger, R. L., and Nazy, J. R. (1959), *J. Amer. chem. Soc.* **81**, 3013.

Levens, E., and Washburn, R. M. (1959), *U.S.P.* 2,875,236.

Levitin, N. E., Westrum, E. F., and Carter, J. C. (1959), *J. Amer. chem. Soc.* **81**, 3547.

Levy, H. A., and Brockway, L. O. (1937), *J. Amer. chem. Soc.* **59**, 2085.

Lewis, D. T. (1956), *Analyst* **81**, 531.

Lewis, G. L., and Smyth, C. P. (1940), *J. Amer. chem. Soc.* **62**, 1529.

Lieber, W. (1957), *Zement-Kalk-Gips.* **10**, 61.

Lincoln, B. H., and Byrkit, G. D. (1947), *U.S.P.* 2,413,718.

Lindeman, L. P., and Wilson, M. K. (1956), *J. chem. Phys.* **24**, 242.

Linnett, J. W. (1940), *Trans. Faraday Soc.* **36**, 1123.

Lippincott, S. B. (1953), *U.S.P.* 2,642,453.

Lipscomb, W. N. (1954), *J. chem. Phys.* **22**, 985.

Lipscomb, W. N. (1957), *J. phys. Chem.* **61**, 23.

Lipscomb, W. N., Pitochelli, A. R., and Hawthorne, M. F. (1959), *J. Amer. chem. Soc.* **81**, 5833.

Long, L. H., and Dollimore, D. (1953), *J. chem. Soc.* 3902.

Long, L. H., and Norrish, R. G. W. (1949), *Phil. Trans.* **A241**, 587.

Long, L. H., and Wallbridge, M. G. H. (1959), *Chem. & Ind.* 295.

Longuet-Higgins, H. C. (1946), *J. chem. Soc.* 139.

Longuet-Higgins, H. C. (1957), *Quart. Rev. chem. Soc., Lond.* **11**, 121.

Lord, R. C., and Nielsen, E. (1951), *J. chem. Phys.* **19**, 1.

Lyle, R. E., DeWitt, E. J., and Pattison, I. C. (1956), *J. org. Chem.* **21**, 61.

Lynds, L., and Stern, D. R. (1959), *J. Amer. chem. Soc.* **81**, 5006.

Lyons, W. E. (1939), *U.S.P.* 2,156,918.

Lyttle, D. A., Jensen, E. H., and Struck, W. A. (1952), *Analyt. Chem.* **24**, 1843.

Maan, C. J. (1929), *Rec. Trav. chim., Pays-Bas* **48**, 332.

McCarty, L. V., Smith, G. C., McDonald, R. S. (1954), *Analyt. Chem.* **26**, 1027.

McCarty, L. V., and Di Giorgio, P. A. (1951), *J. Amer. chem. Soc.* **73**, 3138.

McCoy, R. E., and Bauer, S. H. (1956), *J. Amer. chem. Soc.* **78**, 2061.

McCusker, P. A. (1958), *U.S.P.* 2,820,830.

McCusker, P. A. (1958), *B.P.* 801,707.

McCusker, P. A., and Glunz, L. J. (1955), *J. Amer. chem. Soc.* **77**, 4253.

McCusker, P. A., Ashby, E. C., and Makowski, H. S. (1957), *J. Amer. chem. Soc.* **79**, 5179, 5182, 5185.

McCusker, P. A., Hennion, G. F., and Ashby, E. C. (1957), *J. Amer. chem. Soc.* **79**, 5192.

McCusker, P. A., and Ostdick, T. (1958), *J. Amer. chem. Soc.* **80**, 1103.

McCusker, P. A., and Kilzer, S. M. L. (1960), *J. Amer. chem. Soc.* **82**, 372.

McDaniel, D. H. (1957), *Science* **125**, 545.

McDowell, W. J., and Keenan, C. W. (1956), *J. Amer. chem. Soc.* **78**, 2065, 2069.

McGrath, J. S., Stack, G. G., and McCusker, P. A. (1944), *J. Amer. chem. Soc.* **66**, 1263.

McKone, L. J., and Lyons, W. E. (1939), *U.S.P.* 2,149,856.

Makishima, S., Yoneda, Y., and Tajima, T. (1957), *J. phys. Chem.* **61**, 1618.

Mancera, O., Ringold, H. J., Djerazsi, C., Rosenkranz, G., and Sondheimer, F. (1953), *J. Amer. chem. Soc.* **75**, 1286.

Mann, D. E., and Fano, L. (1957), *J. chem. Phys.* **26**, 1665.

Margrave, J. L. (1954), *J. phys. Chem.* **58**, 258.

Margrave, J. L. (1957), *J. phys. Chem.* **61**, 38.

Mark, H., and Pohland, E. (1925), *Z. Kristallogr.* **62**, 103.

Marquardt, P., and Vogg, G. (1952), *Hoppe-Seyl. Z.* **291**, 143.

Martin, D. R. (1944), *Chem. Rev.* **34**, 461.

Martin, D. R. (1948), *Chem. Rev.* **42**, 581.

Martin, D. R., and Mako, L. S. (1951), *J. Amer. chem. Soc.* **73**, 2674.

Martin, G. R. (1950), *Canadian P.* 466,183.

Marvel, C. S., and Dennoon, C. E., Jr. (1938), *J. Amer. chem. Soc.* **60**, 1045.

Massoth, F. E. (1956), *Dissertation Abstr.* **16**, 1074, *Microfilm*, Publ. No. 16,657.

Matsui, M., Miyano, M., and Tomita, K. J. (1956), *Bull. agr. chem. Soc. Japan* **20**, 139.

Matteson, D. S. (1959), *J. Amer. chem. Soc.* **81**, 5004.

Mattiello, J. J. (1946), "Protective and Decorative Coatings", Vol. V, p. 77, Wiley, New York.

Mattraw, H. C., Erickson, C. E., and Laubengayer, A. W. (1956), *J. Amer. chem. Soc.* **78**, 4901.

Maxwell, L. R. (1940), *J. opt. Soc. Amer.* **30**, 374.

May, F. H. (1957), *U.S.P.* 2,808,424.

May, F. H. (1958), *U.S.P.* 2,855,427.

May, F. H., Levasheff, V. V., and Hammar, H. N. (1958), *U.S.P.* 2,833,623.

Mead, E. J. (1955), *Dissertation Abstr.* **15**, 971.

Meerwein, H. (1948), *Angew. Chem.* **A60**, 78.

Meerwein, H., and Bersin, T. (1929), *Liebigs Ann.* **476**, 113.

Meerwein, H., and Hinz, G. (1931), *Liebigs Ann.* **484**, 1.

Meerwein, H., and Maier-Hüser, H. (1932), *J. prakt. Chem.* **134**, 51.

Meerwein, H., and Pannwitz, W. (1934), *J. prakt. Chem.* **141**, 123.

Meerwein, H., Hinz, G., Majert, H., and Sönke, H. (1936), *J. prakt. Chem.* **147**, 226.

Meerwein, H., and Sönke, H. (1936), *J. prakt. Chem.* **147**, 251.

Meerwein, H., Bock, B., Kirschnick, B., Lenz, W., and Migge, A. (1936), *J. prakt. Chem.* **147**, 211.

Meerwein, H., Battenberg, E., Gold, H., Pfeil, E., and Willfang, G. (1939), *J. prakt. Chem.* **154**, 83.

Melnikov, N. N. (1936), *J. gen. Chem.*, Moscow (*English Translation*) **6**, 636.

Melnikov, N. N., and Rokitskaya, M. S. (1938), *J. gen. Chem.*, Moscow (*English Translation*) **8**, 1768.

Metal Hydrides Inc., Beverley, Mass. Technical Bulletin, 502c.

Meulenhoff, J. (1925), *Z. anorg. Chem.* **142**, 373.

Michaelis, A. (1894), *Ber. dtsch. chem. Ges.* **27**, 244.

Michaelis, A. (1901), *Liebigs Ann.* **315**, 19.

Michaelis, A., and Becker, P. (1880), *Ber. dtsch. chem. Ges.* **13**, 58.

Michaelis, A., and Becker, P. (1882), *Ber. dtsch. chem. Ges.* **15**, 180.

Michaelis, A., and Behrens, M. (1894), *Ber. dtsch. chem. Ges.* **27**, 244.

Michaelis, A., and Luxembourg, K. (1896), *Ber. dtsch. chem. Ges.* **29**, 710.

Michaelis, A., and Hillringhaus, F. (1901), *Liebigs Ann.* **315**, 41.

Michaelis, A., and Richter, E. (1901), *Liebigs Ann.* **315**, 26.

Mikhailov, B. M., and Aronovich, P. M. (1954), *C. R. Acad. Sci. U.R.S.S.* **98**, 791.

Mikhailov, B. M., and Vaver, V. A. (1955), *C. R. Acad. Sci. U.R.S.S.* **102**, 531.

Mikhailov, B. M., and Aronovich, P. M. (1955), *Bull. Acad. Sci. U.R.S.S. Classe sci. chim.*, 946; *English Translation* 859.

Mikhailov, B. M., and Shchegoleva, T. A. (1955), *Bull. Acad. Sci. U.R.S.S., Classe sci. chim.*, 1124.

Mikhailov, B. M., and Vaver, V. A. (1956), *C. R. Acad. Sci. U.R.S.S.* **109**, 94.

Mikhailov, B. M., and Aronovich, P. M. (1956), *Bull. Acad. Sci. U.R.S.S., Classe sci. chim.*, 322; *English Translation*, 311.

Mikhailov, B. M., and Vaver, V. A. (1956), *Proc. Acad. Sci. U.S.S.R., Sect: Chem.* **109**, 341.

Mikhailov, B. M., and Fedotov, N. S. (1956), *Bull. Acad. Sci. U.R.S.S., Classe sci. chim.* 375.

Mikhailov, B. M., and Kostroma, T. V. (1956), *Bull. Acad. Sci. U.R.S.S., Classe sci. chim.* 376.

Mikhailov, B. M., and Vaver, V. A. (1956), *Bull. Acad. Sci. U.R.S.S., Classe sci. chim.* 451, 508; *English Translation*, 441.

Mikhailov, B. M., and Shchegoleva, T. A. (1956), *C. R. Acad. Sci. U.R.S.S.* **108**, 481.

Mikhailov, B. M., and Shchegoleva, T. A. (1956), *Bull. Acad. Sci. U.R.S.S., Classe sci. chim.* 508.

Mikhailov, B. M., Kozminskaya, T. K., Blakhina, A. N., and Shchegoleva, T. A. (1956), *Bull. Acad. Sci. U.R.S.S., Classe sci. chim.* 692; *English Translation* 703.

Mikhailov, B. M., and Kostroma, T. V. (1956), *Bull. Acad. Sci. U.R.S.S., Classe sci. chim.* 1144.

Mikhailov, B. M., and Fedotov, N. S. (1956), *Bull. Acad. Sci. U.R.S.S., Classe sci. chim.* 1511.

Mikhailov, B. M., Kostroma, T. V., and Fedotov, N. S. (1957), *Bull. Acad. Sci. U.R.S.S., Classe sci. chim.* 589.

Mikhailov, B. M., and Kostroma, T. V. (1957), *Bull. Acad. Sci. U.R.S.S., Classe sci. chim.* 646.

Mikhailov, B. M., and Vaver, V. A. (1957), *Bull. Acad. Sci. U.R.S.S., Classe sci. chim.* 812, 989.

Mikhailov, B. M., and Shchegoleva, T. A. (1957), *Bull. Acad. Sci. U.R.S.S.*, *Classe sci. chim.* 1080.

Mikhailov, B. M., and Aronovich, P. M. (1957), *Bull. Acad. Sci. U.R.S.S.*, *Classe sci. chim.* 1123.

Mikhailov, B. M., and Kostroma, T. V. (1957), *Bull. Acad. Sci. U.R.S.S.*, *Classe sci. chim.* 1125.

Mikhailov, B. M., and Tutorskaya, F. B. (1958), *C. R. Acad. Sci. U.R.S.S.* **123**, 479.

Mikhailov, B. M., and Kozminskaya, T. K. (1958), *C. R. Acad. Sci. U.R.S.S.* **121**, 656.

Mikhailov, B. M., and Fedotov, N. S. (1958), *Bull. Acad. Sci. U.R.S.S.*, *Classe sci. chim.* 857.

Mikhailov, B. M., and Shchegoleva, T. A. (1958), *Bull. Acad. Sci. U.R.S.S.*, *Classe sci. chim.* 860.

Mikhailov, B. M., Blokhina, A. N., and Fedotov, N. S. (1958), *Bull. Acad. Sci. U.R.S.S.*, *Classe sci. chim.* 891.

Mikhailov, B. M., and Bazhenova, A. V. (1959), *Bull. Acad. Sci. U.R.S.S.*, *Classe sci. chim.* 76.

Mikhailov, B. M., and Kozminskaya, T. K. (1959), *Bull. Acad. Sci. U.R.S.S.*, *Classe sci. chim.* 80.

Mikhailov, B. M., and Bulinov, Yu. N. (1959), *Bull. Acad. Sci. U.R.S.S.*, *Classe sci. chim.* 172.

Mikheeva, V. I., and Fedneva, E. M. (1955), *C. R. Acad. Sci. U.R.S.S.* **101**, 99.

Mikheeva, V. I., and Fedneva, E. M. (1956), *Zhur. Neorg. Khim.* **1**, 894.

Mikheeva, V. I., and Fedneva, E. M. (1956), *Bull. Acad. Sci. U.R.S.S.*, *Classe sci. chim. (English Translation)*, 925.

Mikheeva, V. I., and Markina, V. Yu. (1956), *Zhur. Neorg. Khim.* **1**, 2700.

Mikheeva, V. I., and Fedneva, E. M. (1957), *Zhur. Neorg. Khim.* **2**, 604.

Miller, J. J., and Hawthorne, M. F. (1959), *J. Amer. chem. Soc.* **81**, 4501.

Milone, M. (1936), *Atti Accad. Torino* **71**, 391.

Milone, M. (1938), *Gazz. chim. ital.* **68**, 582.

Mitchell, J. S. (1958), *U.S.P.* 2,855,382.

Mitra, S. M. (1938), *Indian J. Phys.* **12**, 9.

Moeller, C. W., and Wilmarth, W. K. (1959), *J. Amer. chem. Soc.* **81**, 2638.

Moissan, H. (1891), *C. R. Acad. Sci. Paris* **112**, 717.

Moissan, H. (1891), *C. R. Acad. Sci. Paris* **113**, 624.

Moissan, H. (1892), *C. R. Acad. Sci. Paris* **114**, 622.

Moissan, H. (1895), *Ann. Chim. (Phys.)* (7), **6**, 313.

Montequi, R. (1957), *Inform. quim. anal. Suppl. Ion (Madrid)*, **11**, 8.

Montequi, R., Doadrio, A., and Serrano, C. (1956), *Publ. Inst. Quim. Barba* **10**, 183.

Moore, E. B., Jr., and Lipscomb, W. N. (1956), *Acta cryst.* **9**, 668.

Morgan, G. T., and Tunstall, R. B. (1924), *J. chem. Soc.* **125**, 1963.

Mosher, W. A., and Beachell, H. C. (1958), *Quart. Report*, pp. 1–5. University of Delaware.

Muetterties, E. L. (1957), *Z. Naturf.* **12b**, 264, 265.

Muetterties, E. L. (1957), *U.S.P.* 2,782,233.

Muetterties, E. L. (1957), *J. Amer. chem. Soc.* **79**, 6563.

Muetterties, E. L. (1958), *J. Amer. chem. Soc.* **80**, 4526.

Muetterties, E. L. (1958), *U.S.P.* 2,840,590.

Muetterties, E. L. (1959), *J. Amer. chem. Soc.* **81**, 2597.

Muetterties, E. L., and Rochow, E. G. (1953), *J. Amer. chem. Soc.* **75**, 490.

Mukherji, A. K., and Sant, B. R. (1959), *Analyt. Chem.* **31**, 608.

Mulliken, R. S. (1947), *Chem. Rev.* **41**, 207.

Muraca, R. F. (1954), *Chem. Anal.* **43**, 69.

Muraca, R. F., Collier, H. E., Bonsack, J. P., and Jacobs, E. S. (1954), *Chem. Anal.* **43**, 102.

Muraire, M. (1957), *Chim. anal.* **39**, 184.

Musgrave, O. C. (1956), *J. chem. Soc.* 4305.

Musgrave, O. C. (1957), *Chem. & Ind.* 1152.

Musgrave, O. C., and Park, T. O. (1955), *Chem. & Ind.* 1552.

Nason, H. B. (1857), *Liebigs Ann.* **104**, 126.

Naylor, R. E., Jr., and Wilson, E. B. (1957), *J. chem. Phys.* **26**, 1057.

Nesmeyanov, A. N., and Sazonova, V. A. (1955), *Bull. Acad. Sci. U.R.S.S., Classe sci. chim. (English Translation)*, 187.

Nesmeyanov, A. N., Sazonova, V. A., Liberman, G. S., and Emelyanova, L. I. (1955), *Bull. Acad. Sci. U.R.S.S., Classe sci. chim.*, 48.

Nesmeyanov, A. N., Sazonova, V. A., and Drozd, V. N. (1959), *Bull. Acad. Sci. U.R.S.S., Classe sci. chim.* 163.

Nespital, W. (1932), *Z. phys. Chem.* **B16**, 153.

Neu, R. (1954), *Z. anal. Chem.* **142**, 335.

Neu, R. (1954), *Z. anal. Chem.* **143**, 30, 254.

Neu, R. (1954), *Chem. Ber.* **87**, 802.

Neu, R. (1955), *Chem. Ber.* **88**, 1761.

Neu, R. (1955), *Fette u. Seif.* **57**, 568.

Neu, R. (1956), *Z. anal. Chem.* **151**, 328.

Neu, R. (1956), *Mikrochem. Acta* 1169.

Neu, R. (1958), *Chem. Anal.* **47**, 106.

Neu, R. (1958), *Naturwissenschaften* **45**, 311.

Neu, R. (1959), *Archiv. der Pharm.* **64**, 437.

Niedenzu, K., and Dawson, J. W. (1959), *J. Amer. chem. Soc.* **81**, 3561.

Nielsen, D. R., McEwen, W. E., and VanderWerf, C. A. (1957), *Chem. & Ind.* 1069.

Nielsen, D. R., and McEwen, W. E. (1957), *J. Amer. chem. Soc.* **79**, 3081, 3681.

Nikitina, A. N., Galanin, M. D., Aronovich, P. M., Shchegoleva, T. A., and Mikhailov, B. M. (1958), *Bull. Acad. Sci. U.R.S.S., Sér. phys.* **22**, 12.

Normant, H., and Braun, J. (1959), *C. R. Acad. Sci. Paris* **248**, 828.

Norton, F. J. (1950), *J. Amer. chem. Soc.* **72**, 1849.

Norymberski, J. K., and Woods, G. F. (1954), *Chem. & Ind.* 518.

Norymberski, J. K., and Woods, G. F. (1955), *J. chem. Soc.* 3426.

Nutten, A. J. (1954), *Industr. Chem. Mfr.* **30**, 29, 57.

Nyilas, E., and Soloway, A. H. (1959), *J. Amer. chem. Soc.* **81**, 2681.

Nystrom, R. F., Chaikin, S. W., and Brown, W. G. (1949), *J. Amer. chem. Soc.* **71**, 3245.

O'Connor, G. L., and Nace, H. R. (1955), *J. Amer. chem. Soc.* **77**, 1578.

Oetschlager, H. (1955), *Arch. Pharm., Berl.* **288**, 102.

Oliveto, E. P., and Hershberg, E. B. (1953), *J. Amer. chem. Soc.* **75**, 488.

Onak, T., Landesman, H., and Shapiro, I. (1958), *J. phys. Chem.* **62**, 1605.

Orth, G. O., Jr. (1957), *U.S.P.* 2,809,949.

Osthoff, R. C., and Brown, C. A. (1952), *J. Amer. chem. Soc.* **74**, 2378.

Otto, M. M. (1935), *J. Amer. chem. Soc.* **57**, 1476.

Owen, W. R., and Sutherland, M. D. (1956), *J. Sci. Fd. Agric.* **7**, 88.

Pace, E. (1929), *Atti Accad. Lincei.* **10**, 193.

Palazzo, F. C. (1914), *R. C. Soc. chim. ital.* (2), **5**, 17.

Parry, R. W., and Bissot, T. C. (1956), *J. Amer. chem. Soc.* **78**, 1524.

Parry, R. W., Schultz, D. R., and Girardot, P. R. (1958), *J. Amer. chem. Soc.* **80**, 1.

Parsons, T. D. (1954), *Dissertation Abst.* **14**, 469.

Parsons, T. D., and Ritter, D. M. (1954), *J. Amer. chem. Soc.* **76**, 1710.

Parsons, T. D., Silverman, M. B., and Ritter, D. M. (1957), *J. Amer. chem. Soc.* **79**, 5091.

Pastour, P., and Maréchal, E. (1956), *C. R. Acad. Sci. Paris* **243**, 166.

Pastureau, P., and Veiler, M. (1936), *C. R. Acad. Sci. Paris* **202**, 1683.

Patein, G. (1891), *C. R. Acad. Sci. Paris* **113**, 85.

Paul, R., and Joseph, N. (1953), *Bull. Soc. chim. Fr.* **20**, 758.

Pauling, L. (1947), *Chem. Engng News* **25**, 2970.

Paushkin, Ya. M., and Osipova, L. V. (1955), *C. R. Acad. Sci. U.R.S.S.* **103**, 439.

Pecsok, R. L. (1953), *J. Amer. chem. Soc.* **75**, 2862.

Perkins, G. T., and Crowell, T. I. (1956), *J. Amer. chem. Soc.* **78**, 6013.

Perrine, J. C., and Keller, R. N. (1958), *J. Amer. chem. Soc.* **80**, 1823.

Petry, R. C., and Verhoek, F. H. (1956), *J. Amer. chem. Soc.* **78**, 6416.

Pflaum, R. T., and Howick, L. C. (1956), *Analyt. Chem.* **28**, 1542.

Pictet, A., and Geleznoff, A. (1903), *Ber. dtsch. chem. Ges.* **36**, 2219.

Pictet, A., and Karl, G. (1908), *Bull. Soc. chim. Fr.* (4), **3**, 1114.

Pitzer, K. S. (1945), *J. Amer. chem. Soc.* **67**, 1126.

Platt, J. R. (1954), *J. chem. Phys.* **22**, 1033.

Ploquin, J. (1956), *Bull. Soc. Pharm. Bordeaux* **95**, 13.

Ploquin, J. (1957), *Bull. Soc. Pharm. Bordeaux* **96**, 153.

Ploquin, J. (1958), *Bull. Soc. Pharm. Bordeaux* **97**, 145.

Porter, R. P. (1957), *J. phys. Chem.* **61**, 1260.

Povlock, T. P., and Lippincott, W. T. (1958), *J. Amer. chem. Soc.*, **80**, 5409.

Prescott, R. F., Dosser, R. C., and Sculati, J. J. (1941), *U.S.P* 2,260,336; 2,260,337; 2,260,338; 2,260,339.

Prescott, R. F., Dosser, R. C., and Sculati, J. J. (1942), *U.S.P.* 2,300,006.

Price, W. C., Fraser, R. D. B., Robinson, T. S., and Longuet-Higgins, H. C. (1950), *Disc. Faraday Soc.* **9**, 131.

Procházka, J. (1955), *Czech. P.* 84,283.

Pryde, E. H. (1955), *U.S.P.* 2,715,057.

Quelet, R. (1927), *Bull. Soc. chim. Fr.* (4), **41**, 933.

Quill, L. L. (1956), *129th Meeting, Amer. chem. Soc. Div. of Org. Chem.*, Dallas.

Raff, P., and Brotz, W. (1951), *Z. anal. Chem.* **133**, 241.

Ramsay, W., and Hatfield, H. S. (1901), *Proc. chem. Soc.*, Lond. 152.

Ramsden, H. E., Leebrick, J. R., Rosenberg, S. D., Miller, E. H., Walburn, J. J., Balint, A. E., and Cserr, R. (1957), *J. org. Chem.* **22**, 1602.

Ramser, H., and Wiberg, E. (1930), *Ber. dtsch. chem. Ges.* **63**, 1136.

Rath, W., and Bauer, W. H. (1956), *J. phys. Chem.* **60**, 639.

Ray, N. H. (1954), *U.S.P.* 2,670,333.

Razuvaev, G. A., and Brilkina, T. G. (1952), *C. R. Acad. Sci. U.R.S.S.* **85**, 815.

Razuvaev, G. A., and Brilkina, T. G. (1953), *C. R. Acad. Sci. U.R.S.S.* **91**, 861.

Razuvaev, G. A., and Brilkina, T. G. (1954), *J. gen. Chem., Moscow* **24**, 1415; *English Translation* **24**, 1397.

Reddy, J. van der M., and Lipscomb, W. N. (1959), *J. Amer. chem. Soc.* **81**, 754.
Reedy, A. J. (1957), *Dissertation Abstr.* **17**, 1225; *Microfilm* Publ. No. 20886.
Reeves, R. E. (1949), *J. Amer. chem. Soc.* **71**, 212, 215, 1737, 2116.
Reichard, O. (1953), *Z. anal. Chem.* **140**, 188.
Reid, W. E., Bish, J. M., and Brenner, A. (1957), *J. Electrochem. Soc.* **104**, 21.
Remick, A. E. (1953), "Electronic Interpretations of Organic Chemistry", p. 66. John Wiley & Sons, Inc., New York.
Renner, T. (1957), *Angew. Chem.* **69**, 478.
Rice, B., Barredo, J. M. G., and Young, T. F. (1951), *J. Amer. chem. Soc.* **73**, 2306.
Rice, B., Galiano, R. J., and Lehmann, W. J. (1957), *J. phys. Chem.* **61**, 1222.
Rick, C. E. (1957), *U.S.P.* 2,809,124.
Rideal, S. (1889), *Ber. dtsch. chem. Ges.* **22**, 992.
Rippere, R. E., and La Mer, V. K. (1943), *J. phys. Chem.* **47**, 204.
Rojahn, C. A. (1933), *Chem. Zbl.* **II**, 2704.
Rojahn, C. A. (1933), *Ger. P.* 582,149.
Romain, P., Merland, R., and Laubie, H. (1954), *Bull. Soc. Pharm. Bordeaux* **92**, 131.
Rondestvedt, C. S., Jr., Scribner, R. M., and Wulfman, C. E. (1955), *J. org. Chem.* **20**, 9.
Rose, H. (1856), *Ann. Phys. Lpz.* **98**, 245.
Rosenblum, I. (1934), *U.S.P.* 1,969,761.
Rosenblum, L. (1955), *J. Amer. chem. Soc.* **77**, 5016.
Rosenheim, A., Loewenstamm, W., and Singer, L. (1903), *Ber. dtsch. chem. Ges.* **36**, 1833.
Rothstein, E., and Saville, R. W. (1952), *J. chem. Soc.* 2987.
Rozendaal, H. M. (1951), *Arch. industr. Hyg.* **4**, 257.
Rubia Pacheco, J. de la, and Blasco López-Rubio, F. (1955), *Inform, quím. anal. Suppl. Ion (Madrid)* **9**, 1.
Rüdorff, W., and Zannier, H. (1952), *Z. anal. Chem.* **137**, 1.
Rüdorff, W., and Zannier, H. (1954), *Angew. Chem.* **66**, 638.
Rüdorff, W., and Zannier, H. (1953), *Z. anal. Chem.* **140**, 1.
Rüdorff, W., and Zannier, H. (1953), *Z. Naturf.* **8b**, 611.
Ruigh, W. L., Erickson, C. E., Gunderley, F. C., Sedlak, M., Van der Meulen, P. A., Olin, A. D., and Steinberg, N. G. (1955), *W.A.D.C.* Tech. Report, 55/26, Parts I to IV.
Ryschkewitsch, G. E., Harris, J. J., and Sisler, H. H. (1958), *J. Amer. chem.* **80**, 4515.
Sabatier, P. (1891), *C. R. Acad. Sci. Paris* **112**, 865.
Sanderson, R. T. (1953), *J. chem. Phys.* **21**, 571.
Santucci, L., and Gilman, H. (1958), *J. Amer. chem. Soc.* **80**, 193.
Sazonova, V. A., and Kronrod, N. Ya. (1956), *J. gen. Chem., Moscow*, 1876.
Scattergood, A., Miller, W. H.. and Gammon, J. (1945), *J. Amer. chem. Soc.* **67**, 2150.
Schabacher, W., and Goubeau, J. (1958), *Z. anorg. Chem.* **294**, 183.
Schäfer, H., and Braun, O. (1952), *Naturwissenschaften* **39**, 280.
Schaeffer, G. W., and Anderson, E. R. (1949), *J. Amer. chem. Soc.* **71**, 2143.
Schaeffer, G. W., Schaeffer, R., and Schlesinger, H. I. (1951), *J. Amer. chem. Soc.* **73**, 1612.
Schaeffer, G. W., and Emilius, M. (1954), *J. Amer. chem. Soc.* **76**, 1203.
Schaeffer, G. W., and Basile, L. J. (1955), *J. Amer. chem. Soc.* **77**, 331.

Schaeffer, G. W., Adams, M. D., and Koenig, F. J. (1956), *J. Amer. chem. Soc.* **78**, 725.

Schaeffer, G. W., Roscoe, J. S., and Stewart, A. C. (1956), *J. Amer. chem. Soc.* **78**, 729.

Schaeffer, G. W., Kolski, T. L., and Ekstedt, D. L. (1957), *J. Amer. chem. Soc.* **79**, 5912.

Schaeffer, R. (1957), *J. Amer. chem. Soc.* **79**, 1006.

Schaeffer, R., Steindler, M.. Hohnstedt, L., Smith, H. S., Eddy, L. B., and Schlesinger, H. I. (1954), *J. Amer. chem. Soc.* **76**, 3303.

Schechter, W. H. (1953), *U.S.P.* 2,629,732.

Schechter, W. H. (1954), *U.S.P.* 2,689,259.

Schechter, W. H. (1955), *B.P.* 736,820.

Schechter, W. H. (1957), *U.S.P.* 2,774,136; 2,787,329.

Schiff, H. (1867), *Liebigs Ann., Suppl.* **5**, 158.

Schlesinger, H. I., and Burg, A. B. (1931), *J. Amer. chem. Soc.* **53**, 4321.

Schlesinger, H. I., and Walker, A. O. (1935), *J. Amer. chem. Soc.* **57**, 621.

Schlesinger, H. I., Horvitz, L., and Burg, A. B. (1936), *J. Amer. chem. Soc.* **58**, 407, 409.

Schlesinger, H. I., and Burg, A. B. (1937), *J. Amer. chem. Soc.* **59**, 780.

Schlesinger, H. I., and Burg, A. B. (1938), *J. Amer. chem. Soc.* **60**, 290.

Schlesinger, H. I., Ritter, D. M., and Burg, A. B. (1938), *J. Amer. chem. Soc.* **60**, 1296, 2297.

Schlesinger, H. I., Sanderson, R. T., and Burg, A. B. (1939), *J. Amer. chem. Soc.* **61**, 536.

Schlesinger, H. I., Flodin, N. W., and Burg, A. B. (1939), *J. Amer. chem. Soc.* **61**, 1078.

Schlesinger, H. I., Sanderson, R. T., and Burg, A. B. (1940), *J. Amer. chem. Soc.* **62**, 3421.

Schlesinger, H. I., and Brown, H. C. (1940), *J. Amer. chem. Soc.* **62**, 3429.

Schlesinger, H. I., and Burg, A. B. (1942), *Chem. Rev.* **31**, 1.

Schlesinger, H. I.. Brown, H. C., and Schaeffer, G. W. (1943), *J. Amer. chem. Soc.* **65**, 1786.

Schlesinger, H. I., and Brown, H. C. (1949), *U.S.P.* 2,461,661; 2,461,662; 2,461,663.

Schlesinger, H. I., and Schaeffer, G. W. (1950), *U.S.P.* 2,494,267; 2,528,454.

Schlesinger, H. I.. and Brown, H. C. (1950), *U.S.P.* 2,534,533.

Schlesinger, H. I., and Brown, H. C. (1951), *U.S.P.* 2,543,511.

Schlesinger, H. I., Brown, H. C., Abraham, B., Bond, A. C., Davidson, N., Finholt, A. E., Gilbreath, J. R., Hoekstra, H. R., Horvitz, L., Hyde, E. K., Katz, J. J., Knight, J., Lad, R. A., Mayfield, D. L., Rapp, L., Ritter, D. M., Schwartz, A. M., Sheft, I., Tuck, L. D., Walker, A. O. (1953), *J. Amer. chem. Soc.* **75**, 186.

Schlesinger, H. I., Brown, H. C., Abraham, B., Davidson, N., Finholt, A. E., Lad, R. A., Knight, J., and Schwartz, A. M. (1953), *J. Amer. chem. Soc.* **75**, 191.

Schlesinger, H. I., Brown, H. C., Gilbreath, J. R., and Katz, J. J. (1953), *J. Amer. chem. Soc.* **75**, 195.

Schlesinger, H. I., Brown, H. C., Hoekstra, H. R., and Rapp, L. R. (1953), *J. Amer. chem. Soc.* **75**, 199.

Schlesinger, H. I., Brown, H. C., and Finholt, A. E. (1953), *J. Amer. chem. Soc.* **75**, 205.

Schlesinger, H. I., Brown, H. C., and Hyde, E. K. (1953), *J. Amer. chem. Soc.* **75**, 209.

Schlesinger, H. I., Brown, H. C., Mayfield, D. L., and Gilbreath, J. R. (1953), *J. Amer. chem. Soc.* **75**, 213.

Schlesinger, H. I., Brown, H. C., Finholt, A. E., Gilbreath, J. R., Hoekstra, H. R., and Hyde, E. K. (1953), *J. Amer. chem. Soc.* **75**, 215.

Schlesinger, H. I., and Brown, H. C. (1953), *J. Amer. chem. Soc.* **75**, 219.

Schlesinger, H. I., Brown, H. C., Horvitz, L., Bond, A. C., Tuck, L. D., and Walker, A. O. (1953), *J. Amer. chem. Soc.* **75**, 222.

Schmidt, H. (1955), *Chem. Ber.*, **88**, 459.

Schuele, W. J., Hazel, J. F., and McNabb, W. M. (1956), *Analyt. Chem.* **28**, 505.

Schützenberger, M. (1861), *C. R. Acad. Sci. Paris* **53**, 538.

Schultz, O. E., and Mayer, G. (1952), *Dtsh. ApothZtg.* **92**, 358.

Schultz, O. E., and Goerner, H. (1953), *Dtsh. ApothZtg.* **93**, 585.

Schultz, O. E., and Goerner, H. (1955), *Arch. Pharm., Berl.* **288**, 520.

Schumb, W. C., Gamble, E. L., and Banus, M. D. (1949), *J. Amer. chem. Soc.* **71**, 3225.

Schupp, L. J., and Brown, C. A. (1955), *Abstract Amer. chem. Soc.* 127th Meeting, Ohio, 48R.

Sciarra, J. J., Autian, J., and Foss, N. E. (1958), *J. Amer. pharm. Ass.* **47**, 144.

Scott, L. B., and Morris, R. C. (1958), *U.S.P.* 2,821,463.

Scudi, J. V., Bastedo, W. A., and Webb, T. J. (1940), *J. biol. Chem.* **136**, 399.

Seaman, W., and Johnson, J. R. (1931), *J. Amer. chem. Soc.* **53**, 711.

Secci, M. (1958), *Ann. Chim. Roma* **48**, 1183.

Seidel, C. F., and Stoll, M. (1957), *Helv. chim. acta* **40**, 1990.

Sekiguchi, K. (1958), *J. pharm. Soc. Japan* **78**, 965, 970.

Servoss, R. R., and Clark, H. M. (1957), *J. chem. Phys.* **26**, 1179.

Shamim-Ahmad, S. (1952), *J. Indian chem. Soc.* **29**, 880.

Shapiro, I., and Ditter, J. F. (1957), *J. chem. Phys.* **26**, 798.

Shapiro, I., Weiss, H. G., Schmich, M., Skolnik, S., and Smith, G. B. L. (1952), *J. Amer. chem. Soc.* **74**, 901.

Shapiro, I., and Keilen, B. (1954), *J. Amer. chem. Soc.* **76**, 3864.

Shechter, H., Ley, D. E., and Zeldin, L. (1952), *J. Amer. chem. Soc.* **74**, 3664.

Shechter, H., Ley, D. E., and Roberson, E. B., Jr. (1956), *J. Amer. chem. Soc.* **78**, 4984.

Shepp, A., and Bauer, S. H. (1954), *J. Amer. chem. Soc.* **76**, 265.

Sheppard, N. (1955), *Trans. Faraday Soc.* **51**, 1465.

Shoolery, J. N. (1955), *Disc. Faraday Soc.* **19**, 215.

Shore, S. G., and Parry, R. W. (1955), *J. Amer. chem. Soc.* **77**, 6084.

Siebert, H. (1952), *Z. anorg. Chem.* **268**, 13.

Siegel, B., Mack, J. L., Lowe, J. U. Jr., and Callaghan, J. (1958), *J. Amer. chem. Soc.* **80**, 4523.

Silbiger, G., and Bauer, S. H. (1946), *J. Amer. chem. Soc.* **68**, 312.

Sirotina, I. A., and Alimarin, I. P. (1957), *J. anal. Chem., Moscow* **12**, 367.

Skinner, H. A., and Tees, T. F. S. (1953), *J. chem. Soc.* 3378.

Skinner, H. A., and Smith, N. B. (1953), *J. chem. Soc.* 4025.

Skinner, H. A., and Smith, N. B. (1954), *J. chem. Soc.* 2324, 3930.

Smalley, J. H., and Stafiej, F. S. (1959), *J. Amer. chem. Soc.* **81**, 582.

Smith, J. E., and Kraus, C. A. (1951), *J. Amer. chem. Soc.* **73**, 2751.

Snyder, H. R., Kuck, J. A., and Johnson, J. R. (1938), *J. Amer. chem. Soc.* **60**, 105.

Snyder, H. R., and Weaver, C. (1948), *J. Amer. chem. Soc.* **70**, 232.

Snyder, H. R., and Wyman, F. W. (1948), *J. Amer. chem. Soc.* **70**, 234.

Snyder, H. R., Konecky, M. S., and Lennarz, W. J. (1958), *J. Amer. chem. Soc.* **80**, 3611.

Snyder, H. R., Reedy, A. J., and Lennarz, W. J. (1958), *J. Amer. chem. Soc.* **80**, 835.

Soldate, A. M. (1947), *J. Amer. chem. Soc.* **69**, 987.

Solms, J., and Deuel, H. (1957), *Chimia* **11**, 311.

Solomon, I. J., Klein, M. J., Hatton, K. (1958), *J. Amer. chem. Soc.* **80**, 4520.

Soloway, A. H. (1958), *Science*, **128**, 1572.

Soloway, A. H. (1959), *J. Amer. chem. Soc.* **81**, 3017.

Souchay, P., and Lourijsen, Mme (1956), *Bull. Soc. chim. Fr.* 893.

Sowa, F. J., and Nieuwland, J. A. (1933), *J. Amer. chem. Soc.* **55**, 5052.

Sowa, F. J., and Nieuwland, J. A. (1937), *J. Amer. chem. Soc.* **59**, 1202.

Spanner, S., and Boscott, R. J. (1956), *Scand. J. clin. Lab. Invest.* **250**, 8.

Spier, H. W. (1952), *Biochem. Z.* **322**, 467.

Sporek, K. F. (1956), *Analyst* **81**, 540.

Sporek, K. F., and Williams, A. F. (1955), *Analyst* **80**, 347.

Stafiej, F. S. (1959), *U.S.P.* 2,862,951.

Steele, B. D., and Mills, J. E. (1930), *J. chem. Soc.* 74.

Steindler, M. J., and Schlesinger, H. I. (1953), *J. Amer. chem. Soc.* **75**, 756.

Stewart, J. E. (1955), *J. chem. Phys.* **23**, 2204.

Stewart, J. E. (1956), *J. Res. nat. Bur. Stand.* **56**, 337.

Stock, A. (1921), *Ber. dtsch. chem. Ges.* **54A**, 142.

Stock, A. (1927), *Ber. dtsch. chem. Ges.* **60B**, 1039.

Stock, A. (1932), *Bull. Soc. chim. Fr.* **51**, 697.

Stock, A. (1933), "Hydrides of Boron and Silicon", Cornell University Press, Ithaca, New York.

Stock, A. (1937), *Naturwissenschaften* **25**, 417.

Stock, A. (1943), *Suomen Kemistilehti* **16A**, 75.

Stock, A., and Blix, M. (1901), *Ber. dtsch. chem. Ges.*, **34**, 3039. (This paper deals with "Borimidine" $B_2(NH)_3$.)

Stock, A., and Holle, W. (1908), *Ber.* **41**, 2095.

Stock, A., and Massenez, C. (1912), *Ber. dtsch. chem. Ges.* **45**, 3539.

Stock, A., and Friederici, K. (1913), *Ber. dtsch. chem. Ges.* **46**, 1959.

Stock, A., Friederici, K., and Priess, O. (1913), *Ber. dtsch. chem. Ges.* **46**, 3353.

Stock, A., and Kuss, E. (1914), *Ber. dtsch. chem. Ges.* **47**, 810.

Stock, A., Kuss, E., and Priess, O. (1914), *Ber. dtsch. chem. Ges.* **47**, 3115.

Stock, A., and Zeidler, F. (1921), *Ber. dtsch. chem. Ges.* **54**, 531.

Stock, A., and Kuss, E. (1923), *Ber. dtsch. chem. Ges.* **56B**, 789.

Stock, A., and Siecke, W. (1924), *Ber. dtsch. chem. Ges.* **57B**, 562.

Stock, A., Brandt, A., and Fischer, H. (1925), *Ber. dtsch. chem. Ges.* **58**, 643.

Stock, A., and Pohland, E. (1925), *Ber. dtsch. chem. Ges.* **58B**, 657.

Stock, A., and Pohland, E. (1926), *Ber. dtsch. chem. Ges.* **59B**, 2210, 2215, 2223.

Stock, A., and Pohland, E. (1929), *Ber. dtsch. chem. Ges.* **62B**, 90.

Stock, A., Wiberg, E., and Martini, H. (1930), *Z. anorg. Chem.* **188**, 32.

Stock, A., Wiberg, E., and Martini, H. (1930), *Ber. dtsch. chem. Ges.* **63B**, 2927.

Stock, A., Wiberg, E., Martini, H., and Nicklas, A. (1931), *Z. Phys. Chem.*, *Bodenstein Festband*, 93.

Stock, A., and Wierl, R. (1931), *Z. anorg. Chem.* **203**, 228.

Stock, A., and Wiberg, E. (1932), *Ber. dtsch. chem. Ges.* **65B**, 1711.

Stock, A., Martini, H., and Sütterlin, W. (1934), *Ber. dtsch. chem. Ges.* **67B**, 396.

Stock, A., and Sütterlin, W. (1934), *Ber. dtsch. chem. Ges.* **67B**, 407.

Stock, A., Sütterlin, W., and Kurzen, F. (1935), *Z. anorg. Chem.* **225**, 225.

Stock, A., Kurzen, F., and Laudenklos, H. (1935), *Z. anorg. Chem.* **225**, 243.

Stock, A., and Mathing, W. (1936), *Ber. dtsch. chem. Ges.* **69B**, 1456.

Stock, A., and Laudenklos, H. (1936), *Z. anorg. Chem.* **228**, 178.

Stockmeyer, W. H., Rice, D. W., and Stephenson, C. C. (1955), *J. Amer. chem. Soc.* **77**, 1980.

Stoll, M. (1957), *U.S.P.* 2,803,662.

Stoll M., Hinder, M., and Willhalm, B. (1956), *Helv. chim. acta* **39**, 200.

Stone, F. G. A. (1955), "Chemistry of the Boron Hydrides", *Quart. Rev. chem. Soc., Lond.* **9**, 174.

Stone, F. G. A., and Emeléus, H. J. (1950), *J. chem. Soc.* 2755.

Stone, F. G. A., and Burg, A. B. (1954), *J. Amer. chem. Soc.* **76**, 386.

Stone, F. G. A., and Graham, W. A. G. (1955), *Chem. & Ind.* 1181.

Stout, L. E., and Chamberlain, D. F. (1952), *W.A.D.C. Technical Report* 52/192.

Strecker, W. (1910), *Ber. dtsch. chem. Ges.* **43**, 1131.

Strizhevskii, I. I., and Asinovskaya, G. A. (1955), *Trud. vsesoyuz.nauch. isseld. Inst. Avtogen. Obrabotki Metal* No. 11, 140; (1956), *Referat. Zhur., Met.* 1469.

Sugden, S., and Waloff, M. (1932), *J. chem. Soc.* 1492.

Sugihara, J. M., and Bowman, C. M., (1958), *J. Amer. chem. Soc.* **80**, 2443.

Sujishi, S., and Witz, S. (1954), *J. Amer. chem. Soc.* **76**, 4631.

Sun, K. H. (1956), *Nucleonics* **14**, No. 7, 46.

Sykes, A. (1955), *Industr. Chem. Mfr.* **31**, 308.

Sykes, A. (1956), *Industr. Chem. Mfr.* **32**, 164.

Szmant, H. H., and Irwin, D. A. (1956), *J. Amer. chem. Soc.* **78**, 4386.

Taft, R. W., Jr. (1953), *J. Amer. chem. Soc.* **75**, 4231.

Tahara, A. (1956), *J. Org. chem.* **21**, 442.

Talalay, L., Talalay, J. A., and Bush, T. F. (1956), *U.S.P.* 2,758,980.

Taylor, M. D., Grant, L. R., and Sands, C. A. (1955), *J. Amer. chem. Soc.* **77**, 1506.

Taylor, R. C., and Cluff, C. L. (1958), *Nature* **182**, 390.

Thielens, G. (1958) *Appl. Sci. Res., Hague* **78**, 87.

Thielens, G. (1958), *Naturwissenschaften* **45**, 543.

Thomas, J. R., and Harle, O. L. (1957), *U.S.P.* 2,795,548.

Thomas, L. H. (1946), *J. chem. Soc.* 820, 823.

Thomson, R. T. (1893), *J. Soc. chem. Ind., Lond.* **12**, 432.

Thomson, T., and Stevens, T. S. (1933), *J. chem. Soc.* 556.

Tollin, B. C., Schaeffer, R., and Svec, H. J. (1957), *J. Inorg. & Nuclear Chem.* **4**, 273.

Ton, J. C. (1957), *U.S.P.* 2,802,018.

Topchiev, A. V., Prokhorova, A. A., Paushkin, Y. M., and Kuraschev, M. V. (1958), *Bull. Acad. Sci. U.R.S.S., Classe sci. chim.* 370.

Topchiev, A. V., Zavgorodii, S. V., and Paushkin, Ya. M. (1959), "Borontrifluoride and its Compounds as Catalysts in Organic Chemistry", Translated from the Russian by J. T. Greaves. Pergamon Press.

Torssell, K. (1954), *Acta. chem. scand.* **8**, 1229, 1779.

Torssell, K. (1955), *Acta chem. scand.* **9**, 239, 242.

Torssell, K. (1957), *Arkiv. Kemi.* **10**, 497.

Tranchant, J., and Marvillet, L. (1956), *Mémor. Poud.* **38**, 337.

Trautmann, C. E. (1957), *U.S.P.* 2,813,830.

Trevoy, L. W., and Brown, W. G. (1949), *J. Amer. chem. Soc.* **71**, 1675.

Tronov, B. V., and Petrova, A. M. (1953), *J. gen. Chem. Moscow* **23**, 1019.

Turner, H. S., and Warne, R. J. (1958), *Chem. & Ind.* 526.

Tyson, G. N. (1959), *U.S.P.* 2,884,440.

Ulich, H., and Nespital, W. (1931), *Z. Elektrochem.* **37**, 559.

Ulmschneider, D., and Goubeau, J. (1957), *Chem. Ber.* **90**, 2733.

Ulmschneider, D., and Goubeau, J. (1958), *Z. physik. Chem.* (*Frankfurt*) **14**, 56.

Upson, R. W. (1950), *U.S.P.* 2,511,310; 2,517,944; 2,517,945.

Upson, R. W. (1952), *U.S.P.* 2,599,144.

Urry, G., Wartik, T., and Schlesinger, H. I. (1952), *J. Amer. chem. Soc.* **74**, 5809.

Urry, G., Wartik, T., Moore, R. E., and Schlesinger, H. I. (1954), *J. Amer. chem. Soc.* **76**, 5293.

Urry, G., Kerrigan, J., Parsons, T. D., and Schlesinger, H. I. (1954), *J. Amer. chem. Soc.* **76**, 5299.

Vaughn, T. H. (1937), *U.S.P.* 2,088,935.

Veiss, A., and Ieviņš, A. (1959), *J. anal. Chem.*, *Moscow* **14**, 143.

Venkataramaraj Urs, S., and Gould, E. S. (1952), *J. Amer. chem. Soc.* **74**, 2948.

Vol'pin, M. E., Zhdanov, S. I., and Kursanov, D. N. (1957), *C. R. Acad. Sci. U.R.S.S.* **112**, 264.

Warburton, D. M. (1952), *U.S.P.* 2,594,370.

Wartik, T., Moore, R. E., and Schlesinger, H. I. (1949), *J. Amer. chem. Soc.* **71**, 3265.

Wartik, T., and Schlesinger, H. I. (1953), *J. Amer. chem. Soc.* **75**, 835.

Wartik, T., and Pearson, R. K. (1955), *J. Amer. chem. Soc.* **77**, 1075.

Wartik, T., and Apple, E. F. (1958), *J. Amer. chem. Soc.* **80**, 6155.

Watt, W. J. (1956), *Dissertation Abstr.* **16**, 650., *Univ. Microfilm.* Publ. No. 16266.

Webster, S. H., and Dennis, L. M. (1933), *J. Amer. chem. Soc.* **55**, 3233.

Wechter, W. J. (1959), *Chem. & Ind.* 294.

Weiss, H. G., and Shapiro, I. (1953), *J. Amer. chem. Soc.* **75**, 1221.

Weiss, H. G., and Shapiro, I. (1959), *J. Amer. chem. Soc.* **81**, 6167.

Weissbach, A. (1958), *J. org. Chem.* **23**, 329.

Wendlandt, W. W. (1957), *Analyt. chim. acta* **16**, 216.

Wendlandt, W. W. (1957), *Chem. Anal.* **46**, 38.

Werner, R. L., and O'Brien, K. G. (1955), *Australian J. Chem.* **8**, (2) 355.

Werner, R. L., and O'Brien, K. G. (1956), *Australian J. Chem.* **9**, 137.

Whatley, A. T., and Pease, R. N. (1954), *J. Amer. chem. Soc.* **76**, 835.

Wheeler, C. M., and Sandstedt, R. A. (1955), *J. Amer. chem. Soc.* **77**, 2024.

Whelan, W. J., and Morgan, K. (1955) *Chem. & Ind.* 1449.

Wiberg, E. (1928), *Z. anorg. Chem.* **173**, 199.

Wiberg, E. (1929), *Z. anorg. Chem.* **179**, 309.

Wiberg, E. (1930), *Z. anorg. Chem.* **187**, 362.

Wiberg, E. (1930), *Z. anorg. Chem.* **191**, 43, 49.

Wiberg, E. (1931), *Z. anorg. Chem.* **195**, 288.

Wiberg, E. (1935), *Z. anorg. Chem.* **225**, 262.

Wiberg, E. (1936), *Ber. dtsch. chem. Ges.* **69B**, 2816.

Wiberg, E. (1948), *Naturwissenschaften* **35**, 182, 212.

Wiberg, E. (1949), *F.I.A.T. Rev. Germ. Sci. Inorg. Chem.* **1**, 129.

Wiberg, E. (1955), *Angew. Chem.* **67**, 711.

Wiberg, E. (1957), *Experimentia Suppl.* 183.

Wiberg, E., and Sütterlin, W. (1931), *Z. anorg. Chem.* **202**, 1, 22, 31, 37.

Wiberg, E., and Schuster, K. (1933), *Z. anorg. Chem.* **213**, 77, 89, 94.

Wiberg, E., and Schuster, K. (1934), *Ber. dtsch. chem. Ges.* **67B**, 1805.

Wiberg, E., and Sütterlin, W. (1935), *Z. anorg. Chem.* **222**, 92.

Wiberg, E., and Heubaum, U. (1935), *Z. anorg. Chem.* **222**, 98.

Wiberg, E., and Smedsrud, H. (1935), *Z. anorg. Chem.* **225**, 204.

Wiberg, E., and Heubaum, U. (1935), *Z. anorg. Chem.* **225**, 270.

Wiberg, E., and Sütterlin, W. (1935), *Ber. dtsch. chem. Ges.* **68B**, 296.

Wiberg, E., and Mathing, W. (1937), *Ber. dtsch. chem. Ges.* **70B**, 690.

Wiberg, E., and Ruschmann, W. (1937), *Ber. dtsch. chem. Ges.* **70B**, 1393, 1583.

Wiberg, E., and Bolz, A. (1938), *Angew. Chem.* 396.

Wiberg, E., and Bolz, A. (1940), *Ber. dtsch. chem. Ges.* **73B**, 209.

Wiberg, E., and Hertwig, K. (1947), *Z. anorg. Chem.* **255**, 141.

Wiberg, E., Hertwig, K., and Bolz, A. (1948), *Z. anorg. Chem.* **256**, 177.

Wiberg, E., Bolz, A., and Buchheit, P. (1948), *Z. anorg. Chem.* **256**, 285.

Wiberg, E., and Karbe, K. (1948), *Z. anorg. Chem.* **256**, 307.

Wiberg, E., and Bolz, A. (1948), *Z. anorg. Chem.* **257**, 131.

Wiberg, E., and Hertwig, K. (1948), *Z. anorg. Chem.* **257**, 138.

Wiberg, E., Bolz, A., and Buchheit, P. (1948), unpublished work cited by J. Goubeau in *F.I.A.T. Rev. Germ. Sci.: Inorg. Chem.* Vol. I, 218.

Wiberg, E., and Buchheit, P. (1948), unpublished work, cited by J. Goubeau in *F.I.A.T. Rev. Germ. Sci.: Inorg. Chem.* Vol. I, 218.

Wiberg, E., Buchheit, P., and Hertwig, K. (1948), unpublished work, cited by J. Goubeau in *F.I.A.T. Rev. Germ. Sci.: Inorg. Chem.* Vol. I, 218.

Wiberg, E., and Bauer, R. (1950), *Z. Naturf.* **5B**, 397.

Wiberg, E., and Horeld, G. (1951), *Z. Naturf.* **6B**, 338.

Wiberg, E., and Bauer, R. (1952), *Z. Naturf.* **7B**, 58.

Wiberg, E., and Henle, W. (1952), *Z. Naturf.* **7B**, 575, 582.

Wiberg, E., and Krüerke, U. (1953), *Z. Naturf.* **8B**, 608, 609, 610.

Wiberg, E., and Sturm, W. (1953), *Z. Naturf.* **8B**, 529, 530, 689.

Wiberg, E., and Michaud, H. (1954), *Z. Naturf.* **9B**, 497.

Wiberg, E., and Sturm, W. (1955), *Z. Naturf.* **10B**, 108.

Wiberg, E., and Hartwimmer, R. (1955), *Z. Naturf.* **10B**, 290, 291, 294 295.

Wiberg, E., Nöth, H., and Hartwimmer, R. (1955), *Z. Naturf.* **10B**, 292.

Wiberg, E., and Sturm, W. (1955), *Angew. chem.* **67**, 483.

Wiberg, E., and Gösele, W. (1956), *Z. Naturf.* **11B**, 485.

Wiberg, E., and Jahn, A. (1956), *Z. Naturf.* **11B**, 489.

Wiberg, E., Nöth, H., and Usón Lacal, R. (1956), *Z. Naturf.* **11B**, 490.

Wiberg, E., and Hartwimmer, R. (1956), *Ger. P.* 945,625.

Wiberg, E., and Modritzer, K. (1956), *Z. Naturf.* **11B**, 747, 748, 750, 751, 753, 755.

Wiberg, E., Dittmann, O., Nöth, H., and Schmidt, M. (1957), *Z. Naturf.* **12B**, 59.

Wiberg, E., Dittmann, O., and Schmidt, M. (1957), *Z. Naturf.* **12B**, 61.

Wiberg, E., Edward, J., Evans, F., and Nöth, H. (1958), *Z. Naturf.* **13B**, 263, 265.

Wilson, C. O., Jr. (1959), *U.S.P.* 2,880,227.

Winternitz, P. F. (1959), *Abs.* 135th Meeting, *Amer. chem. Soc.* 19M.

Wittig, G., and Keicher, G. (1947), *Naturwissenschaften* **34**, 216.

Wittig, G., Keicher, G., Rückert, A., and Raff, P. (1949). *Liebigs Ann.* **563**, 110.

Wittig, G., and Rückert, A. (1950), *Liebigs Ann.* **566**, 101.

Wittig, G., Meyer, F. J., and Lange, G. (1951), *Liebigs Ann.* **571**, 167.

Wittig, G., and Raff, P. (1951), *Liebigs Ann.* **573**, 195.

Wittig, G., and Schloeder, H. (1955), *Liebigs Ann.* **592**, 38.

Wittig, G., and Haag, W. (1955), *Ber. Chem.* **88**, 1654.

Wittig, G., and Herwig, W. (1955), *Ber. Chem.* **88**, 962.

Wittig, G., and Stilz, W. (1956), *Liebigs Ann.* **598**, 93.

Wittig, G., and Polster, R. (1956), *Liebigs Ann.* **599**, 13.

Wittig, G., Stilz, W., and Herwig, W. (1956), *Liebigs Ann.* **598**, 85.

Wittig, G., and Wittenberg, D. (1957), *Liebigs Ann.* **606**, 1.

Wohler, F., and Deville, H. St. C. (1858), *Ann. Chim. (Phys.)* (3), **52**, 84.

Wolfrom, M. L., and Wood, H. B. (1951), *J. Amer. chem. Soc.* **73**, 2933.

Wolfrom, M. L., and Solms, J. (1956), *J. org. Chem.* **21**, 815.

Wuyts, H., and Duquesne, A. (1939), *Bull. Soc. chim. Belge.* **48**, 77.

Yabroff, D. L., and Branch, G. E. K., (1933) *J. Amer. chem. Soc.* **55**, 1663.

Yabroff, D. L., Branch, G. E. K., and Almquist, H. J. (1933), *J. Amer. chem. Soc.* **55**, 2935.

Yabroff, D. L., Branch, G. E. K., and Bettman, B. (1934), *J. Amer. chem. Soc.* **56**, 1850.

Zachariasen, W. H. (1934), *Z. Kristallogr.* **88**, 150.

Zachariasen, W. H. (1954), *Acta cryst.* **7**, 305.

Zakharkin, L. I., and Okhlolystin, O. Yu (1959), *Bull. Acad. Sci. U.R.S.S., Classe sci. chim.* 181.

Zakrzewski, K., May, Z., and Murawski, K. (1956), *Biokhimiya*, **21**, 596.

Zeidler, L. (1952), *Hoppe-Seyl. Z.* **291**, 177.

Zhigach, A. F., Kazakova, E. B., and Kigel, R. A. (1956), *C. R. Acad. Sci. U.R.S.S.* **106**, 69.

Zhigach, A. F., Kazakova, E. B., and Krongauz, E. S. (1956), *C. R. Acad. Sci. U.R.S.S.* **111**, 1029.

Zief, M., and Woodside, R. (1959), *J. org. Chem.* **24**, 1338.

Ziegler, K. (1956), *Angew. Chem.* **68**, 721.

Ziegler, K., Gellert, H. G., Martin, H., Nagel, K., and Schneider, J. (1954), *Liebigs Ann.* **589**, 91.

Zittle, C. A. (1951), *Advanc. Enzymol.* **12**, 493.

Zvonkova, Z. V. (1958), *Kristallografiya* **3**, 564.

Zvonkova, Z. V., and Ghishkova, V. P. (1958), *Kristallografiya* **3**, 539.

Zwahlen, K. D., Horton, W. J., and Fujimoto, G. I. (1957), *J. Amer. chem. Soc.* **79**, 3131.

AUTHOR INDEX

Where a page number is marked with an asterisk (*) the author is not specifically mentioned in the text but papers by him are relevant to the subject matter.

A

Abdel-Akher, M., 148
Abdine, H., 111
Abel, E. W., 12, 42, 60, 69, 70, 74, 87, 89, 226
Abraham, B., 6, 7, 8, 120, 121
Abraham, M. H., 100, 115
Accascina, F., 110
Adams, M. D., 161
Adams, R. M., 165
Ahmad, T., 5, 52
Ainley, A. D., 67, 77, 79, 80, 116
Aklin, O., 112
Alimarin, I. P., 112
Allen, E. C., 210
Allen, S., 32*
Almquist, H. J., 67
Alton, E. R., 164
Altshuller, A. P., 130*
Amin, A. M., 111
Ananthakrishnan, R., 11
Anderson, E. R., 160, 162, 164, 167, 177
Anderson, H. C., 220*
Anderson, J. R., 7*
Anderson, T. F., 122*
Anderson, W. E., 122*
Angyl, S. J., 18
Antikainen, P. J., 20
Appel, F. J., 7*
Apple, E. F., 196
Arbuzov, B. A., 10
Arimoto, F. S., 76
Arnold, H. R., 58
Aronovich, P. M., 72, 76, 82, 88, 94
Armour, A. G., 116
Arnold, J. R., 124
Asahara, T., 11*
Ashby, E. C., 83, 85, 95, 102, 155
Ashikari, N., 100, 101
Asinovskaya, G. A., 7
Atoji, M., 214
Atteberry, R. W., 128*
Auten, R. W., 85
Autian, J., 21

B

Badin, E. J., 157
Balacco, F., 69*
Balint, A. E., 75
Ballard, S. A., 5
Baltimore Paint and Varnish Production Club, 8
Bamford, C. H., 96, 97, 114
Bannister, W. J., 5, 7
Banus, J., 85, 169
Banus, M. D., 131, 138, 158, 220
Barbaras, G. D., 130*
Barbaras, G. K., 140, 165
Barker, E. F., 122*
Barker, S. A., 16,* 17,* 18*
Barnes, R. F., 7
Barnes, R. P., 128
Barredo, J. M. G., 164
Bartholomay, L. H., Jr., 93, 165, 180
Bartleson, J. D., 20
Bashkirov, A. N., 8
Basile, L. J., 166
Bastedo, W. A., 18
Battenberg, E., 211
Bauer, S. H., 11, 59,* 69, 120, 122,* 124, 126,* 130,* 174, 176*
Bauer, W. H., 122*
Bawn, C. E. H., 101
Bax, C. M., 164
Beach, J. Y., 11, 69, 130*
Beachell, H. C., 124, 128, 129
Bean, F. R., 67, 69
Becher, H. J., 11, 96, 165, 170, 171, 181, 211, 218, 226, 229, 230
Beck, F. F., 18
Becker, P., 58, 60, 67, 69, 80
Behrens, M., 80, 86
Bell, R. P., 120, 174
Bellamy, L. J., 223, 225, 226, 227, 231
Belz, L. H., 220*
Bennett, H., 19
Bent, H. E., 108*
Benton, F. L., 219
Bera, B. C., 148

Beran, F., 76*
Berkhout, H. W., 111*
Berl, W. G., 159
Berneis, H. L., 140, 165
Bersin, T., 15
Berzelius, J. J., 22
Besson, A., 220, 221
Bethke, G. W., 96*
Bettman, B., 67, 69, 202
Bihari, M., 21
Birr, K. H., 222
Bish, J. M., 130*
Bissot, T. C., 164, 167, 189
Blakhina, A. N., 72, 82, 88, 94
Blasco López-Rubio, F., 109*
Blau, E. J., 96
Blau, J. A., 11, 32, 231
Blum, E., 122*
Bock, B., 83, 97, 114
Böeseken, J., 16, 18
Bogaty, H., 160
Böhm, H., 76*
Böhm, U., 8, 69, 165, 210
Böhme, H., 108*
Boll, E., 108*
Bolz, A., 161, 163, 165, 166, 167, 168, 169, 174, 176, 177
Bond, A. C., 6, 7, 8, 120, 121, 130*
Bonner, T. G., 22*
Bonner, W. H., 140, 165
Bonsack, J. P., 111*
Boone, J. L., 164, 196
Booth, H. S., 209
Booth, R. B., 83, 84, 85, 182
Borax Consolidated Ltd., 8*
Borisov, A. E., 81
Boscott, R. J., 134*
Bouquet, 22, 36
Bourne, E. J., 16,* 17,* 18,* 32*
Bowden, S. T., 10, 11
Bowlus, H., 183, 185
Bowman, C. M., 72,* 206
Boyd, A. C., 133, 135, 137, 139, 147, 165
Boyer, J. H., 134*
Bradley, M. J., 182, 224, 229, 230
Bragdon, R. W., 6, 138, 158
Bragg, J. K., 124
Branch, G. E. K., 67, 69, 202
Brandenberg, W., 11*
Braun, J., 75
Braun, O., 16,* 17,* 18*
Bremer, C., 16,* 17,* 18*
Brendel, G., 189
Brenner, A., 130*
Brewster, J. F., 17
Brilkina, T. G., 165

Brinckman, F. E., 59
Brindley, P. B., 60, 63, 69, 71, 72, 73, 74, 84
Brockway, L. O., 22,* 96, 161,* 218*
Broida, H. P., 159
Brokow, R. S., 157
Brotz, W., 111
Brown, A. E., 160
Brown, C. A., 165, 170, 173,* 176, 180, 181
Brown, E. H., 170
Brown, H. C., 6, 7, 8, 39, 93, 97, 98, 102, 109, 114, 119, 120, 121, 130, 133, 135, 137, 139, 140, 141, 142, 147, 153, 155, 156, 157, 158, 165, 180, 183, 209
Brown, J. F., 180
Brown, R. D., 164
Brown, W. G., 134,* 142, 144, 150
Bruin, de, J. A., 16, 18
Brumberger, H., 189
Bryusova, L. Ya., 48
Buchheit, P., 83, 163, 165, 166, 167, 168, 169, 174, 176, 177
Bues, W., 96, 226
Bujwid, Z. J., 59, 218, 219
Buls, V. W., 83, 107
Burg, A. B., 69, 84, 85, 119, 120, 122,* 157, 161, 163, 164, 165, 166, 167, 168, 169, 170, 172, 174, 176, 177, 187, 189, 196
Burkhardt, L., 11,* 129
Burns, J. J., 136
Bush, T. F., 160
Buzzel, A., 183
Byrkit, G. D., 8*

C

Cahours, A., 22*
Callaghan, J., 129
Callery Chem. Co., 161*
Calvert, R. P., 7*
Cambi, L., 15
Campbell, D. H., 164
Campbell, G. W., 164
Cardon, S. Z., 165
Carpmael, A., 19
Carpenter, R. A., 100
Carr, C. J., 18
Carter, J. C., 164
Caujolle, F., 76*
Cazes, J., 64
Ceron, P., 213, 214, 217
Chaikin, S. W., 142, 144, 150
Chainani, G. R., 14, 193
Challenger, F., 67, 77, 79, 80, 116
Chamberlain, D. F., 19,* 20

Charnley, T., 11
Chatt, J., 98,* 99,* 152,* 153*
Chaudhuri, T. C., 7*
Cheney, L. C., 22
Cherbuliez, E., 15
Chu, T. L., 109
Clark, H. M., 6
Clark, M. M., 10
Clark, S. L., 104
Clarke, R. P., 124
Clear, C. G., 67
Clifford, J., 107
Cluff, C. L., 164
Cluley, H. J., 111*
Coates, G. E., 167
Coever, H. J., 183
Coffin, K. P., 59,* 176*
Cohn, G., 5, 15
Colclough, T., 5, 7, 12, 34, 43
Collier, H. E., 111*
Commerford, J. D., 9
Consden, R., 18
Cook, H. G., 13, 51, 52, 211
Coolidge, A. S., 108*
Cooper, S., 49
Cooper, S. S., 111*
Copaux, H., 15
Core, A. F., 122*
Cornwell, C. D., 122*
Councler, C., 6, 22
Cowan, R. D., 122*
Cowley, E. G., 11
Cowie, W. P., 65
Coyle, T. D., 196
Crane, F. E., Jr., 112*
Crawford, B. L., Jr., 122,* 228*
Crawhall, J. C., 152
Criner, G. X., 145
Crowell, T. I., 15
Csapo, F., 21
Cserr, R., 75
Cueilleron, J., 219
Curran, C., 22,* 96, 183
Currell, B. R., 4, 29

D

Dandegaonker, S. H., 60, 61, 62, 69, 70,
  74, 87, 199, 201, 205, 225
Darling, S. M., 20, 76
Das, T. P., 96
Davidson, J. M., 110
Davidson, N., 6, 7, 8, 120, 121
Davies, A. G., 100, 107, 115, 116, 117
Davies, T., 109*
Davis, M., 137, 151

Davis, O. L., 83, 107
Davis, W. D., 132*
Dawson, J. W., 182
Dehmelt, H. G., 96
Dennis, L. M., 6
Dennoon, C. E., Jr., 19
Denson, C. L., 15
Deuel, H., 21
Deville, H. St.C., 220
Dewar, M. J. S., 197, 198
DeWitt, E. J., 95
Diamond, H., 7
Dickens, P. G., 123
Dickerson, R. E., 124
Dietz, R., 198
Di Giorgio, P. A., 122*
Dillard, C., 130*
Dillon, T. E., 219
Dilthey, W., 57*
Dimroth, O., 51
Ditter, J. F., 123
Dittmar, P., 92, 97, 108, 114, 165
Djerazsi, C., 134*
Doadrio, A., 109*
Dodson, R. M., 143*
Dodson, V. H., 114, 135, 156, 165
Dollimore, D., 85, 93, 209, 220,* 221
Domash, L., 102, 139, 141, 142, 147, 153,
  158, 165, 209
Dorfman, M., 108*
Dornow, A., 171
Dosser, R. C., 5
Drozd, V. N., 109*
Dürst, J., 112
Duncanson, L. A., 11, 231
Dupire, A., 6
Dupont, J. A., 128
Duppa, B. F., 92, 97
Duquesne, A., 5, 7
Dworkin, A. S., 69
Dzurus, M., 163, 176, 177

E

Easterbrook, E. K., 79
Ebelman, 22, 36
Eberhardt, W. H., 122*
Ebling, I. N., 20
Eddy, L. B., 176
Edmonds, J. T., 20
Edsall, J. T., 228*
Edward, J., 126, 132
Edwards, J. O., 17
Edwards, J. D., 28, 40, 42, 60, 74, 87
Edwards, L. J., 163
Ehrlich, P., 6

Ekhoff, E., 96, 226
Ekstedt, D. L., 128
Elliott, D. F., 152
Ellis, C., 19
Ellzey, S. E., Jr., 134*
Elson, R. E., 124
Emeléus, H. J., 85,* 120, 152, 174, 176*
Emelyanova, L. I., 109*
Emi, K., 21
Emilius, M., 176, 177
Engelmann, F., 109*
Epple, R., 107
Erickson, C. E., 58, 69, 182, 206, 230
Etherington, T. L., 122*
Etridge, J. J., 6, 8, 10
Evans, F., 126, 132
Evans, H. E., 18

F

Fajans, E. W., 8
Fano, L., 214
Farkas, L., 122*
Faust, T., 51
Fay, P. S., 20, 159
Fedneva, E. M., 121, 128, 163
Fedorova, L. S., 100
Fedotov, N. S., 72, 82, 88, 94
Ferguson, G. A., Jr., 96
Ferguson, R. P., 10, 96
Fetter, N. R., 129
Fields, P. R., 7
Finch, A., 213, 214, 217
Fine, S. D., 10
Finholt, A. E., 6, 7, 8, 120, 121, 130*
Fischer, E. O., 109*
Fischer, R., 112
Fitch, S. J., 128
Flaschka, H., 111
Fletcher, E. A., 165
Flodin, N. W., 120, 165
Florin, R. E., 189
Fornaseri, M., 17
Foss, N. E., 21
Foster, A. B., 18, 148
Fowler, D. L., 108, 165
Frankland, E., 7, 92, 97, 114, 221
Fraser, R. D. B., 122, 230
Frazer, M. J., 4, 14, 29, 49, 52, 54, 190, 191, 192, 209
French, C. M., 110
French, F. A., 96
French, H. E., 10
French, R. H. V., 105
Frey, J., 213, 214, 217
Frick, S., 229, 230

Friederici, K., 118
Frohnsdorff, R. S. M., 160
Frush, H. L., 17, 148
Fuchs, O., 136, 138
Fuchs, R., 147, 150
Fujimoto, G. I., 8
Fuoss, R. M., 110
Furukawa, G. T., 122*
Furukawa, J., 100

G

Gabor, V., 136, 138
Gakle, P. S., 122*
Galanin, M. D., 76
Galat, A., 11*
Galhac, E., 8*, 135
Galiano, R. J., 164
Gamble, E. L., 120, 220
Gammon, J., 5, 13
Gardner, D. M., 187*
Garner, P. J., 20
Garrett, E. R., 142
Gasselin, V., 210
Gates, N. V., 8
Gattermann, L., 37
Gautier, J. A., 112*
Gautier, M., 183
Gayel, P., 76*
Gayhart, E. L., 159
Gebauhr, W., 109
Gehrt, H. H., 171
Geilmann, W., 109
Geleznoff, A., 51, 52
Geller, S., 183
Gellert, H. G., 153
Gel'Perin, N. I., 8
George, P. D., 11
Gerrard, W., 2, 3, 4, 5, 7, 11, 12, 13, 14, 19, 23, 24, 25, 26, 27, 28, 29, 31, 32, 34, 35, 37, 39, 40, 42, 44, 47, 48, 49, 51, 52, 53, 54, 57, 59, 60, 61, 62, 63, 64, 69, 70, 71, 72, 73, 74, 81, 84, 85, 87, 89, 91, 105, 172, 173, 176, 180, 181, 182, 183, 185, 186, 189, 190, 191, 192, 193, 195, 206, 209, 211, 212, 218, 219, 223, 225, 226, 227, 228, 229, 230, 231
Gerstein, M., 93, 165, 180, 183
Ghishkova, V. P., 64, 195
Gibbs, T. R. P., 131, 138, 158
Gilbreath, J. R., 6, 7, 8, 120, 121, 140, 165
Gilde, M., 21
Gillette Industries Ltd., 160
Gilman, H., 66, 67, 76, 202, 203
Gilmont, P., 120

Gintis, D., 102, 139, 141, 142, 147, 153, 158, 165, 209
Girardot, P. R., 161
Gloss, G. H., 111*
Glunz, L. J., 69, 70, 102, 114, 155
Goerner, H., 108, 112
Gold, H., 211
Good, C. D., 164, 196
Goodspeed, N. C., 130*
Gorin, P. A. J., 148
Goubeau, J., 8, 69, 84,* 85, 96, 99, 107, 165, 170, 181, 210, 218, 219, 222, 226
Gould, E. S., 12, 165
Gould, J. R., 159
Graber, F. M., 164, 196
Graham, J. H., 128
Graham, W. A. G., 86, 99, 152, 158
Grant, L. R., 160, 163, 189
Graves, G. D., 7*
Grayson, M., 140, 165
Green, L. G., 132*
Greenwood, N. N., 96, 211
Griffel, F., 165, 170
Griffey, P. F., 189, 190
Grim, E. C., 21
Grimley, J., 40
Groszos, S. J., 182
Grummit, O., 69, 78, 80, 83, 93, 97, 114, 115, 221
Gudriniece, E., 109*
Gunderley, F. C., 58, 182, 206, 230
Gunn, S. R., 132*
Gustavson, G., 51*
Guter, G. A., 128*

H

Haag, W., 109
Haber, R. G., 36
Hänsel, R., 20
Haider, S. Z., 5, 11, 52
Hamilton, J. K., 148, 149
Hamilton, S. B., 199, 200, 201, 205
Hamilton, W. C., 123
Hamlen, R. P., 123
Hammar, H. N., 6
Hare, D. G., 107, 115, 116, 117
Harle, O. L., 20
Harman, A., 122*
Harris, J. J., 166,* 182
Harris, P. M., 132*
Hartmann, H., 222
Hartwimmer, R., 170, 194
Hatfield, H. S., 118
Hatton, K., 107
Hauser, C. R., 207*

Havir, J., 111*
Haworth, D. T., 182
Hawthorne, M. F., 122, 127, 128, 155, 205
Hayter, R. G., 52
Hazel, J. F., 219
Hedberg, K., 126*
Hein, F., 11*
Hendrickson, A. R., 67, 79
Henkel, 7*
Hennion, G. F., 83, 85, 102, 156
Henry, M. C., 219
Hermans, P. H., 16, 17, 18
Hershberg, E. B., 150
Hertwig, K., 86, 163, 165, 166, 167, 168, 169, 174, 177, 179
Herwig, W., 108, 109, 112
Herzberg, G., 122*
Herzog, H. L., 150
Hewitt, F., 163
Heyl, W., 109*
Hill, W. H., 98
Hillringhaus, F., 34, 58, 86
Hinckley, A. A., 138
Hinder, M., 8
Hinz, G., 83, 97, 101, 114
Hipple, J. A., Jr., 122*
Hirao, N., 15
Hoard, J. L., 183
Hochstein, F. A., 144
Hoekstra, H. R., 6, 7, 8, 120, 121, 130*
Hörhammer, L., 20
Hoffenberg, D. S., 207*
Hoffmann, A. K., 205
Hohnstedt, L. F., 182
Holasek, A., 111
Holle, W., 170
Holliday, A. K., 40, 48, 163, 213, 215
Holmes, R. R., 102, 133, 139, 141, 142, 147, 153, 158, 165, 209
Holt, N. B., 17
Holzbecher, Z., 67,* 79
Honeycutt, J. B., 66, 67, 76, 203
Horeld, G., 86, 178
Horii, Z., 134*
Horner, L., 108*
Hornig, H. C., 124
Horowitz, R. H., 133, 135, 137, 139, 147, 165
Horton, W. J., 8
Horvitz, L., 6, 7, 8, 84, 120, 121, 163, 177
Hough, L., 148
Hough, W. V., 163, 176, 177
Howe, B. K., 13, 25
Howick, L. C., 111*
Hsami, Y., 20
Hu, J. H., 123
Hudson, H. R., 181, 182, 189, 225

Hughes, E. C., 20, 159
Hughes, E. W., 161*
Hunter, D. L., 8
Hurd, D. T., 95, 152
Hyde, E. K., 6, 7, 8, 120, 121
Hyman, M., 19

I

Ievins, A., 109,* 112
Iffland, D. C., 145, 146
I.G. Farbenind, A–G, 19
Ilett, J. D., 13, 51, 52, 211
Iliceto, A., 143
Inglis, J., 220
Ingold, C. K., 12*
Inoi, T., 134*
Inoue, S., 100
Ioffe, M. L., 48
Irany, E. P., 19
Irwin, D. A., 211*
Isbell, H. S., 17, 148
Ishibashi, M., 21

J

Jablonski, E. M., 51*
Jablonski, F. E., 96
Jackson, A. H., 65
Jacobs, E. S., 111*
Jacobson, R. A., 214
Jander, G., 218
Jansons, E., 112
Jaulmes, P., 8,* 135
Jensen, E. H., 130,* 132*
Jevnik, M. A., 150
Jira, R., 109*
Joannis, A., 170, 221
Joglekar, M. S., 11
Johannesen, R. B., 140, 165, 183
Johnson, A. R., 169, 183
Johnson, F. W., 8
Johnson, J. R., 67, 69, 78, 80, 83, 93, 97, 114, 221
Johnson, S., 165
Jolly, W. L., 124
Jones, J. K. N., 148
Jones, J. R., 104
Jones, J. V., 4, 29
Jones, F., 118
Jones, L. C., 51
Jones, R. C., 170, 171, 178, 180
Jones, W. J., 10, 11
Jong, W. E. van R. de, 16, 18
Joseph, N., 153
Josien, M. L., 225
Julius, A., 18

K

Kaesz, H. D., 196
Kahovec, L., 6, 11, 52
Kanabu, K., 11*
Karawia, M. S., 112
Karbe, K., 163, 165, 166, 167, 168, 169, 174, 176, 177
Karl, G., 51, 52
Karrer, P., 18
Kaspar, J. S., 132*
Katritzky, A. R., 164
Katroka, H., 21
Katz, J. J., 6, 7, 8, 120, 121, 130,* 140, 165
Katz, J. R., 11
Kaufman, J. J., 123
Kazakova, E. B., 129, 177
Keenan, C. W., 170, 221
Kehiaian, H. V., 15
Keicher, G., 108, 110
Keil, T., 6
Keilen, B., 123
Keller, H., 8, 69, 165, 210
Keller, R. N., 69*
Keller, W., 112*
Kemula, W., 111*
Kent Wilson, M., 96*
Keough, A. H., 69, 72, 79
Kerrigan, J., 213, 214, 217
Khan, I. A., 35
Khotinskii, E. S., 7
Khotinsky, E., 62
Khundkar, M. H., 5, 11, 52
Kigel, R. A., 129, 177
Kilzer, S. M. L., 212
King, G. W., 122*
Kinney, C. R., 22, 69, 170, 171, 178, 179, 180
Kinoshita, Y., 20
Kirkpatrick, W. H., 8
Kirschnik, B., 83, 97, 114
Kitadani, S., 21
Klein, M. J., 107
Klemm, L., 122*
Klemm, W., 122*
Klimentova, N. V., 100
Klingel, A. R., Jr., 20
Knight, J., 6, 7, 8, 120, 121
Koenig, F. J., 161
König, W., 67, 69, 86
Köster, R., 95, 96, 121, 157, 163
Kohler, M., 111
Koike, H., 20
Kolbezen, M. J., 170, 171, 179
Kolesnikov, G. S., 100

Kollonitsch, J., 136, 138
Kolski, T. L., 128, 136
Konecky, M. S., 200, 203
Kornacki, J., 111*
Koski, W. S., 123
Kostroma, T. V., 72, 82, 88, 94
Kozminskaya, T. K., 72, 82, 88, 94
Kraffczyk, K., 218
Krantz, J. C., 18
Kraus, C. A., 83, 84, 85, 92, 93, 108, 165, 170, 182
Krause, E., 69, 92, 97, 102, 108, 114, 165
Kreshkov, A. P., 8*
Krongauz, E. S., 129, 177
Kronrod, N. Ya., 63
Krüerke, U., 85, 108,* 194
Kubba, V. P., 197
Küchlin, A. T., 16, 18
Kuck, J. A., 21, 69, 78, 114
Kudryasheva, N. V., 8
Kuemmel, D. F., 36
Kuivila, H. G., 67, 69, 72, 78, 79, 116
Kuljian, E. S., 168
Kuraschev, M. V., 93
Kuriyama, S., 20
Kursanov, D. N., 109*
Kury, J. W., 124
Kurz, P. F., 159
Kurzen, F., 118
Kuskov, V. K., 8, 35, 54
Kuss, E., 118
Kynaston, W., 181

L

Lad, R. A., 6, 7, 8, 120, 121
Ladd, J. R., 11
La Mer, V. K., 16,* 17,* 18*
Landesman, H., 218
Lane, T. J., 22*
Lange, G., 109, 165
Lange, K. R., 124, 129
Langer, S. H., 20
Lappert, M. F., 5, 7, 10, 11, 12, 13, 17, 23, 26, 27, 28, 31, 32, 34, 37, 39, 40, 42, 44, 47, 54, 59, 60, 61, 62, 63, 69, 70, 71, 72, 73, 74, 81, 84, 85, 87, 89, 91, 96, 172, 173, 176, 181, 183, 185, 186, 195, 212, 218, 223, 224, 225, 226, 227, 228, 230, 231
Laubie, H., 138*
Laudenklos, H., 118
Lauterbur, P. C., 123
Lawbengayer, A. W., 10, 52, 69, 96, 128, 176, 183
Lawesson, S. O., 64, 90

Lawrence, F. I. L., 7
Lazier, W. A., 71*
Lebas, J. M., 225
Leber, J. P., 15
Ledwith, A., 101
Leebrick, J. R., 75
Leets, K. V., 8
Lehmann, W. J., 164
Lennarz, W. J., 200, 203
Lenz, W., 83, 97, 114
Le Roi Nelson, K., 140, 165
Letsinger, R. L., 71, 72, 81, 84, 87, 90, 96, 199, 200, 201, 205, 224
Levasheff, V. V., 6
Levens, E., 7
Levi, D. L., 96, 97, 114
Levinskas, G. J., 98
Levitin, N. E., 164
Levy, H. A., 96
Lewin, J. U., 160
Lewis, D. T., 21
Lewis, E. S., 128
Lewis, G. L., 11
Ley, D. E., 144, 147
Liberman, G. S., 109*
Lieber, W., 111*
Lincoln, B. H., 8*
Lindeman, L. P., 226
Lindsay, M., 14, 85, 189
Link, R., 8, 69, 165, 210
Linnett, J. W., 122, 127
Lippincott, S. B., 5
Lippincott, W. T., 86
Lipscomb, W. N., 122, 124, 128, 214
Loewenstamm, W., 51*
Long, L. H., 85, 93, 96, 125, 209, 220,* 221
Longuet-Higgins, H. C., 122, 134, 230
Lord, R. C., 122*
Lourijsen, Mme., 16,* 17,* 18*
Lowe, J. U., Jr., 129
Lücke, K. E., 96, 226
Lyle, R. E., 95
Lynds, L., 217
Lyons, W. E., 8,* 20
Lyttle, D. A., 132,* 142

M

Maan, C. J., 16
McCarty, L. V., 122,* 124, 132*
McCoskey, R. E., 122*
McCoy, R. E., 126*
McCusker, P. A., 22,* 69, 70, 83, 85, 96, 102, 114, 155, 209,* 212
McDaniel, D. H., 123

Macdonald, C. G., 18
McDonald, R. S., 122*
McDowell, W. J., 170, 221
McElroy, A. D., 163
McEwen, W. E., 64, 72, 79, 95, 126
McGrath, J. S., 209*
McHugh, D. J., 18
Mack, J. L., 129
McKee, W. E., 164, 167, 170, 172
Macklen, E. D., 2, 24, 39, 48, 49, 61, 211
McKone, L. J., 8*
McNabb, W. M., 219
MacWood, G. E., 123
Mahoney, C. L., 170, 171, 179
Maier-Hüser, H., 52
Majert, H., 83, 97, 114
Makishima, S., 6
Mako, L. S., 22
Makowski, H. S., 83, 85, 96, 102
Maksimova, Z. G., 8
Mancera, O., 134*
Mann, D. E., 214
Marcus, R. A., 189
Maréchal, E., 152
Margerison, D., 100
Margrave, J. L., 123
Mark, H., 122*
Markina, V. Yu., 121, 128, 163
Marquardt, P., 112
Marra, J. V., 102
Martin, D. R., 22, 209, 218,* 220*
Martin, G. R., 98,* 99*
Martin, H., 153
Martin, K. J., 136
Martin, R. L., 211
Martini, H., 118, 176
Marvel, C. S., 19
Marvillet, L., 111*
Mason, L. S., 132*
Massenez, C., 118
Massey, A. G., 213, 215
Massoth, F. E., 126*
Mathing, W., 85, 92, 93, 118, 215
Matsui, M., 134*
Matteson, D. S., 75
Mattiello, J. J., 19
Mattraw, H. C., 69
Maxwell, L. R., 122*
May, F. H., 6
May, Z., 20
Mayer, G., 112
Mayfield, D. L., 6, 7, 8, 120, 121
Mead, E. J., 15, 16, 102, 114, 133, 135,
    137, 139, 140, 141, 142, 147, 153, 156,
    158, 165, 209
Meeker, T. R., 124, 129

Meerwein, H., 15, 16, 52, 83, 97, 101, 114,
    210, 211, 212
Meibohm, E. P., 132*
Melamed, M., 62
Mellon, M. G., 36
Melnikov, N. N., 79, 86, 87
Merland, R., 138*
Merrill, J. M., 98
Metal Hydrides Inc., 130*
Meulenhoff, J., 16, 17
Meyer, F. J., 109, 165
Michaelis, A., 34, 58, 60, 67, 69, 80, 86,
    92, 202
Migge, A., 83, 97, 114
Mijs, J. A., 16, 18
Mikhailov, B. M., 72, 76, 82, 88, 94
Mikheeva, V. I., 121, 128, 163
Miller, E. H., 75
Miller, J. J., 128
Miller, R. R., 176
Miller, W. H., 5, 13
Mills, J. E., 118*
Milone, M., 11
Mincer, A. M. A., 2, 24, 39, 49, 61, 211
Mistry, S. N., 48, 49
Mitchell, J. S., 35
Mitra, S. M., 11
Miyano, M., 134*
Moeller, C. W., 97
Mohler, F. L., 189
Moissan, H., 220, 221
Montequi, R., 109*
Moodie, R. B., 115, 116, 117
Mooney, E. F., 52, 59, 180, 181, 182, 189,
    211,225
Moore, E. B., Jr., 214
Moore, H. B., 136
Moore, L. O., 76, 202
Moore, R. E., 213, 217
Morgan, G. T., 51
Morgan, K., 149
Morris, J. H., 96
Morris, R. C., 76
Morrison, G. C., 17
Moscarella, C., 76*
Mosher, W. A., 128
Mountfield, B. A., 11, 35, 231
Muetterties, E. L., 54, 59, 185, 186
Mukherji, A. K., 112
Mulligan, B. W., 96
Mulliken, R. S., 122*
Muraca, R. F., 111*
Muraire, M., 111*
Murawski, K., 20
Murray, K., 153, 155, 156, 157, 165
Musgrave, O. C., 64, 65, 71, 219

N

Nace, H. R., 10
Nagel, K., 153
Nason, H. B., 7
Naylor, R. E., Jr., 211
Nazy, J. R., 199, 200, 201, 205
Nechvatal, A., 105
Nesmeyanov, A. N., 109*
Nespital, W., 39, 183
Neu, R., 86, 91, 108, 112
Newitt, D. M., 96, 97, 114
Newkirk, A. E., 10, 96, 132*
Nicklas, A., 118
Niedenzu, K., 182
Nielsen, D. R., 64, 72, 79, 95, 126
Nielsen, E., 122*
Nieuwland, J. A., 183, 185
Nikitina, A. N., 76
Nikuni, Z., 36
Nitsche, R., 92, 108, 114
Nobbe, P., 97, 102, 114, 165
Nobile, A., 150
Normant, H., 75
Norrish, R. G. W., 96
Norton, F. J., 122,* 124
Norymberski, J. K., 149
Nöth, H., 126, 132, 170
Nutten, A. J., 111*
Nyilas, E., 206, 208
Nystrom, R. F., 150

O

O'Brien, K. G., 7,* 11, 223
O'Conner, G. L., 10
Oetschlager, H., 207*
Okamoto, M., 20
Okhlolystin, O. Yu, 95
Olin, A. D., 58, 182, 206, 230
Oliveto, E. P., 150
Olsen, H. L., 159
Olson, B., 111*
Onak, T., 218
Onyszchuk, M., 85*
Orth, G. O., 20
Osipova, L. V., 207*
Ostdick, T., 102
Osthoff, R. C., 165, 170, 173,* 181
Otto, M. M., 11
Owen, T. B., 183
Owen, W. R., 8
Ozols, J., 109*

P

Pace, E., 58, 126, 206
Palazzo, F. C., 64*

Pannwitz, W., 210, 212
Park, R. P., 122*
Park, T. O., 71
Parry, R. W., 161, 164, 167, 189
Parsons, T. D., 85, 93, 102, 104, 114, 213, 214, 217
Partington, J. R., 11
Pastour, P., 152
Pastureau, P., 16,* 17,* 18*
Patein, G., 183
Patel, J. K., 14, 190, 191, 192, 209
Patron, G., 143
Pattison, I. C., 95
Paul, R., 153
Pauling, L., 122*
Paushkin, Ya. M., 93, 207,* 209
Pearce, C. A., 172, 173, 225
Pearsall, H., 93, 165
Pearson, R. K., 130
Pease, R. N., 124, 152, 157
Pecsok, R. L., 131*
Pellerin, F., 112*
Perkins, G. T., 15
Perlin, A. S., 148
Perlman, P. L., 150
Perrine, J. C., 69*
Petrova, A. M., 8*
Petrucci, S., 110
Petry, R. C., 115
Pettit, R., 197
Pfeil, E., 211
Pflaum, R. T., 111*
Phillips, J. R., 14
Pictet, A., 51, 52
Pilyavskaya, A. I., 8
Pitochelli, A. R., 122
Pitzer, K. S., 122
Platt, J. R., 124
Ploquin, J., 11, 54
Podall, H., 102, 139, 141, 142, 147, 153, 158, 165, 209
Pohland, E., 118, 122,* 161, 163, 174, 176, 213
Pohorilla, M. J., 7
Polack, H., 165
Polster, R., 108, 112
Pontz, D. F., 22, 69
Porter, R. P., 6
Povlock, T. P., 86
Prescott, R. F., 5
Prey, V., 76*
Price, W. C., 122, 230
Priess, O., 118
Pritchard, E. H., 10, 11
Prochazka, J., 36
Prokhorova, A. A., 93

Pryde, E. H., 131*
Pupko, S. L., 7
Pyszora, H., 11, 183, 185, 231

Q

Quelet, R., 65
Quill, L. L., 9*
Quinn, E., 189

R

Raff, P., 108, 110, 111, 165
Rahtz, M., 181, 218
Ramsay, W., 118
Ramsden, H. E., 75
Ramser, H., 26, 37
Ramsey, W. J., 124
Ranck, R. O., 66, 67, 76, 202, 203
Randolph, C. L., 163, 164, 167, 168, 170, 172
Rapp, L., 6, 7, 8, 120, 121
Rasmussen, R. S., 96
Rath, W., 122*
Ray, N. H., 8*
Razuvaev, G. A., 165
Reddy, J. van der M., 128
Reedy, A. J., 67,* 200, 203
Reeves, R. E., 16,* 17,* 18*
Reichard, O., 111*
Reid, W. E., 130*
Reilly, M. L., 122*
Reinert, K., 96
Remes, N., 71, 72, 81, 84, 87, 90, 96, 224
Remick, A. E., 102*
Renault, J., 112*
Renner, T., 220
Reuter, F. H., 7*
Rice, B., 164
Rice, D. W., 131*
Richards, E. L., 148
Richards, O. V., 77
Richardson, N. M., 100
Richter, D. E., 165, 170
Richter, E., 34, 58, 86
Rick, C. E., 7
Rideal, S., 178
Ringold, H. J., 134*
Rippere, R. E., 16,* 17,* 18*
Ritter, D. M., 6, 7, 8, 85, 93, 102, 114, 120, 121, 140, 161, 163, 165, 168, 176
Roberson, E. B., Jr., 147
Robinson, T. S., 230
Rochow, E. G., 185, 186
Rohwedder, K. H., 107
Rojahn, C. A., 11*

Rokitskaya, M. S., 79, 86, 87
Romain, P., 138*
Rondestvedt, C. S., Jr., 98, 108
Roscoe, J. S., 161
Rose, H., 7
Rosenberg, S. D., 75
Rosenblum, I., 19
Rosenblum, L., 97, 157
Rosenheim, A., 51*
Rosenkranz, G., 134*
Ross, V. F., 17
Roth, L. E., 136
Rothstein, E., 81, 84
Roux, G., 76*
Rozendaal, H. M., 122*
Rozenoer, A. A., 8
Rubia Pacheco, J. de la, 109*
Rückert, A., 108, 110
Rüdorff, W., 111, 112
Ruigh, W. L., 58, 182, 206, 230
Ruschmann, W., 85, 92, 93, 215
Rutkowski, A. J., 102
Ryschkewitsch, G. E., 182, 224, 229, 230

S

Sabatier, P., 118
Sachsee, H., 122*
Sakai, T., 134*
Salmon, O. N., 183
Salzberg, P. L., 71*
Sanderson, R. T., 120, 122,* 130,* 157, 165
Sands, C. A., 160, 163
Sandstedt, R. A., 211*
Sant, B. R., 112
Santucci, L., 66, 67, 76, 203
Saunders, B. C., 13, 51, 52, 211
Saville, N. M., 32
Saville, R. W., 81, 84
Sazonova, V. A., 63, 109*
Scattergood, A., 5, 13
Schabacher, W., 219
Schäfer, H., 16,* 17,* 18*
Schaeffer, G. W., 120, 128,* 130, 136, 160, 161, 162, 163, 164, 166, 167, 176, 177
Schaeffer, R., 96,* 128, 177
Schar, W. C., 129
Scharrnbeck, W., 67, 69, 86
Schechter, W. H., 6, 9, 20
Scherf, K., 108*
Schiff, H., 6, 7, 8, 15
Schlesinger, H. I., 6, 7, 8, 69, 84, 119, 120, 121, 124, 130, 140, 157, 161, 163, 165, 166, 168, 169, 174, 176, 177, 213, 214, 215, 217

Schloeder, H., 109
Schmich, M., 123
Schmid, H., 36
Schmidt, H., 20
Schneider, J., 153
Schuele, W. J., 219
Schuler, K. E., 159
Schultz, D. R., 161
Schultz, J. W., 17
Schultz, O. E., 108
Schumb, W. C., 220
Schupp, L. J., 180
Schuster, K., 169, 171
Schützenberger, M., 51*
Schwartz, A. M., 6, 7, 8, 120, 121
Sciarra, J. J., 21
Scott, L. B., 76
Scribner, R. M., 98, 108
Scudi, J. V., 18
Sculati, J. J., 5
Seaman, W., 67, 69
Sears, D. S., 183
Secci, M., 93
Sedlak, M., 58, 182, 206, 230
Seidel, C. F., 8
Sekiguchi, K., 20
Selman, J., 11
Sen, D., 35
Serrano, C., 109*
Servoss, R. R., 6
Seus, D., 109*
Shafferman, R., 11, 14, 54, 81, 91, 225, 231
Shamim-Ahmad, S., 96
Shapiro, I., 123, 133, 218
Shchegoleva, T. A., 72, 76, 82, 94
Shechter, H., 144, 147
Sheft, I., 6, 7, 8, 120, 121, 140, 165
Sheiman, B. M., 8, 35
Shepp, A., 122*
Sheppard, N., 226
Shoaf, C. J., 102, 139, 141, 142, 147, 153, 158, 165, 209
Shoolery, J. N., 122
Shore, S. G., 161*
Shumeiko, A. K., 8
Siddiqullah, Md., 5, 11
Siebert, H., 96
Siecke, W., 118
Siegel, B., 129
Silbiger, G., 130*
Silver, H. B., 31, 44, 47
Silverman, M. B., 93, 114
Singer, L., 51*
Sirotina, I. A., 112
Sisler, H. H., 182, 229, 230
Sitteri, P. K., 79

Skinner, H. A., 11, 83, 229
Skoog, I. H., 71, 72, 81, 84, 87, 90, 96, 224
Slack, S. C., 79
Smalley, J. H., 182
Smith, F., 148, 149
Smith, G. C., 122*
Smith, J. E., 93
Smith, N. B., 83, 229
Smith, R. K., 7
Smith, S. H., 176
Smyth, C. P., 11
Snyder, H. R., 69, 77, 78, 80, 83, 93, 97, 114, 200, 203, 221
Soboczenski, E. J., 69, 72, 79
Soboleva, T. A., 100
Soldate, A. M., 132
Sollman, P. B., 143*
Solms, J., 21, 148, 206
Solomon, I. J., 107
Solopenkov, K. N., 8
Soloway, A. H., 201, 206, 208
Sondheimer, F., 134*
Sönke, H., 83, 97, 114
Souchay, P., 16,* 17,* 18*
Sowa, F. T., 185
Sowler, J., 48
Spanner, S., 134*
Spier, H. W., 111
Sporek, K. F., 111*
Stacey, G. J., 13, 51, 52, 211
Stacey, M., 18, 148
Stack, G. G., 209*
Stafiej, F. S., 7, 182
Stainier, W. M., 18
Staveley, L. A. K., 109*
Steele, B. D., 118*
Stegeman, G., 132*
Stehle, P. F., 114, 156, 165
Steinberg, N. G., 58, 182, 206, 230
Steindler, M. J., 166, 169
Stephenson, C. C., 131*
Stern, D. R., 217
Stevens, T. S., 109*
Stewart, A. C., 161, 163, 176, 177
Stewart, J. E., 96, 164
Stiff, J. F., 120
Stilz, W., 108, 112
Stock, A., 92, 97, 118, 119, 120, 121, 161, 165, 169, 170, 174, 176, 213
Stockmeyer, W. H., 131*
Stoll, M., 8
Stone, F. G. A., 59, 84, 86, 99, 120, 152, 158, 166, 174, 187, 196
Stosick, A. J., 126*
Stout, L. E., 19,* 20
Strecker, W., 62*

Strickson, J. A., 14, 85
Strizhevskii, I. I., 7
Struck, W. A., 132*
Sturm, W., 85, 170, 194
Subba Rao, B. C., 102, 114, 133, 135, 137, 139, 141, 142, 147, 153, 155, 156, 157, 158, 165, 209
Sugden, S., 6, 8, 10, 185, 186, 210
Sugihara, J. M., 206
Sujishi, S., 97, 99, 165
Sun, K. H., 21
Sutherland, M. D., 8
Sütterlin, W., 11, 22, 26, 37, 118, 169
Sutton, L. E., 164
Svec, H. J., 96*
Swayampati, D. R., 66, 67, 76, 202, 203
Sykes, A., 111*
Szabo, L. S., 20, 159
Szmant, H. H., 211*
Sztanko, E., 21

T

Taft, R. W., Jr., 99
Tahara, A., 143
Tajima, T., 6
Talalay, J. A., 160
Talalay, L., 160
Tannenbaum, S., 122*
Taylor, M. D., 93, 128, 160, 163, 165, 180
Taylor, R. C., 164
Taylor, R. L., 118
Tees, T. F. S., 83
Teh-Fu, Yen, 146
Thatte, V. N., 11
Thielens, G., 66
Thomas, D. L., 7*
Thomas, J. R., 20
Thomas, L. H., 5, 6, 10, 11
Thomas, R. I., 83, 107
Thomas, W. M., 205
Thompson, H. T., 22
Thompson, P. G., 52
Thomson, R. T., 16
Thomson, T., 109*
Tierney, P. A., 39, 114, 135, 153, 155, 156, 157, 165
Tollin, B. C., 96*
Tomita, K. J., 134*
Ton, J. C., 6
Topchiev, A. V., 93, 209
Torssell, K., 71, 90, 93, 96, 115, 200, 201
Tranchant, J., 111*

Trautmann, C. E., 7
Trevoy, L. W., 134*
Tronov, B. V., 8*
Tsuruta, T., 100
Tuck, L. D., 6, 7, 8, 120, 121
Tunstall, R. B., 51
Tupa, R. C., 159
Turner, H. S., 179, 181, 182
Tyson, G. N., 7

U

Ulich, H., 39
Ulmschneider, D., 84*
Ulrich, A. M., 15
Upson, R. W., 182*
Urry, G., 213, 214, 217

V

van Artsdalen, E. R., 69
van Campen, M. G., 69, 78, 80, 83, 93, 97, 114, 221
van der Meulen, P. A., 58, 182, 206, 230
Vanderwerf, C. A., 64, 72, 79, 95, 126, 147, 150
Vaughn, T. H., 7
Vaver, V. A., 72, 82, 88, 94
Veiler, M., 16,* 17,* 18*
Veiss, A., 112
Veloric, H. S., 124, 129
Venkataramaraj Urs, 12, 165
Verhoek, F. H., 115
Vermaas, N., 16, 18
Videla, G. H., 176*
Vinogradova, V. S., 10
Vogg, G., 112
Vol'pin, M. E., 109*
von Grosse, A., 108
Vrestal, J., 111*

W

Wagner, R. I., 84, 85, 166, 169, 187
Walburn, J. J., 75
Walker, A. O., 6, 7, 8, 84, 120, 121, 124
Wall, L. A., 189
Wallbridge, M. G. H., 125
Wallis, J. W., 183, 185, 186, 227, 228, 229
Waloff, M., 185, 186
Warburton, D. M., 218,* 220*
Warne, R. J. 179, 181, 182
Wartik, T., 130,* 196, 213, 217
Washburn, R. M., 7
Watt, W. J., 20
Weaver, C., 77
Webb, T. J., 18
Webster, S. H., 6

Wechter, W. J., 160
Weismann, T. J., 109
Weiss, F., 112*
Weiss, H. G., 133
Weissbach, A., 18
Wendlandt, W. W., 112
Werner, R. L., 11, 223
Werntz, J. H., 7*
West, C. D., 19
Westrum, E. F., 164
Whatley, A. T., 152
Wheatley, P. J., 214
Wheelans, M. A., 2, 51, 52, 53
Wheeler, C. M., 211*
Whelan, W. J., 149
White, R. F. M., 107, 115, 116, 117
Wiberg, E., 11, 16, 22, 26, 37, 69, 81, 83,
    85, 86, 92, 93, 118, 120, 126, 132, 161,
    163, 165, 166, 167, 168, 169, 170, 171,
    174, 176, 177, 178, 179, 194, 215
Wierl, R., 118
Willfang, G., 211
Willhalm, B., 8
Williams, A. F., 111*
Williams, R. L., 223, 225, 226, 227, 231
Williams, R. M., 69, 72, 79
Wilmarth, W. K., 97
Wilson, B. M., 105
Wilson, C. O., Jr., 7
Wilson, E. B., 211
Wilson, M. K., 226
Wilzbach, K. E., 130*
Winternitz, P. F., 96
Wittig, G., 108, 109, 110, 112, 165
Wittmeier, H. W., 96, 226
Witz, S., 99, 165
Wohler, F., 220
Wolfrom, M. L., 148, 206
Wood, H. B., 148

Woodhead, A. H., 4
Woodrow, H. W., 85, 169
Woods, G. F., 149
Woodside, R., 112
Wulfman, C. E., 98, 108
Wuyts, H., 5, 7
Wyman, F. W., 77
Wyvill, P. L., 2, 24, 25, 39, 49, 61, 211

Y

Yabroff, D. L., 67, 69, 202
Yagi, S., 15
Yogi, K., 36
Yoneda, Y., 6
Young, T. F., 164
Yur'eva, L. P., 8, 35

Z

Zachariasen, W. H., 18
Zakharkin, L. I., 95
Zakrzewski, K., 20
Zalkin, A., 124
Zannier, H., 111, 112
Zappel, A., 165, 170
Zavgorodii, S. V., 209
Zeidler, F., 92, 97, 118, 165
Zeidler, L., 112*
Zeldin, L., 144
Zhdanov, S. I., 109*
Zhigach, A. F., 129, 177
Zhukova, V. A., 8, 35
Zief, M., 112
Ziegler, K., 96, 121, 153
Zittle, C. A., 18
Zvonkova, Z. V., 64, 195
Zwahlen, K. D., 8
Zweifel, G., 153, 155, 156, 157, 165

SUBJECT INDEX

A

Acetamide, reaction with boron trifluoride, 185

Acetamide–boron trifluoride, preparation of, 185
  reaction with alcohols, 185
  amines, 185

Acetic acid esters, reaction with boron trichloride, 51
  mechanism of reaction, 53

Acetonitrile, reaction with boron trichloride ether complex, 184
  boron dichlorodiethyl ether complex, 184
  n-butyl chloroborinate, 183
  relative donor strength of, 184

Acetonitrile–boron trichloride, preparation of, 183
  reaction with di-n-butyl sulphide, 184
  o-nitrophenol, 184
  pyridine, 184
  tetrahydrofuran, 184

p-(β-acetovinyl)-phenyl boronic acid, 204

Acetoxy derivatives of boron, 231

Acetoxy groups, effect of acetoxy groups on bonding to boron, 231

Acetyl chloride, reaction with n-butyl borate, 15

Acrylonitrile polymerization, using tri-n-butylboron, 100
  using triethylboron, 100

Alcohols, isolation as borates, 6
  purification as borates, 6
  reaction with boron trichloride, 22, 23, 24, 25
  phenylboron dihalides, 60
  sodium borohydride, 158
  trialkylborons, 97

Aldehydes, reaction with boron trichloride, 54, 55
  trialkylborons, 97
  reduction with sodium borohydride, 142

Aliphatic primary amines, reaction with boron trichloride, 179

Alkaloids, determination of, using tetraphenylboron, 112

Alkenyl alcohols, reaction with boron trichloride, 31

Alkenylborons, oxidative stability of, 114

Alkoxyboron compounds containing amino groups, 170

Alkoxydecaboranes, 128

Alkoxysilanes, reaction with boron trichloride, 14

Alkylaluminium sesquihalides, for preparation of trialkylborons, 95

Alkylaminohalogenoborinates, preparation of, 173

Alkylammonium chloride, reaction with boron trichloride, 181

Alkylated boranes, 84, 124–126

Alkyl borates, chemical properties, 11
  co-ordination of, 12
  hydrolysis of, 12
  I.R. spectra of, 223
  kinetics of solvolysis, 15
  physical properties, 10
  use of, as catalysts, 101
    as heat exchange media, 11

N-Alkylborazoles, 177

Alkylborinic acids, diethanolamine esters, 90

Alkylboron dihalides, alcoholysis of, 60
  hydrolysis of, 59
  oxidative stability of, 114
  preparation of, 58, 104
  stability of, 104
  use of, in preparation of alkylated diboranes, 125, 126

Alkylboronic acids, esters of, 114
  formation of, 125
  stability of, 78

Alkylboronic halogen esters, 73

Alkyl chlorides, formation from mixed ethers, 61
  trialkylphosphate and boron trichloride, 190
  reaction with boron halides, 95

Alkyl chloroalkyl ethers, reaction with boron trichloride, 42

Alkyl chloroboronates, see Di-alkyl chloroboronates

Alkyl decaboranes, preparation by Grignard reaction, 129

Alkyl derivatives of borazole, 176, 177

297

Alkyl dichloroborinates, preparation, 23, 26, 38
reactions of, 28, 29, 32
Alkyl diethylborinates, 83
Alkyl hydroperoxides, reaction with boron trichloride, 115
Alkyllithium, reaction with n-butyl di-n-butylborinate, 94
Alkyl metaborates, 9
I.R. spectra of, 224
Alkyl phenylchloroborinates, I.R. spectra of, 226
Alkyl phosphinoboranes, 187
Alkyl phosphates, reaction with boron halides, 190
Alkyl phosphites, reaction with boron trichloride, 189
Alkynyl alcohols, reaction with boron trichloride, 31
o-Allylphenol from allyl phenyl ether, 47
Aluminium borohydride, reaction with boron trichloride, 134, 157
Aluminium trichloride with boron compounds, 70
as Lewis acid catalyst, 29, 30
Amide-boron trichloride, alcoholysis of, 185
hydrolysis of, 185
I.R. spectra of, 186
N.M.R. measurements, 186
point of co-ordination in, 186
preparation of, 185
pyrolysis of, 185
structure of, 186
Amides, complexes with boron trihalide, 185
reaction with boron trihalide, 185
reduction with lithium borohydride, 151
tetraphenyl boron derivatives of, 112
Amine-boranes, 155, 164
Amine-boron trichloride, 180
Amines, reaction with alkyl and aryl boron halides, 178, 179
boron trihalides, 178, 179, 180
trimethylboron complexes, 98
Aminoborazoles, 159
Aminodialkylborons, 84
2-Aminoethyl esters of aryl-boronic acids, and diarylborinic acids, 71, 90
2-Aminoethyl diphenyl borinate, I.R. spectrum of, 224
Ammonia, reaction with dialkyl boron halides, 84
diborane, 161
methylboronic anhydride, 70
tetramethylborane, 165

Ammonium borohydride, decomposition of, 161
Ammonium chloride, reaction with diammonia diborane, 162
n-Amylisobutyl-t-butylboron, 104
Aniline, reaction with boron trichloride, 178
Aniline–boron trichloride, 178
p-Anisidine, reaction with boron trichloride, 179
p-Anisidine hydrochloride, reaction with boron trichloride, 179
Anisole, reaction with boron trichloride, 39
p-Anisylaminoboron dichloride, 179
Antimony trifluoride, reaction with di-n-propylboron iodide, 85
tetrachlorodiboron, 217
Antimony trihalides, halogen exchange reagents, 209
l-Arabinose, reaction with phenylboronic acid, 206
Arylamine boron dichlorides, I.R. spectra of, 227
Arylamine boron trichloride, 180
Arylated boranes, preparation of, 126
pyridine complexes of, 127
reactivity of, 128
Arylation of boron, 108
Aryl borates, chemical properties of, 11, 35
physical properties of, 10
preparation of, 34
Arylboron dihalides, 58, 59
Arylboronic halogeno esters, 73
Arylborons, oxidative stability of, 115
Aryl dichlorophosphinates, complexes with boron halides, 192
Arylmagnesium bromide, reaction with tri-isobutyl borate, 86
tri-n-butyl borate, 86
trimethoxyboroxine, 86
Aryl phosphates, reaction with boron halides, 190
Aryl phosphites, reaction with boron halides, 191
Arylthallium halides, 80
Autoxidation of boronic acids, 78

B

B- and F-strain theory, 165
Back–co-ordination, 12
B-aminoborazoles, 159
Basic strength sequences, 165
Benzamide, reaction with boron trifluoride, 185
Benzene, reaction with diborane, 95

Benzyl acetate, reaction with boron tri-
chloride, 53
Biological activity of boron heterocycles,
197, 207, 208
Bis-aminoborons, 169
Bis-2,1-borazaronaphthyl ether, 198
Bis-2-chlorovinylboron chloride, 81
Bis(dialkylboron)acetylene, 222
Bis-1,2-dichloroborylethane, 213, 214
Bis(diethylamine)boron chloride, I.R.
spectrum of, 226
Bis-dihydrobenzoboradiazole, 199
Bis-dimethylaminoborane, 126
Bis-dimethylaminoboric acid, 126
Bis-dimethylaminoboron halides, 169,
170
Bis-dimethylaminoboronic acid, 169
Bis-monomethylamine boron bromide,
169
Borate ion, 17
Borazanes, 166
Borazenes, addition to, 168
monomeric forms of, 168
effect of heat on, 166
reactivity of, 167
redistribution in, 168
reaction with hydrogen halides, 167,
168
Borazines, 174
Borazole, 174, 175, 176
Borazole polymers, 182
Borazoles, I.R. spectra of, 224
Boric esters, conversion into ether, 56
co-ordination, 11
electrophilic power of oxygen in, 14
I.R. spectra of, 224
mechanism of hydrolysis of, 11
of polyhydric alcohols, 19
preparation from boric oxide, 5
from borax, 7
from sodium borohydride, 159
reaction with carboxylic acid, 15
lithium alkyls and aryls, 63
metal alkoxides, 141
oxyacids, 15
thermal stability of, 11
uses of, 7
Boric oxide, reaction with sodium hydride,
132
for preparation of borates, 5
Borimidazolones, preparation, 206
use in cancer therapy, 208
Borinates, I.R. spectra of, 223
Borinic anhydrides, I.R. spectra of, 225
proton magnetic resonance, 107
redistribution of, 106

Borohydride ion, dependence of stability
on electron affinity of associated
metal ion, 134
nucleophilic attack by, 147
Borole, 194
Boron, B—H frequencies, 231
B—N frequencies, 229
B—Cl frequencies, 226
analysis of, 21
arylation of, 108
B-methyl frequencies, 226
boron–aryl bonds, 225
boron–oxygen bonding, back–co-ordination in, 223
dealkylation of, 81
detection in glasses, 6
Boron acetate, 51
Boronates, alkyl-oxy exchange, 72
n-butyl, 63
hydrolysis of, 71
I.R. spectra of, 223
thermal stability of, 71
Boron–boron bonds, in triphenylboron
sodium dimer, 109
in B_2Cl_4, 213
Boron–carbon bonds, cleavage of, 77, 78,
79, 102, 114, 202
Boron compounds, biological activity of,
66, 76, 197, 207, 208
luminescence of, 66
Boron-containing ores, preparation of
borates from, 7
Boron dihalides, from boronic anhydrides
and $AlCl_3$, 70
from boronic anhydrides and BCl_3, 69
using Grignard reagent, 62
Boron heterocycles, hydrolytic stability,
200
preparation of, 197
Boron hydride ion, existence of, 162
Boron hydrides, adsorption in gas phase,
123
historical, 118
preparation of, 119, 121
use of, as fuels, 158
as additives to conventional fuels,
158
Boronic acids, acid strength, 67
anhydrides, 69
trimeric form, 69
for preparation of boron dihalides,
70
applications, 75
biological activity, 76
crystal structure, 64
dehydration, 69

Boronic acids (*cont.*)
  esterification, 71
  halogen esters, 73
  preparation of cyclic esters, 72
    preparation from Grignard reagents, 62
  purification, 65
  styrylboronic acids, preparation, 64
Boronic anhydrides, esterification, 69
  formation of, 66
Boronic esters, containing functional group in α-position, 75
  cyclic, preparation of, 72
  preparation from trialkylborons, 114
Boron modified polymers, 19
Boron–nitrogen bond length, in borazole, 175
  in trisdimethylamino boron, 171
Boron–nitrogen–carbon bond angle in trisdimethylaminoboron, 171
Boron nitrogen compounds, 161
Boron peroxy compounds, 115, 116
Boron phosphate, 190
Boron phosphorus bonds in triphenylboron triphenylphosphorus, 109, 187
Boron phosphorus compounds, 187
  polymeric forms, 187
Boron polymers, 58, 62, 182
Boron sulphur compounds, 194
Boron tribromide, complex with dioxan, 49
  conductiometric titrations with quinoline, picoline and *iso*quinoline, 219
  ether fission by, 219
  interaction of, with amides, 185
  reaction with alcohols, 218
    aryl phosphites, 14
    diborontetrachloride, 213
    lithium ethoxide, 218
    monomethyl amine, 169
    trimethylboroxine, 218
    triphenylphosphite, 191
Boron trichloride, complex with amines, 180
  aniline, 178
  dialkyl alkylphosphonates, 192
  nitrobenzene, 180
  p-phenylene diamine, 180
  phosphorus oxychloride, 14
  trialkyl phosphites, 191
  triphenyl phosphite, 190
  dipole moment of, 39
  exchange with trialkyl borons, 106
  fission of ethers with, 46
  I.R. spectra of complexes of, 227
  nitroaniline, lack of reaction with, 180

Boron trichloride (*cont.*)
  reaction with acetates, 53
    alcohols, 22, 23, 24, 25
    aldehydes and mechanism, 54, 55
    aliphatic primary amines, 180
    alkenyl alcohols, 31
    alkoxysilanes, 14
    alkylammonium chloride, 31
    alkyl borates, 14
    alkyl chloroalkyl ethers, 42
    alkyl phosphites, 14
    alkynyl alcohols, 31
    aminostyrene, 198
    amides, 185
    aniline, 178
    anisole, 39
    aryl phosphites, 14
    benzyl acetate, 53
    n-butyl hydroperoxide, 115
    tert-butyl magnesium chloride, 102
    cyclic ethers, 40, 41
    cyclohexylamine, 181
    diamines, 180
    ethers, 36, 37
    ethyl acetate, 53
    ethylene glycol, 32, 33
    hydroxyesters, 52
    ketones, 57
    lithium aluminium hydride, 121
    methyl amine, 178
    (+)-2-octanol, 27
    phenols, 34, 35, 36
    phenyl acetate, 53
    primary aromatic amines, 180
    trialkylborons, 98
    trialkyl phosphites, 191
    tri-*n*-butylboron, 83
    trifluoroacetic acid, 54
    triphenyl phosphite, 190
    unsaturated ethers, 44, 45
    zinc alkyls, 92
  as reaction catalyst, 47
Boron trifluoride, catalysis of Claisen rearrangement, 48
  complex with ether, use of, 109
  complexes with alcohols, 211
  reaction with acetamide, 185
    benzamide, 185
    n-butyl magnesium bromide, 18
    Grignard reagents, 92
    lithium hydride, 121
    propionamide, 185
    secondary alcohols, 212
    sodium methoxide, 210
    tri-*n*-butyl borate, 212
    trimethyl borate, 210, 211

Boron trifluoride acetamide complex, reaction with alcohols, 185
Boron trifluoride 1 : 1 amide complexes, effect of heat on, 185
hydrolysis, 185
Boron trihalides, complexes with aryl phosphates, 192
  amides, 185
  ethers, 48
  nitriles, 183
  order of acceptor strength, 192, 209
  reaction with alkyl chlorides, 95
    amides, 185
    amines, 178
    di-*n*-butylborinic anhydride, 82
    diphenylborinic anhydride, 87
    diphenylborinic esters, 87
    diphenylboron halides, 89
    Grignard reagents, 62, 92
    pyridinium halides, 181
    trialkylamine, 169
    trialkylborons, 83
    zinc alkyls, 81
Boron triiodide complex with ammonia, 221
  reaction with alcohols and ethers, 221
Boron oxalate, 54
Boroxole (boroxine), 70
Bromal, reaction with triethylboron, 83
Bromine, reaction with tri-*n*-butylboron, 83
Bromoborosulfole, 194
α-Bromonitrocycloalkanes, debromination using sodium borohydride, 146
p-Bromophenylboronic acid, 64
N-Bromosuccinimide, 146
  reaction with arylboronic acids, 203
ω-Bromo-*p*-tolueneboronic acid, 203
Butadiene, reaction with triphenyl methyl sodium, 108
  with B_2Cl_4, 214
n-Butylamine, reaction with *n*-butyl-boron dichloride, 182
t-Butyl alcohol, reaction with boron tri-chloride, 25
n-Butyl-bis-diethylamino borinate, 172
  reaction with hydrogen chloride, 172, 173
n-Butyl borate, physical properties of, 10
  reaction with acetyl chloride, 15
    phosphorus chlorides, 14
    silicon tetrachloride, 13
    thionyl chloride, 13
  solubility of hydrogen chloride in, 13
  use of in preparation of 2,2-tolandi-boronic acid, 199

n-Butylboron dichloride, reaction with liquid ammonia, 182
n-Butyl boronimine, 182
n-Butyl dialkylaminochloroborinate, 174
t-Butyl di-*n*-amyl boron, 104
n-Butyl dibromoborinate, 218
n-Butyl dibutenylborinate, 83
n-Butyl di-*n*-butylborinate, 82
  reaction with alkyl lithium, 94
    phosphorus pentachloride, 82
n-Butyl dichloroborinate, 27
  reaction with acetonitrile, 183
iso-Butyl dichloroborinate, 28
n-Butyl difluoroborinate, 212
n-Butyl hydroperoxide, reaction with boron trichloride, 115
t-Butyl hydroperoxide, reaction with boron trichloride, 115
t-Butyl hypochlorite, reaction with tri-*n*-butylboron, 83
t-Butyl magnesium bromide, reaction with boron trifluoride, 81
t-Butyl magnesium chloride, reaction with boron trichloride, 103
n-Butyl di-*n*-octylborinate, 84
n-Butyl phenylbromoborinate, 74

C

Carbon[14] labelled alkylborons, 93
Carbon heterocycles containing boron, 95, 96, 197
Carbonium ions, 14, 38, 61, 74
Carbonyl compounds, reduction with metal borohydrides, 150
Carbonyl group, co-ordination of, to boron, 91, 231
Carboxylic acids, reaction with borates, 15
  trialkyl borons, 83, 97
m-Carboxyphenylboronic acid, 201
p-Carboxyphenylboronic acid, 201
Catechol dioxygen ethers, solubility of hydrogen chloride in, 50
Chelate orthoborate, reaction with Grignard reagents, 84
Chelation, 29, 34, 91
Chloretone, reference to boron trichloride, 25
2-Chloro-2,1-borazaronaphthalene, 198
Chloroborosulfole, 194
β-Chlorovinylboron dichloride, 58
Cholesterol, reaction with diborane, 160
Complex hydride anion, 134
Conjugated nitroalkenes, reduction of with metal borohydrides, 147
Covalent borohydrides, 134

Cyclic boron–carbon compounds, 95, 96, 197

Cyclic ethers, reaction with boron trichloride, 40, 41

Cyclododeca-1,5,9-triene, hydroboration of, 96

Cyclohexylamine, reaction with boron trichloride, 181

Cyclohexylaminoboron dichloride, 181

Cyclohexyl metaborate, 10

Cyclohexyloxyboroxine, 10

D

Debromination reduction, 145

Decaborane, acid function of, 128
  deuterium exchange with diborane, 123
  preparation of complexes of, 128, 129
  reaction with alcohols, 129

Dechlorination, of substituted halogeno acetophenones using metal borohydrides, 147

Dialkoxyborohydrides, 140

Dialkyl alkanephosphonates, reaction with boron trihalides, 192

Dialkylamines, reaction with boron trichloride, 169

Dialkylamino boronates, 173

Dialkylboron halides, 84
  reactions with metal alkenyls, 85
  in preparation of alkylated diboranes, 125, 126

Dialkylboronic acids, 82, 125

Dialkyl chloroboronates, 23, 30

Dialkyl phenylboronates, I.R. spectra, 226

Diamines, reaction with boron trichloride, 180

Diammonia diborane, 161, 162, 174

Diaryl boranes, pyridine complexes of, 127, 128

Diazomethane, reaction catalysed by alkyl borates, 101

o,o'-Dibenzyldilithium, reaction with butyl borate, 90

Diborane, complex with phosphine, 187
  co-ordination compounds derived from, 161
  deuterium exchange with decaborane, 123
  preparation, 119, 121
  properties, 122
  pyrolysis of, 124
  reaction with ammonia, 161
    cholesterol, 160
    dimethylamine, 126
    dimethylphosphine, 187

Diborane (cont.)
  reaction with ethylene, 152
    hydrazine, 166
    lithium hydride, 130
    methanol, 126
    N-methyl-hydroxylamine, 164
    olefins, 152
    potassium methoxyborohydride, 131
    sodium methoxyborohydride, 131
    sym-dimethyl-hydrazine, 166
    trialkylborons, 124
  sorption of, 124

Diborane diammoniate, 161

Diboronic acids, 65, 66

Diboron tetrachloride, 213
  reaction with dimethyl sulphide, 196

Di-n-butylborinic anhydride, 82
  reaction with boron trichloride, 82
    optically active octanol, 82
    phosphorus pentachloride, 82

Di-n-butylboron derivatives, 81

Di-n-butylboronic acid, 82

Di-n-butyl chloroboronate, I.R. spectrum of, 226

Di-n-butyl dibutylaminoboronate, 173

Di-n-butyl diethylaminoboronate, back–co-ordination in, 173
  hydrolysis of, 173
  reaction with n-butanol, 173
    di-n-butylamine, 173
    hydrogen chloride, 173

Di-n-butyl phenylboronate, reaction with phenylboron dichloride, 60

Di-n-butyl sulphide, reaction with boron trichloride, 195

Diethanolamine arylboronates, 72

Diethylaminoboron dichloride complex with hydrogen chloride, 181

Diethylammonium tetrachloroborate, 181

Diethylene glycol dimethyl ether (diglyme) complex with sodium borohydride, 39, 135
  use as solvent for sodium borohydride, 135

Diethyl ethylphosphonate complex with boron trichloride, 193
  reaction with boron trichloride, 193

Diethyl phenylboronate, reaction with lithium aluminium hydride, 127

Dimethoxyborane, 126

Dimethylamine borane, 164

Dimethylamineboron difluoride, 169

Dimethylamine, boron trichloride complex reaction with pyridine, 180
    triethylamine, 179

Dimethylaminoborane, 126, 169

Dimethylaminoboron dichloride complex
with HCl, 180
Dimethylborazene, 167, 168
Dimethylborazole, 176
Dimethylboron fluoride, 81
Dimethylborophosphine, 188
Dimethylboron iodide, 222
Dimethyl fluoroboronate, 210
Dimethylphosphine, reaction with di-
borane, 187
Dimethylphosphinoborane, 187, 188
Dimethyl sulphide, reaction with diboron
tetrachloride, 196
Diols, effect on acid strength of boric acid,
16
Dioxygen cyclic ethers, reaction with
boron trihalides, 48
Diphenylborinic acid, I.R. spectrum of,
226
Diphenylborinic anhydride, I.R. spectrum
of, 225
Diphenylboron halides, I.R. spectra of,
225, 226
Diphenylboronium perchlorate, reaction
with α, α′-dipyridyl, 110
Diphenyldiborane, 126
pyridine complex, 127
use as reducing agent, 128
Diphenyl ether, boron trichloride system,
42
α, α′-Dipyridyldiphenylboronium perchlorate, as example of boron containing
cation, 110
Di(trifluoromethyl)phosphinoborine, 189

E

Electron density, disposition of in 1 : 1
amine boron trichloride complexes, 180
of alcoholic oxygen atoms, 1
Esters, reduction of, with metal borohydrides, 147
Ethanolamine esters, 90
Ethers, reaction with boron trichloride,
36, 37
2-Ethoxycarbonyl-1-methylvinyl di-n-
butylborinate, aromatic character of,
91
stability by chelation, 91
2-Ethoxycarbonyl-1-methylvinyl diphenylborinate, aromatic character of, 91
stability by chelation, 91
Ethylene, polymerization of, using tri-n-
butyl boron, 100, 101
reaction with diborane, 152
diboron tetrachloride, 214

Ethylene chloroboronate, 32, 33
Ethylenediamine, complex with trimethylboron, 99
reaction with trimethylboron, 170
Ethylene glycol, reaction with boron
trichloride, 32, 33
Ethyl acetate, reaction with boron
trichloride, 53
Ethyl acetoacetate, reaction with dialkylboron chloride, 91
Ethylaluminium sesquichloride, reaction
with trimethyl borate, 95
Ethyl borate, 7, 10
Ethyl metathioborate, 195
p-Ethylphenylboronic anhydride, reaction
with N-bromosuccinimide, 203

F

Ferric chloride, effect of, on decomposition
of alkyl dichloroborinates, 29
of n-butyl phenylchloroborinate, 73
of dialkyl chloroboronates, 30
on reaction between n-butyl borate
and acetyl chloride, 15
n-butyl borate and phosphorus
pentachloride, 14
n-butyl borate and phosphorus
trichloride, 13
n-butyl borate and silicon tetrachloride, 13
n-butyl borate and thionyl chloride,
13
F-strain, 180
Four-centre broadside mechanism, 3, 14
Furan, reaction with boron trichloride, 46
Furylboronic acid, 78, 79, 80

G

Glyceryl borate, 52
Glyceryl diborate, I.R. spectrum of, 225
Glycol borates, 19
Grignard reagents, bifunctional, in preparation of diboronic acids, 65
reaction with boron halides, 62, 92
chelateorthoborate, 84
trialkylborons, 94
use of in preparation of alkylboronic
acids, 63
arylboronic acids, 62
diboronic acids, 64

H

Halogens, reaction with trialkylborons, 83
Halogeno acetophenones, dehalogenation
of, 147

Heterocycles containing boron, 197
Hexachloro*iso*propanol, reaction with boron trichloride, 13
Hexamethylsiloxane, reaction with dimethylboron bromide, 85
Hexamine, reaction with *o*-bromomethyl phenylboronic acid, 204
B—*N*—Hexamethylborazole, 177
High-energy fuels, alkylated boranes, 130
    boron hydrides, 130
Hydration of olefins, anti-Markownikoff, 154
Hydrazine, reaction with diborane, 166
Hydrido compounds of boron, 118
    polymers from, 159
    suggested application, 158
    use as fuels, 158
Hydroboration of olefins, 152
Hydrogenation, non-catalytic method, 154
Hydrogen chloride, reaction with alcohols, 2
    diammonia diborane, 162
    trialkylborons, 83
    tri-*n*-butylboron, 83
    tris-aminoborons, 171
    tris-phenylaminoboron, 171, 179
    solubility in borates, 13
Hydrogen halides, reaction with B—O—C links, 13
    diphenylborinic esters, 88
    trialkylborons, 83
Hydrolysis of, boronic esters, 71
    diphenylboron halides, 88
    diphenylboronic acid ester, 88
Hydroxyesters, reaction with boron trichloride, 52
o-Hydroxyphenylboronic acid and anhydride, C—B cleavage, 202
    reaction with benzene diazonium chloride, 203
Hyperconjugation, 96, 97

I

Inductive effect, in alkyl dichloroborinates, 29
    in dialkyl chloroboronates, 30
Infra-red spectra, 223
Inorganic benzene, *see* Borazole
Isomerization, in trialkylborons, 103, 156

K

Ketones, reaction with boron trichloride, 57
    reduction with sodium borohydride, 143

Keto-groups, selective reduction of, with sodium borohydride, 143, 149

L

Lewis acids, strength sequence, 140, 165, 209
    use as catalysts, 13, 29, 30
Lithium alkyls and aryls, reaction with borates, 63
Lithium aluminium hydride, reaction with boron trichloride, 121
    dialkylboron halides, 125, 126
    diethyl phenylboronate, 127
    phenylboron dichloride, 126
    trialkyl boroxines, 127
    triaryl boroxines, 127
Lithium borohydride, reducing ability of, 136
    reaction with methylamine hydrochloride, 178
    trialkylborons, 125
    reduction of amides with, 151
    of trichloroborazole with, 177
Lithium chloride, reaction with sodium borohydride, 132, 137
Lithium hydride, reaction with $BF_3.Et_2O$, 121
    diborane, 130
Lithium tetraphenylboron, dephenylation of, 110
    reaction with metal chlorides, 110
Lithium triphenylborohydride, 109

M

d-Mannitol, tris-(phenylboronate) of, 206
l-Menthyloxyboroxine, 10
Mesomerism, 12, 39, 42
Metaboric acid esters, preparation of, 8, 115
    recent work on, 10
Metal alkoxyboron compounds, 15
Metal borohydrides, applications of, 141, 160
    covalency in, 134
    dependence of reducing ability on metal ion, 134
    mechanism of reduction by, 141
    preparation of, 121, 130
    reaction with olefins, 153
        trimethylamine hydrochloride, 164
    use to reduce:
        acid chlorides, 144
        carbonyl compounds, 150
        conjugated nitroalkenes, 147

Metal borohydrides (*cont.*)
  use to reduce: esters, 150
    nitroketones, 144
    substituted halogeno acetophenones, 147
    selective reduction of keto groups, 143
Metal ions, effect of, on reducing ability of sodium borohydride, 134
3-Methallyl chloride, polymerization of, in presence of boron trichloride, 46
Methanol, in preparation of borate, 6
  reaction with diborane, 126
    trinaphthylboron, 98
    triphenylboron, 98
Methanolysis, 44
Methylamine, reaction with boron trichloride, 178
Methylamine hydrochloride, reaction with lithium borohydride, 178
Methylated diboranes, pyrolysis of diammonia compounds of, 177
Methyl borate, co-ordination compounds of, 12
  physical properties of, 6, 10
  preparation of, 7
  structure of, 12
B-Methyl borazoles, reaction with hydrogen chloride, 176
    water, 176
N-Methyl borazoles, reaction with water, 175
Methylboronic acid, 70, 176
Methylboronic anhydride trimer, 69, 175
Methyl dichloroborinate, 26
Methyldivinylboron, 114
Methyl phosphinoborine, 188
p-Methoxybenzonitrile, complex with boron trichloride, 184
p-Methoxyphenylboron dichloride, 62
Monoaminoborons, *see* Borazenes
Monomethylamine, reaction with boron tribromide, 169
Monomethylborazoles, 176
Mutual replacement, of OR and halogen or R by R', 33, 60, 94, 102, 103, 211
  see also Redistribution tendencies

N

α-Naphthyl borate, 10
*N*eopentyl alcohol, 13, 25
Nitric acid, reaction with borates, 15
Nitrile–boron trichloride complex, 183
Nitrile–boron trihalide complex, 183
Nitriles, 1 : 1 complexes with boron trihalides, 183

Nitroalkanes, synthesis of, from conjugated nitroalkenes, 147
Nitroalkenes, reduction of, using metal borohydrides, 147
Nitrobenzene, complex with boron trichloride, 180
Nitrocarboxyphenylboronic acids, preparation of, 201
Nitro*cyclo*alkanes, preparation of, 145
o-Nitrophenol, reaction with acetonitrile boron trichloride complex, 184
o-Nitrophenylchloroboron compounds, reaction with butyl hydroperoxides, 115
o-Nitrophenyl dichloroborinate, 34, 184
p-(β-Nitrovinyl)-phenylboronic acid, 204
Nomenclature of boron ring compounds, 198
N-trialkylborazoles, 178, 179
Nucleophilic power, of alcoholic or phenolic, oxygen atoms, 3

O

2-Octanol, reaction with di-*n*-butyl boronic anhydride, 82
(+)-2-Octanol, reaction with boron trichloride, 27
n-Octyl diphenylborinate, 89
n-Octyllithium, reaction with tri-*n*-butyl borate, 84
n-Octyl magnesium bromide, reaction with tri-*n*-butyl borate, 84
Olefins, hydration of, 154
  hydroboration of, 152, 153
  hydrogenation of, 154
  reaction with diborane, 152
    sodium borohydride, 153
Optically active compounds, 27, 38, 71, 73, 88

P

Pentaborane, reaction with alcohols, 129
  ketones, 129
  structure, 123
Perhydro-9b-boraphenalene, 96
Peroxy-boron compounds, 115
Phenols, reaction with boron trichloride, 34, 35, 36
Phenyl acetate, reaction with boron trichloride, 53
Phenylaminodialkylboron, I.R. spectra, 229
Phenylaminodiarylboron, I.R. spectra, 229
Phenylborane, 95
  pyridine complex, 128
  triarylamine complex, 127

Phenyl borate, chemical properties, 12
  co-ordination compounds of, 12
  physical properties, 10
  preparation from triphenyl phosphite and boron trichloride, 191
2-Phenyl-2,1-borazaronaphthalene, 198
Phenylboron dibromide, reaction with ethers, 60
  preparation, 59
Phenylboron dichloride, preparation, 58
  reaction with alcohols, 74
    ammonia, 182
    di-n-butyl phenylboronate, 60, 72
    ethers, 61
    lithium aluminium hydride, 95
  use of in preparation of boron heterocycles, 197
Phenylboron dihalide, reaction with alcohols, 60
  water, 59, 60
Phenylboronic acid, acid strength of, 67
  oxidation of, 116
2-Phenyl-2,3-borothiaindole, 198
2-Phenyl-2,3-borothianaphthene, 197
2-Phenyl-2,3-boroxocoumarone, 197
Phenyl dichloroborinate, intermediate formation of, 53
  mechanism of decomposition of, 73
Phenyl diphenylborinate, 98
p-Phenylenediamine boron trichloride complex, 180
o-Phenylene chloroboronate, 35
o-Phenylene hydrogen borate, 35
p-Phenylenediboronic acid, 65
Phenyl lithium, reaction with triphenylboron, 110
1-Phenylethanol, reaction with boron trichloride, 25
Phosphine, complex with diborane, 187
Phosphino-borane (-borine), 187
Phosphinodimethylboron, 84
Phosphites, reaction with boron trichloride, 14
Phosphorus oxychloride, reaction with boron trichloride, 14
  n-butyl borate, 14
Phosphorus pentachloride, reaction with n-butyl borate, 14
  n-butyl di-n-butylborinate, 82
  di-n-butylborinic anhydride, 82
Phosphorus pentahalide, reaction with diphenylborinic anhydride, 87
Phosphorus trichloride, in reaction of triphenyl phosphite and boron trichloride, 191
  reaction with borates, 13

α-Picoline, complex with trimethylboron, 98
Polarizability, of B—Cl and B—Br bonds, 219
Polyhydroxy boric acid polymers, 19
Polymeric boron phosphorus compounds, applications, 187
  hydrolysis, 187
  preparation, 187, 193
  properties, 193
Polymerization, by trialkylboron catalysts, 100
Polymers, boron modified, 19
Potassium, estimation of, using tetraphenylboron, 111
Potassium borohydride, 136
Potassium fluoroborate, use of in formation of styryl boronic acid, 64
Potassium methoxyborohydride, reaction with diborane, 131
Propenyl lithium, reaction with dialkylboron halides, 85
Propionamide, reaction with boron trifluoride, 185
n-Propyl borate, physical properties, 10
Protonated double-bond theory in diborane, 122
Pyridine, complex with trimethylboron, 98
  with decaborane, 129
  function in removal of HCl, 4, 26
  reaction with boron trichloride dimethylamine complex, 180
Pyridine borane, 128, 160, 164

Q

Quaternary ammonium, 158
Quinoline borane, 128
8-Quinolineboronic acid, 201

R

Redistribution (Disproportionation) tendencies in boron compounds, 173
  order of in trialkylborons, 104
Resins, boronated, 19
Ring frequency, of B—N ring system, 230
  of polymeric borazoles, 230
Ring systems containing boron, 197

S

Silanols, dehydration of, 70
Silicon, electrophilic power of, 3
Silicon hydrides, presence of, in borohydrides, 118
Silicon tetrachloride, reaction with borates, 13

Sodium, reaction with di-*n*-butylboron chloride, 85
triphenylboron, 108
Sodium borohydride, decomposition of, to diborane, 131
effect of alkoxy substituents on reducing properties of, 134
presence of polyvalent metal halides on reducing properties of, 138
hydrolysis of, 132
industrial preparation of, 131
properties of, 132
reaction with alcohols, 133, 135
ketones, 135
lithium halides, 131, 137
potassium hydroxide, 136
tetraalkylammonium hydroxide, 158
trialkylborons, 125, 126
in reduction in presence of salts, 136, 137
of aliphatic ketoximes, 146
of steroidal compounds, 149
of sugars, 148
solubility of, in organic solvents, 133
use of, in preparation of alkoxy-decaboranes, 130
Sodium decaborane, use of, in preparation of alkoxydecaboranes, 130
Sodium hydride, 131
reaction with boric oxide, 132
sodium methoxyhydride, 131
trimethyl borate, 140
Sodium methoxide, reaction with boron trifluoride, 210
Sodium tetraalkoxyborons, 15, 16, 141
Sodium tetraphenylboron, 109
Sodium trialkoxyborohydrides, 140
reducing properties of, 141
Sodium triethoxyborohydride, 140
Sodium triphenylborohydride, 109
Sodium triphenylcyanoboron, 108
Sodium triphenylhydroxyboron, 108
Steric effects, in amine–trialkylboron complexes, etc., 12, 13, 98, 99, 160, 165
Steroidal compounds, reduction of, with sodium borohydride, 149
Styrene, polymerization of, using trialkylboron catalysts, 101, 102
Styrylboronic acid, 64
Substitution in arylboronic acids, 67
Sugars, reduction of, with metal borohydrides, 148
Sulphur compounds, molecular addition compounds of, with boron compounds, 195
Sulphuric acid, reaction with borates, 15

T

Tetraacetyl diborate, 51, 54
Tetraalkoxy boron compounds, of lithium, sodium, potassium, calcium, zinc, and thorium, 141
Tetraalkylammonium hydroxide, reaction with sodium borohydride, 158
Tetraborane, 119, 123
reaction with trimethylamine, 166
Tetrachloroborate, 181
Tetrahaloborate anion, 181
Tetramethyl diborane, 120
reaction with ammonia, 165
Tetraoctyl diborane, 155
Tetraphenyl tin, use in preparation of phenylboron dichloride, 59, 207
Tetravinyl tin, use of, in preparation of vinylboron dichloride, 59
Thienylboronic acid, 78, 79, 80
Thionyl chloride, reaction with borates, 13
Three-centre bond, in diborane, 122
p-Toluidine, complex with boron trichloride, 180
reaction with boron trichloride, 180
o-Tolyl borate, 10
p-Tolylborinic anhydride, 86
Trialkoxyboroxines, 8
Trialkylaminoboranes, reaction with olefins, 155
Trialkylborons, 92
co-ordination with nitrogen bases, 164
dealkylation of, 83
isomerization of, 154
mechanism of, 156
oxidation of, 83
preparation by hydroboration, 153, 154
reaction with boron halides, 83
diborane, 124
halogens, 83
hydrogen chloride, 83
lithium aluminium hydride, 125
lithium borohydride, 125, 126
sodium borohydride, 125, 126
Trialkylboroxines, reaction with lithium aluminium hydride, 127
Trialkyl (aryl) phosphites (phosphates), reaction with boron trichloride, 190, 191
N-tri-*p*-anisyl-*B*-trichloroborazole, 179
Triarylborons, 92, 97
co-ordination with nitrogen bases, 164
Tri-α-naphthylboron, 97, 98
B-tri-*n*-butylborazole, 182
Tri-*n*-butylboron, oxidation of, 83
reaction with boron trichloride, 83, 98

Tri-*n*-butylboron (*cont.*)
  reaction with bromine, 83
    Grignard reagents, 94
    hydrogen halides, 83, 98
  B-Trichloroborazole, 175, 176
    reduction with lithium borohydride, 177
  B-Trichloro-*N*-trialkyl borazoles, 179
Triethoxyboroxine, 9
Triethylamine, reaction with dimethyl-
  amine trichloride complex, 179
B-Triethylborazole, 177
Triethylboron, from diborane and ethylene,
  152
  reaction with alcohols, 83
    aldehydes, 83
    bromal, 83
Triethylene diborate, 33
Trifluoroacetic acid, reaction with boron
  trichloride, 54
Trimethoxyboroxine, 9
Trimethoxyboroxole, *see* Trimethoxy-
  boroxine
Trimethylamine, complex with methyl
  boronic anhydride, 70
  reaction with dimethylamine boron
    difluoride, 169
    tetraborane, 164
Trimethylamineborane, 160, 163, 164
Trimethylamine hydrochloride, reaction
  with metal borohydrides, 164
Trimethylamine methylborane, 163
Trimethyl borate, from dimethoxyborane,
  126
  reaction with sodium hydride, 131
Trimethylborazane, 164
Trimethylborazoles, 175
Trimethylboron, reaction with ethylene
  diamine, 170
N-Trimethyl-*B*-trichloroborazole, 178
Triphenylaminoboron, 179
B-Triphenylborazole, 182
Triphenylboroxine, reaction with lithium
  aluminium hydride, 127
N-Triphenyl-*B*-trichloroborazole, 179

Tri-*o*-phenylene bisborate, 36
Tri-*n*-propoxyboroxine, 9
Trisaminoborons, 170
Tris(di-*iso*-butylcarbinyl)borate, 7
Tris(dimethylamino)boron, 171
Tris(methylamino)boron, 170
Tris(phenylamino)boron, 171
Tris-2,2,2,-trichloroethyl borate, 10
Tristrifluoroacetyl borate, 54
Trivinylboron, 114

U

Unsaturated alcohols, reaction with boron
  trichloride, 31
Unsaturated ethers, reaction with boron
  trichloride, 44, 45
Uranium borohydride, 120

V

Vinylboronic acid (ester), 75
Vinyl chloride, polymerization of, using
  trialkylborons as catalysts, 100
Vinyl ether, reaction with boron trichloride, 46
Vinyl ethyl ether, polymerization of,
  using boron trichloride, 48
Vinyl monomers, polymerization of, using
  trialkylborons as catalysts, 100
p-Vinylphenylboronic acid, 205
Vinyl sodium, reaction with dialkylboron
  halides, 85

W

Wagner-Meerwein rearrangements, in alkyl
  dichloroborinates, 28, 29, 38
  in dialkyl chloroboronates, 30

Z

Zinc alkyls, reaction with boron trihalides,
  58, 81